Statistical Studies of Income, Poverty and Inequality in Europe

Computing and Graphics in R Using EU-SILC

Aims and scope

Large and complex datasets are becoming prevalent in the social and behavioral sciences and statistical methods are crucial for the analysis and interpretation of such data. This series aims to capture new developments in statistical methodology with particular relevance to applications in the social and behavioral sciences. It seeks to promote appropriate use of statistical, econometric and psychometric methods in these applied sciences by publishing a broad range of reference works, textbooks and handbooks.

The scope of the series is wide, including applications of statistical methodology in sociology, psychology, economics, education, marketing research, political science, criminology, public policy, demography, survey methodology and official statistics. The titles included in the series are designed to appeal to applied statisticians, as well as students, researchers and practitioners from the above disciplines. The inclusion of real examples and case studies is therefore essential.

Published Titles

Analyzing Spatial Models of Choice and Judgment with R
David A. Armstrong II, Ryan Bakker, Royce Carroll, Christopher Hare, Keith T. Poole, and Howard Rosenthal

Analysis of Multivariate Social Science Data, Second Edition
David J. Bartholomew, Fiona Steele, Irini Moustaki, and Jane I. Galbraith

Latent Markov Models for Longitudinal Data
Francesco Bartolucci, Alessio Farcomeni, and Fulvia Pennoni

Statistical Test Theory for the Behavioral Sciences
Dato N. M. de Gruijter and Leo J. Th. van der Kamp

Multivariable Modeling and Multivariate Analysis for the Behavioral Sciences
Brian S. Everitt

Multilevel Modeling Using R
W. Holmes Finch, Jocelyn E. Bolin, and Ken Kelley

Bayesian Methods: A Social and Behavioral Sciences Approach, Second Edition
Jeff Gill

Multiple Correspondence Analysis and Related Methods
Michael Greenacre and Jorg Blasius

Applied Survey Data Analysis
Steven G. Heeringa, Brady T. West, and Patricia A. Berglund

Informative Hypotheses: Theory and Practice for Behavioral and Social Scientists
Herbert Hoijtink

Statistical Studies of Income, Poverty and Inequality in Europe: Computing and Graphics in R Using EU-SILC
Nicholas T. Longford

Foundations of Factor Analysis, Second Edition
Stanley A. Mulaik

Linear Causal Modeling with Structural Equations
Stanley A. Mulaik

Handbook of International Large-Scale Assessment: Background, Technical Issues, and Methods of Data Analysis
Leslie Rutkowski, Matthias von Davier, and David Rutkowski

Generalized Linear Models for Categorical and Continuous Limited Dependent Variables
Michael Smithson and Edgar C. Merkle

Incomplete Categorical Data Design: Non-Randomized Response Techniques for Sensitive Questions in Surveys
Guo-Liang Tian and Man-Lai Tang

Computerized Multistage Testing: Theory and Applications
Duanli Yan, Alina A. von Davier, and Charles Lewis

Chapman & Hall/CRC
Statistics in the Social and Behavioral Sciences Series

Statistical Studies of Income, Poverty and Inequality in Europe

Computing and Graphics in R Using EU-SILC

Nicholas T. Longford

CRC Press
Taylor & Francis Group
Boca Raton London New York

CRC Press is an imprint of the
Taylor & Francis Group, an **informa** business

A CHAPMAN & HALL BOOK

CRC Press
Taylor & Francis Group
6000 Broken Sound Parkway NW, Suite 300
Boca Raton, FL 33487-2742

First issued in paperback 2022

ISBN 13: 978-1-03-247736-7 (pbk)
ISBN 13: 978-1-4665-6832-7 (hbk)

DOI: 10.1201/b17164

This book contains information obtained from authentic and highly regarded sources. Reasonable efforts have been made to publish reliable data and information, but the author and publisher cannot assume responsibility for the validity of all materials or the consequences of their use. The authors and publishers have attempted to trace the copyright holders of all material reproduced in this publication and apologize to copyright holders if permission to publish in this form has not been obtained. If any copyright material has not been acknowledged please write and let us know so we may rectify in any future reprint.

Publisher's Note
The publisher has gone to great lengths to ensure the quality of this reprint but points out that some imperfections in the original copies may be apparent.

Visit the Taylor & Francis Web site at
http://www.taylorandfrancis.com

and the CRC Press Web site at
http://www.crcpress.com

To the distant memories of Minsk, Amdahl and VAX

Contents

Preface   xiii

List of Figures   xvii

List of Tables   xxi

1   Poverty Rate   1
    1.1   Background   1
       1.1.1   Median and Percentiles   3
       1.1.2   Populations and Samples   4
    1.2   Income Distribution   7
       1.2.1   Poverty-Rate Curves   8
    1.3   Comparisons   12
    1.4   Sampling Weights   14
       1.4.1   Trimming   19
    1.A   Appendix. Programming Notes   20
       1.A.1   Data Input   20
       1.A.2   Estimating a Quantile   23
       1.A.3   Summarising the Weights   25
       1.A.4   Some Auxiliary Functions   27
       1.A.5   Plotting a Set of Curves   29
       1.A.6   Final Touch in a Diagram   31
       1.A.7   Countries in EU-SILC   35

2   Statistical Background   37
    2.1   Replications. Fixed and Random   37
    2.2   Estimation. Sample Quantities   38
       2.2.1   Weighted Sample Median   39
    2.3   Sampling Variation. Bootstrap   40
    2.4   Horvitz–Thompson Estimator   45
    2.5   Fragility of Unbiasedness and Efficiency   47
       2.5.1   Lognormal Distribution   48
    2.A   Appendix   49
       2.A.1   Bootstrap   49
       2.A.2   Moments of the Lognormal Distribution   52

3 Poverty Indices   **53**
  3.1   Poverty Index 53
      3.1.1 Which Kernel? 56
  3.2   Relative and Log-Poverty Gaps 60
  3.3   Lorenz Curve and Gini Coefficient 63
  3.4   Scaled Quantiles 73
      3.4.1 Permutation Test 75
  3.5   Income Inequality. Kernels, Scores and Scaling 75
  3.A Appendix 77
      3.A.1 Negative Values of eHI 77
      3.A.2 Newton Method in R 80
      3.A.3 More on Poverty Indices 82
      3.A.4 Lorenz Curve and Gini Coefficient 87
      3.A.5 Scaled Quantiles 91
      3.A.6 Permutation Test 93

4 Mixtures of Distributions   **97**
  4.1   Introduction 97
  4.2   Fitting Mixtures 100
  4.3   Examples 102
      4.3.1 Exploration of the Fitted Probabilities 104
      4.3.2 Results for Several Countries 110
  4.4   Improper Component 113
  4.5   Components as Clusters 116
      4.5.1 Confusion Matrix 120
  4.A Appendix. Programming Notes 121
      4.A.1 EM Algorithm for Mixtures of Normal Distributions . 121
      4.A.2 Improper Component 131
      4.A.3 Confusion Index 132

5 Regions   **135**
  5.1   Introduction 135
  5.2   Analysis of Regions 137
  5.3   Small-Area Estimation 142
  5.4   Using Auxiliary Information 146
  5.5   Regions of Spain 148
      5.5.1 Composite Estimation of the Poverty Rates 150
  5.6   Regions of France 156
  5.7   Simulations 158
  5.A Appendix 160
      5.A.1 Estimation of Region-Level (Co-)Variances 160
      5.A.2 Report Card for Austria and Its Regions 161
  5.B Programming Notes 161
      5.B.1 Composite Estimation 168
      5.B.2 Multivariate Composition 170

 5.B.3 Graphics 173

6 Transitions **175**
 6.1 Panel Data 175
 6.2 Absolute and Relative Rates of Transition 182
 6.3 Substantial Transitions 186
 6.4 Partial Scoring of Transitions 191
 6.5 Transitions over Several Years 194
 6.6 Imputed Patterns 197
 6.A Appendix. Programming Notes 203
 6.A.1 National Panel Databases 203
 6.A.2 Rates of Transition 207

7 Multivariate Mixtures **215**
 7.1 Multivariate Normal Distributions 215
 7.1.1 Finite Mixtures of Normal Distributions 216
 7.2 EM Algorithm 217
 7.3 Example 218
 7.4 Improper Component 227
 7.5 Mixture Models for the Countries in EU-SILC 230
 7.6 Stability of Income 232
 7.7 Confusion and Separation 234
 7.A Appendix 236
 7.A.1 What Can Go Wrong in Iterations of EM 236
 7.B Programming Notes 237
 7.B.1 Improper Component 245
 7.B.2 Stability of Income 247
 7.B.3 Confusion Index 249

8 Social Transfers **251**
 8.1 Capacity of Social Transfers 251
 8.2 Impact of Social Transfers 254
 8.3 Potential and Effectiveness 261
 8.4 Nonparametric Regression 264
 8.4.1 Smoothing Sequences 270
 8.5 Perils of Indices 272
 8.A Appendix. Programming Notes 274
 8.A.1 Nonparametric Regression 277
 8.A.2 Graphics for Nonparametric Regression 281

9 Causes and Effects. Education and Income **285**
 9.1 Background and Motivation 285
 9.2 Definitions and Notation 287
 9.2.1 Treatment-Assignment Mechanism 288
 9.3 Missing-Data Perspective 290
 9.4 Propensity and Matched Pairs 293

9.4.1   Regression as an Alternative   294
9.5   Application   296
9.5.1   Results for Other Countries   304
9.5.2   Regression of Outcome on Treatment and Background   305
9.5.3   Potential Versions of Variables   307
9.A   Appendix. Programming Notes   308
9.A.1   Second-Level Functions   313
9.A.2   Graphics for the Balance Diagnostics   322

Epilogue   **327**

Bibliography   **331**

Subject Index   **341**

Index of User-Defined R Functions   **353**

Preface

A majority of the population in the established members of the European Union (EU) have over the last few decades enjoyed prosperity, comfort and freedom from existential threats, such as food shortage, various forms of destruction of our lifes, homes and other possessions, judicial excesses or barred access to vital services, such as health care, education, insurance and transportation. New technologies, epitomised by the Internet and the mobile phone, but also micro-surgery and cheap long-distance travel, have transformed the ways we access information, communicate with one another, obtain health care, education, training and entertainment, and how public services and administration operate. Our economies and societies have a great capacity to invent, apply inventions and package them in forms amenable for personal use by the masses. These great achievements have not been matched in one important area, namely, tackling poverty. Poverty is about as widespread in our societies as it was a few decades ago when, admittedly, our standards for what amounts to prosperity were somewhat more modest (Atkinson, 1998).

Yet, there is no shortage of incentives to reduce poverty in our societies. The purely economic ones are that the poor are poor consumers, and much of our prosperity is derived from the consumption by others; the poor are poor contributors to the public funds (by taxes on income, property and consumption), which pay for some of the vital services and developments. More profound concerns are that the poor are a threat to the social cohesion, are more likely to be attracted to criminal and other illegal activities, and represent a threat to all those who are not poor, because we would not like ourselves and those dear to us to live in such circumstances. By definition, regardless of its details, the poor are poorly integrated in societies in which prosperity dominates, and substantial reduction of poverty would be generally hailed as a great social and economic achievement.

Poverty is studied in economics and political sciences, and population surveys are an important source of information about it. The design and analysis of such surveys are principally statistical subject matter and the computer is essential for their data compilation and processing. That motivates the focus of this volume on statistical computing. In the future, a lot of information about poverty is likely to be derived from population registers, but their comparability across the countries and arrangements to maintain confidentiality of the records without restricting access to them are largely unresolved problems. Statistics is likely to, and computing will certainly, play a central role in such a setting.

The European Union and the governments of its member states have made commitments to reduce poverty and social exclusion (Europe 2020 strategy; Maître, Nolan and Whelan, 2012), but the prospects of fulfilling them are not encouraging while the economic crisis strains the government budgets and introduces new poor to the count that is meant to be reduced substantially in a relatively short time. An element of the general strategy is the collection of data from which the levels of poverty can be reliably monitored and compared against set goals, but also across countries and over the years. The European Union Statistics on Income and Living Conditions (EU-SILC) is a programme of annual national surveys which collect data related to poverty and social exclusion by interviews with adult members of randomly selected households. The surveys have several important elements in common, such as the broad features of the sampling design, the questionnaire content, data processing and dissemination. Such harmonisation enables their analysis by established methods, common across the countries, and using well-established software.

This volume presents a set of statistical analyses pertinent to the general goals of EU-SILC. The analyses are conducted with the statistical language and environment R (R Development Core Team, 2011). The combination of EU-SILC and R is chosen because EU-SILC is believed not to be analysed by as many parties and in as much depth as the data collection effort and general importance of the surveys' theme might justify, and R has become over the years the leading statistical software among academic and methodologically oriented analysts. The contents of the volume is biased toward computing and statistics, with reduced attention to economics, political and other social sciences. The emphasis is on methods and procedures as opposed to results, because the data from annual surveys made available since publication and in the near future will degrade the novelty of the data used and the results derived in this volume. The presented methods do not exhaust the range of analyses suitable for EU-SILC, but will hopefully stimulate the search for new and adaptation of established methods that cater for the identified purposes.

The aim of this volume is not to propose any specific methods of analysis, but to open up the analytical agenda and address the elements of the key definitions in the subject of poverty assessment that entail nontrivial elements of arbitrariness. While soundly selected conventions are useful for uniformity and comparability, and for generating a pretense of certainty, simplicity and authority, far too much is at stake to ignore their fragile nature. This is addressed in the volume by exploring alternative definitions and assumptions (sensitivity analysis) and by searching for consensus among the results based on them. Uncertainty is an implacable adversary in our everyday lives as well as in statistical inference. Its eradication is not feasible, so we combat it by its in-depth understanding (and control of the survey design when applicable or feasible) and detailed assessment of its impact. Its denial is in the long term a poor strategy.

Exploration of EU-SILC data or of any similar database is an open-ended process, without a point at which it could be confidently declared that it has been completed. In this respect, the volume and the extensive code associated with it are incomplete, and are meant as a mere taster of what can (and perhaps should) be done in the analysis of EU-SILC and similar collections of population (household) surveys. Another purpose of the volume is to demonstrate the flexibility of R and its good match to the addressed problems. It has become customary to compile R code in packages with standard documentation, which would be easy to apply by other users, especially those not willing to engage in code development. This mode of operation is not well-suited with this volume, because the code is not intended to be applied uncritically, but to be adapted to the analysts' (and their clients') perspectives, goals and priorities, as outlined and illustrated by examples in the narrative. Eurostat (2010) and similar publications in the Eurostat Statistics Books series provide concise background to the EU-wide social policy issues which the EU-SILC data is intended to inform.

The reader is expected to be acquainted with the fundamentals of R, although knowledge of other software with extensive programming capabilities, such as Stata or MATLAB®, is a transferable skill. There are a lot of textbooks and other literature for immersion in R; the level and extent of Dalgaard (2008) or Zuur, Ieno and Meesters (2009) are sufficient for this volume. For more thorough background I recommend Venables and Ripley (2010). Elements of linear algebra and basic calculus are used in the volume without any introduction; they are also essential for understanding how R operates. Some familiarity with elementary statistical methods, including linear models and survey sampling methods, is also expected.

Chapter 1 introduces the general theme of poverty by studying the percentage of those regarded as poor and discussing the context: estimation of population quantities, working with data from surveys and dealing with sampling weights. Chapter 2 reviews some important statistical concepts and methods used in the volume. Chapter 3 defines and explores some commonly used indices of poverty and inequality, including the Lorenz curve and Gini coefficient. The topic of Chapter 4 is mixture distributions as a general vehicle for detailed description of the household income in a population. Chapter 5 deals with estimation of indices for regions (divisions of a country) and applies methods for small-area estimation. Chapter 6 moves beyond the confines of a cross-sectional (single-year) study and explores methods for describing changes (transitions) between the states of poverty and prosperity, first for a pair of years and then for longer sequences of years. Chapter 7 deals with multivariate mixture distributions for sequences of (annual) values of income and highlights the value of multivariate analysis over a sequence of (annual) univariate analyses. The subject of Chapter 8 is social transfers and their presumed role of alleviating poverty. Chapter 9 presents an application of the potential outcomes framework to assessing the value of postsecondary education.

Except for Chapter 2, every chapter contains a few technical elements from statistics or computing, mainly in isolated short sections. The chapters have extensive R programming notes in the appendices, in which some of the code used in the text is explained in detail. In general, the code becomes more complex and more extensive as the chapters proceed, and only some of the functions are documented, with a focus on the key ideas and 'tricks'. The entire code with accompanying notes can be downloaded from www.sntl.co.uk/EUsilc.html. It contains nearly 250 files; 60 of them have R functions, of which there are more than 200, and 85 files have code for graphics, for all the diagrams and illustrations in the volume.

Access to the EU-SILC database is essential for extracting the full value of this volume. Institutions, such as universities, research institutes and (national) central banks, can apply for access to

estat-microdata-access@ec.europa.eu.

Available are annual (cross-sectional) national databases for all EU countries and Norway and Iceland (released in March 2013 for years up to 2011) and longitudinal (four-year rotating panels) national databases for all of these countries except Germany, Ireland and Romania (released in August 2013 for years up to 2011).

I want to express my thanks to Jitka Bartošová and Catia Nicodemo for highly competent assistance with data extraction and orientation in the database. Discussions with Grazia Pittau and her encouragement are acknowledged. Insightful comments and suggestions by anonymous reviewers of earlier drafts of this volume are acknowledged. I am grateful to Taylor and Francis for arranging these reviews and for assistance with typesetting in LaTex.

Barcelona, April 2014.

List of Figures

1.1 Density, distribution function and quantile function of the standard normal distribution. 5

1.2 Histograms of the equivalised household income (eHI) of households in the EU-SILC sample for Austria in 2010 on the linear and log scales. 8

1.3 Estimated annual poverty-rate curves for Austria in 2004–2010 and their odds ratios. 9

1.4 Poverty odds ratios in Austria in 2004–2010, plotted for a restricted range of thresholds. 10

1.5 Annual individual-level poverty rates for the countries in EU-SILC in 2004 and 2010. 11

1.6 Annual individual-level poverty rates for the countries in EU-SILC in 2004 and 2010, related to Austria. 12

1.7 Quantile–quantile plot of the values of eHI in 2004 and 2010 in Austria. 14

1.8 Distributions of the sampling weights in the datasets in EU-SILC in 2004 and 2010. 15

1.9 Means and dispersions of the weights and log-weights in national surveys in EU-SILC in 2010. 17

1.10 Scaled percentiles and standard deviations of the weights; EU-SILC in 2010. 18

2.1 Weighted sample median. 40

2.2 Bootstrap of a sample with unequal sampling probabilities. . 41

2.3 Bootstrap estimation of the poverty rate in Austria in 2010. . 43

2.4 Bootstrap versions of the quantile–quantile curves for eHI in Austria in 2004 and 2010. 44

3.1 National poverty gaps, defined by the identity (linear) kernel, as functions of the poverty threshold (individual level); 2004 and 2010. 55

3.2 Newton method. 58

3.3 National poverty-gap curves for a range of exponents r; year 2010. 59

3.4 National relative poverty-gap curves in 2004 and 2010. 61

3.5   Absolute and relative poverty-gap curves for Austria in 2004–2010. 62

3.6   Mean log-poverty gaps for the countries in EU-SILC in 2004 and 2010. 64

3.7   Lorenz curve for a set of 100 values. 65

3.8   Lorenz curves for the countries in EU-SILC in 2004 and 2010. 67

3.9   Gini coefficients for the countries in EU-SILC in 2004 and 2010. 68

3.10 Transformed Lorenz curves; EU-SILC in 2004 and 2010. . . . 68

3.11 Lorenz curves for Austria in 2010, with data infiltrated by an extreme observation. 69

3.12 Bootstrap replicates of the transformed Lorenz curve for Austria in 2010 and the estimated standard errors. 70

3.13 Lorenz curves and Gini coefficients for lognormal distributions. 72

3.14 Scaled quantiles for the countries in EU-SILC in 2004–2010. 74

4.1   Densities of mixtures of normal distributions. 99

4.2   Mixture model fits to log-eHI for Austrian households in 2010. 105

4.3   Log-eHI and fitted probabilities $\hat{r}_{i,2}$. 106

4.4   Fitted densities with two to five mixture components; Austria in 2010. 107

4.5   Ternary plots of the fitted assignment probabilities for the three- and four-component mixture model fits; Austria in 2010. 109

4.6   Assignment probabilities for alternative mixture models; Austria in 2010. 110

4.7   Fits of the four-component mixture model to the countries in EU-SILC since 2004; households in 2010. 111

4.8   Fitted probabilities $\hat{r}_{i,0(d)}$ as functions of eHI. 115

4.9   Confusion index $r_{A|B}$ of the standard normal distribution ($p = 0.9$) by a normal distribution with different expectation and variance. 119

4.10 Pairs of scaled densities on the borderline of being clusters for $T = 0.05$. 120

5.1   Sample sizes for the countries and their regions in EU-SILC in 2010. 136

5.2   Estimates of the poverty rate functions for Austria and its regions in 2004 and 2010. 138

5.3   Scaled quantiles for Austria and its regions in 2004–2010. . . 139

5.4   Gini coefficients and transformed Lorenz curves for Austria and its regions, 2004–2010; individual level. 140

5.5   Scaled quantiles of eHI for Czech Republic and its regions in 2005–2010. 141

5.6 Bootstrap estimates of the standard errors of the scaled percentiles for the regions of Czech Republic in 2010. 142

5.7 Sample sizes and estimated population sizes of the Spanish autonomous communities and cities; 2004 – 2010. 149

5.8 Direct and univariate composite estimates of the poverty rates in the regions of Spain, 2004 – 2010. 151

5.9 Direct and univariate to seven-variate composite estimates of the poverty rates in the regions of Spain, 2004 – 2010 and estimated standard errors. 153

5.10 Direct estimates of the region-level poverty rates in Spain for 2008 and 2009 against the estimates in 2010. 155

5.11 Sample sizes and estimated population sizes of the regions of France. 156

5.12 Direct and univariate composite estimates of the poverty gap for the regions of France. 158

5.13 Multivariate composite estimates of the poverty gap for the regions of France. 159

6.1 Rotating panel design. 176

6.2 Patterns of presence on the panel for Austria and Belgium; rotational groups 2007 – 2010. 181

6.3 Estimated rates of transition between poverty states in Austria from 2009 to 2010. 183

6.4 Estimated rates of transition between poverty states in Austria from 2007 to 2010; linear and logit scales. 185

6.5 Relative rates of transition between poverty states in countries with panel data from 2007 to 2010; intact households. 187

6.6 Relative equivalised household income of households with different poverty status in 2009 and 2010. 188

6.7 Absolute and relative rates of substantial transition from 2009 to 2010 for centre $\tau = 60\%$ and half-widths in the range $(0, 10)\%$. 190

6.8 Subscores and scores for transitions. 193

7.1 Ternary plot of the fitted probabilities in mixture model with three components; Austria 2007 – 2010. 223

7.2 Random samples from subsets of the intact households that are very likely to belong to the four mixture components; Austria 2007 – 2010. 224

7.3 Fits of mixture models with two to five components; Austria 2007 – 2010. 225

7.4 Fitted probabilities of assignment to the mixture components in the four- and five-component models; Austria 2007 – 2010. 226

7.5   Fitted means, standard deviations and correlations as functions of the improper density for models with three proper components; Austria 2007–2010.   228

7.6   Fitted marginal probabilities of the proper components for the mixture models with two to four components; Austria 2007–2010.   229

7.7   Fitted marginal percentages of the improper component for the mixture models with up to four components; Austria 2007–2010.   230

7.8   Summaries of the fits of the mixture models with four components for the countries with panels in 2007–2010.   231

7.9   Estimated stability curves for the countries with panels since 2007.   234

8.1   Annual national social transfers and disposable income in 2004–2010.   253

8.2   Estimated percentages of households that received more than a given percentage of their disposable income from social transfers; 2004–2010.   255

8.3   Estimated poverty rates in the A- and R-worlds and their odds ratios for Austria, 2004–2010.   256

8.4   Scaled eHI in the A- and R-worlds; Austria 2010.   258

8.5   Absolute and relative rates of success and failure of the social transfers in the EU-SILC countries in 2010.   259

8.6   Potential and effectiveness of the social transfers in EU-SILC countries in 2010.   262

8.7   Nonparametric regression fits to equivalised social transfers for Austria in 2010.   267

8.8   Nonparametric regression fits for countries in eastern Europe in 2010.   268

8.9   Nonparametric regression fits for the Mediterranean countries in 2009 and 2010.   271

9.1   Hot deck for potential outcomes.   292

9.2   Influential values of a background variable.   296

9.3   Histograms of the fitted propensity scores within the treatment groups.   300

9.4   Numbers of households in the propensity groups; Austria 2011.   301

9.5   Balance plot for the categorical background variables used in the propensity model; Austria, 2011.   302

9.6   Balance plot for the categorical background variables used in the propensity model; Austria, 2004–2011, except for 2009. .   303

List of Tables

2.1 Estimates of the ratios of percentiles of eHI for Austria, 2010 vs. 2004. 45

3.1 Record of convergence of the Newton method. 57

4.1 Mixture model fits to log-eHI for Austrian households in 2010. 104
4.2 Fitted assignment probabilities $\hat{r}_{i,k}$ for models with two and three components. 108
4.3 Fit of the model with two normally distributed (proper) components and one improper component with constant value $D = 10^{-d}$. 116

5.1 Report card for Austria and its regions, 2004 – 2010 (individual level). 162

6.1 Household-level sample sizes of the national panels in EU-SILC in the 2007 – 2010 database. 178
6.2 Patterns of presence of households in the EU-SILC rotational groups 2007 – 2010. 180
6.3 Estimated rates of poverty patterns in 2007 – 2010. 196
6.4 Multiple imputation analysis of poverty patterns for Austria, 2007 – 2010; $\tau = 60\%$ and $\nu = 5\%$. 200
6.5 Estimated frequencies and MI standard errors for poverty patterns in 2007 – 2010; $\tau = 60\%$ and $\nu = 10\%$. 201

7.1 Mixture model fits to the values of eHI with multivariate lognormal distributions as the basis; intact households in the Austrian EU-SILC panel 2007 – 2010. 219
7.2 Leading digits of the probabilities of assignment in two- and three-component model fits; Austria, 2007 – 2010. 222

8.1 Linear poverty gap in the A- and R-worlds and the potential and effectiveness of the social transfers in Austria in 2010. . . 261

9.1 Background variables in the propensity model. 298
9.2 Propensity model fit (logistic regression); Austria 2011. . . . 299
9.3 Numbers of households in the propensity groups; Austria 2011. 301

9.4   Diagnostic summaries for age and the sampling weights as background variables in matched-pairs analysis for Austria in 2004–2011.   304

9.5   Estimates of the average treatment effects for a selection of EU countries and years 2004–2011.   305

9.6   Linear regression fit to log-eHI for Austria in 2011.   306

1

Poverty Rate

1.1 Background

This section introduces the primary measure of poverty in a population and reviews the basic statistical concepts relevant to its estimation.

Poverty is defined as the shortage of essentials for everyday life of an individual or a household. The acquisition of many of these essentials, though not of all of them, is mediated by money—by paying for goods and services. That is why poverty is commonly interpreted as shortage of financial resources. Health care, education, clean environment, transportation, telecommunications, legal services, political representation, personal safety and security, and the like, are other essentials. Where they are provided universally with no barriers to access, their availability is taken for granted, and financial poverty is then rightly the principal concern. In everyday discourse, we qualify the term 'poverty' according to its various aspects, including severity (e.g., as absent, mild, moderate, severe or extreme). Yet, when we turn to its more profound characterisation as a social phenomenon ubiquitous in most countries and their regions, we often opt to describe an individual or a household by a dichotomy, as either poor or not poor, which is too coarse for many purposes.

Poverty is a concept closely linked to a standard, called sufficiency, to which we relate the resources of an individual or a household. This standard cannot be defined universally, because people have different needs, perceptions of necessity and material ambitions that they would expect to be satisfied. Establishing a separate standard for every household is beyond the scope of a survey or a similar data collection exercise. Instead, a list (basket) of goods and services may be compiled, their prices established, and a household's status, as poor or not poor, established with respect to this standard. In an alternative approach, a household is classified according to its ownership of goods and access or subscription to services on a suitable list of essentials. The vast variety of needs, perceptions and perspectives, not only in a large country but even in a small community, renders such a standard far from perfect. Prices differ among the regions and localities of a country, and the needs also tend to differ from one region to another.

Much of the research on poverty has set these considerations aside in favour of simpler definitions that are easier to operationalise in national surveys and their analyses (Sen 1976 and 1979; Atkinson 1987). First, income has taken over as the leading (and often the only) determinant of the poverty status. The only alternative considered is expenditure, but it is in general much more difficult to record than income. Next, a single national standard, defined by a poverty threshold, has been adopted. Further, income is considered over a period of one year, either a calendar year or a tax year (e.g., from 6th April of one year to 5th April of the next in the United Kingdom). Income over a shorter period may be subject to substantial deviations from a longer-term average, which can be smoothed over by saving and using various forms of credit. The standard is defined in relation to the median annual household income, suitably pro-rated to a member of the household. As a convention promoted by Eurostat, the Statistical Office of the European Union, a household and all its members are declared as poor if their household income, adjusted for the household composition, falls short of 60% of the national median household income.

The *Laeken indicators*, defined in Marlier *et al.* (2007, Chapter 2), are a comprehensive list of indices associated with poverty and social exclusion. Collecting reliable data about them remains a challenge. Some of our analyses in this and later chapters point to some problems even in the collection of data about household income. A comprehensive review of practices in the collection of data about household income and a set of guidelines, essential for comparability of statistics across countries as well as in other contexts, is given in UNECE (2011).

Poverty is an attribute of a household because its members are assumed to share all their resources, even if their income is uneven. For example, one adult may earn all the income of a household comprising two adults and two children. A household of two adults requires fewer resources than two adults living alone in separate households. Similarly, a typical child requires fewer resources than an adult. (Some parents may contest this statement.)

The *equivalised household size* (eHS) is defined by counting unity for one (adult) member of the household, 0.5 for each subsequent adult and 0.3 for every child in the household. Every member of a household aged 14 or below is classified as a child. Thus, for a household of two adults $(1.0 + 0.5)$ and three children (3×0.3), eHS $= 2.4$. In general, a household with H_A adults and H_C children has

$$\text{eHS} = 1 + c_A (H_A - 1) + c_C H_C \,,$$

where $c_A = 0.5$ and $c_C = 0.3$. We use a notation for these two coefficients because they are set by a convention, and exploring some alternatives breaches no scientific principles. In fact, $c_A = 0.7$ and $c_C = 0.4$ are sometimes applied. According to the formula with $c_A = 0.5$ and $c_C = 0.3$, the needs of a family of two (adult) parents and two children exceed by only 5% the needs of two solitary persons living in separate households (eHS $= 2.1$ vs. 2.0). With $c_A = 0.7$ and $c_C = 0.4$, the difference is 25% (2.5 vs. 2.0).

An argument as to which pair (c_A, c_C) is correct can be constructively resolved only by considering a plausible set of pairs (c_A, c_C), which most likely contains both alternatives proposed above. Even if there is a correct answer, it is beyond our capacity to establish it. This weakness of ours is not as harmful as the authoritarian imposition of a single 'correct' pair (c_A, c_C) may be. We should hedge our bets against such key definitions or settings in case some alternatives that could almost equally well have been adopted yield substantially different conclusions. This cautious attitute is founded on the general ideas of sensitivity analysis (Saltelli *et al.*, 2004). See Longford and Nicodemo (2009) for their application to the European Union Statistics on Income and Living Conditions (EU-SILC).

The total income of a household, denoted by HI, is the sum of the income of all its members, supplemented by the income received by the household that is not connected with any one of its members in particular. In a survey questionnaire, the sources of income are usually classified as employment, self-employment, investments, receipts from rental properties, inheritance, social transfers (including unemployment benefit), pensions, and the like, to make the interview process easier. Losses, in business transactions and investments, are regarded as negative income. The *equivalised household income* (eHI) is defined by pro-rating, as the ratio of the total income, HI, and eHS:

$$eHI = \frac{HI}{eHS}.$$

This is a key outcome variable used in most analyses in this book.

1.1.1 Median and Percentiles

The median of a variable, or of a set of values, is defined as a quantity such that exactly half of the values of the variable are smaller than or equal, and half are greater than or equal to this quantity. Some compromise has to be made when the result of implementing this definition is not unique. When the number of values is even, say $2L$, then, by convention, the median is set to the average of the Lth and $(L+1)$st largest values, even though any quantity in the range delimited by them satisfies the original definition.

The median is a special case of a percentile. The P-percentile ($0 \leq P \leq 100$) of a variable is defined as a quantity such that P percent of the values of the studied variable are smaller than or equal and $100 - P$ percent are greater than or equal to the quantity. Conventions similar to those for the median have to be adopted to make the percentile unique. For $p \in (0, 1)$, the p-quantile is defined as the $(100p)$-percentile. The quantile function for a variable is defined as the function that assigns to each value $p \in (0, 1)$ the corresponding quantile. We denote it by $Q(p)$, or $Q_X(p)$ when we want to indicate the variable involved. The zero-quantile (and percentile) is defined as the minimum, and the one-quantile (and 100th percentile) as the maximum of the variable. The 25th and 75th percentiles are called the lower and upper *quartiles*, respectively. The percentiles 10, 20, $\ldots$, 90, are called *deciles*.

The *distribution* of a variable X is defined as the extent of its values and their relative frequencies. The distribution function of X is the function F, or F_X, which assigns to each value x the probability (frequency in a finite population) of values of X smaller than or equal to x; $F(x) = P(X \leq x)$, $-\infty < x < +\infty$. A distribution defined in an infinite population is said to be absolutely continuous (smooth) if its distribution function is differentiable throughout. The derivative of the function is called the *density*. There are continuous distributions that are not absolutely continuous, but they are of no practical importance. We therefore drop the qualifier 'absolutely' throughout, even though it always applies.

Although we have defined percentiles for a variable, they are defined equivalently for a distribution. The percentiles and quantiles of a distribution are related to the distribution function of a variable X. For most continuous distributions there is no ambiguity in the definitions of quantiles and percentiles. For them, the quantile function is the *inverse* of the distribution function; by applying the quantile function to the distribution function at a value x we recover this value:

$$Q_X\{F_X(x)\} = x.$$

Similarly, $F_X\{Q_X(p)\} = p$ for any probability p. We write $Q_X(p) = F_X^{-1}(p)$.

The normal distribution is the most commonly used continuous distribution; we denote it by $\mathcal{N}(\mu, \sigma^2)$, where μ is the expectation and $\sigma^2 > 0$ the variance. The standard normal distribution is defined as $\mathcal{N}(0, 1)$. From a variable X with a normal distribution $\mathcal{N}(\mu, \sigma^2)$ we obtain a variable with $\mathcal{N}(0, 1)$ by the linear transformation $(X - \mu)/\sigma$, called *standardisation*. The distribution function of $\mathcal{N}(0, 1)$ is denoted by Φ and its density by ϕ. The distribution function of $\mathcal{N}(\mu, \sigma^2)$ is $\Phi\{(x - \mu)/\sigma\}$ and its density is $\phi\{(x - \mu)/\sigma\}/\sigma$. Let q be a quantile of the standard normal distribution. Then $\mu + q\sigma$ is the quantile of $\mathcal{N}(\mu, \sigma^2)$. The three functions derived from $\mathcal{N}(0, 1)$ are displayed in Figure 1.1. We also use the notation $\phi(x; \mu, \sigma^2)$ and $\Phi(x; \mu, \sigma^2)$ for the respective density and distribution function of $\mathcal{N}(\mu, \sigma^2)$.

1.1.2   Populations and Samples

A typical study is concerned with a population, such as all the households in a country. Usually the population has a finite size (finite number of members), but it is large, e.g., comprising several million residents. Such a population is often treated as if its size were infinite. In a population of finite size N, every variable is discrete, because it could not have more than N distinct values. However, when the number of distinct values is large, the values are ordinal and, in principle, any value in an identified range could be realised, then the variable can reasonably be regarded as continuous. For example, the values of income defined in a currency are all integers or have two decimal places, but it is meaningful to regard income as a continuous variable, and seek a continuous distribution for its description.

The resources available to us, or to another party with a stake in the study, are not sufficient to collect the relevant data, the values of a set of

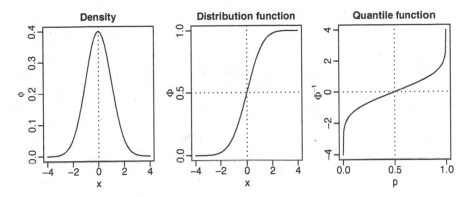

FIGURE 1.1
Density, distribution function and quantile function of the standard normal distribution.

variables, from every member (individual or household) of the population. The compromise made is to collect data from only a sample of the members. Extrapolation from the results obtained for the sample to the population is the principal task of an analysis and one of the main preoccupations of survey methodology (Kish, 1965; Cochran, 1977). The analysis is aided (or made feasible) by a suitable sampling design, a protocol for selecting the members of the population who will form the sample. These members are called *subjects*.

The sampling design that affords the simplest analysis in most contexts is simple random sampling. With this design, the members of the population are treated interchangeably: every member has the same probability of inclusion in the sample, all pairs of members also have identical probabilities of both being included, as do all triplets, and so on. With such a design, the sample can be treated as a faithful miniature of the population. In a general design, the sampling weight of a subject is defined as the reciprocal of the probability of inclusion in the sample. It can be interpreted as the number of members of the population represented in the survey by the subject, and therefore as the importance of the subject in the sample. The total of the weights is an estimator of the size of the population, but if the weights are scaled (multiplied by a positive constant), then this no longer applies. In a sample with a simple random sampling design, every subject has the same weight.

Simple random sampling is often impractical in large human populations and sometimes not efficient for all the intented purposes. Various elements of the agenda of the sponsors of the survey are better served by some more complex designs. Two key devices in survey sampling are *stratification* and *clustering*. Stratification amounts to dividing the domain of the survey (e.g., a country) to several subdomains (e.g., regions), called *strata*, and conducting independently a separate survey in every stratum. In this way, a national survey, a substantial task, is split to several regional surveys, and these smaller tasks are easier to manage.

Clustering can be applied in domains in which a large number of (administrative) units, such as districts, schools or the branches of a company are defined. We apply first a sampling design to these (first-stage) units, selecting some of them into the so-called first-stage sample, and then apply, independently, a sampling design in every unit that has been selected. Clustering amounts to selecting a sample in stages: first we select districts and then subjects within the selected districts. A clustered sampling design can have more than two stages. For example, districts are selected first, then neighbourhoods within the selected districts, and finally households within the selected neighbourhoods. Stratification and clustering can be combined, by selecting a clustered sample in each stratum (region).

As a result of applying a sampling design with unequal probabilities, the subjects of a survey are associated with unequal sampling weights. These weights are adjusted for nonresponse and other imperfections in the conduct of the survey, such as inadequacies of the sampling frame on which the sample selection is based. The adjustment depends on the composition of the sample; see Deville, Särndal and Sautory (1993) for details. Holt and Smith (1979) provide a strong rationale for such adjustment. A different perspective to dealing with missing values is presented by Little and Rubin (2002) and Rubin (2002), and the extensive literature spawn by them. The perspectives are not exclusive and a practical solution may combine elements of both perspectives.

A summary, such as the mean or median, is defined for a population and a sample. In most settings, the population summary is of principal interest but is not available. The purpose of the survey is to enable an analyst to make an inferential statement about such a summary. Estimates, confidence intervals and verdicts of hypothesis tests are common formats used for making inferential statements. They are obtained from the survey data—the values of one or several variables for the subjects.

A sample has two versions. Prior to its drawing, it is a random subset of the population, defined by the sampling design. After the draw, the sample is fixed; to emphasise this, it is referred to as the *realised sample*. An inferential statement, being based on a sample, has two corresponding versions, as a random quantity, best represented by the sampling design and as a formula, computational procedure, or a computer programme to be applied on the anticipated sample, and as a realised statement, the result of applying this procedure on the realised sample. The population quantity is referred to as the *inferential target*.

1.2 Income Distribution

An individual or a household is classified as poor if its value of eHI is below 60% of the national median eHI. The median is evaluated for individuals but, in principle, could also be evaluated for households. The poverty rate of a country is defined as the percentage of individuals classified as poor. In this definition of poverty rate, the median is chosen because it is a more robust measure of the location (central tendency) of a variable. More precisely, the sample median is a robust estimator of the population median, whereas the sample mean is not a robust estimator of the population mean. Note that the population mean and population median differ for a variable with a skewed (asymmetric) distribution. A robust estimator of the population mean could be found and adopted.

It is well-established that the distribution of income in a typical country is asymmetric and highly skewed; usually many households have income (and eHI) near a minimum and there is a long tail of those with (extremely) high income. Log transformation often brings the distribution of income much closer to symmetry. An example is given in Figure 1.2, where the histogram is drawn for the values of eHI for the households in the EU-SILC sample for Austria in 2010 in the left-hand panel, and the log-transformed values are plotted in the right-hand panel. While the assumption of normality for the transformed values may be contested, their distribution is clearly much closer to normality than for the original values.

In Figure 1.2, we omitted the extreme 0.25% of the smallest and largest values, for 16 subjects each, reducing their range to $(2767, 110\,466)$ Euro. Values outside this range are extreme by any standards. The resolution of the histograms would be greatly lowered by their inclusion. Values equal or very close to zero may be erroneous entries or misleading responses. For orientation, log-eHI for 3000 Euro is approximately 8.0 and for 60 000 Euro it is approximately 11.0. Instead of the log transformation, $\log(x)$, we have in fact applied the transformation $\log(1 + x)$, so that zero on the original scale is mapped by the transformation to zero. For more usual values of eHI, of several thousand Euro, adding a single Euro makes negligible difference. On the original scale, the values have a long right-hand tail. After the transformation, the left-hand tail is slightly more pronounced.

The normal distribution with the same (sample) mean and standard deviation as the data, on the original scale or after the transformation, provides a simple, if inadequate, description of the data. Chapter 4 deals with data description in greater detail. In brief, we will describe the values by a few (log-)normal distributions, pretending that they originate from a set of sub-populations which got mixed up and can no longer be identified. Hence the term *mixture* of distributions.

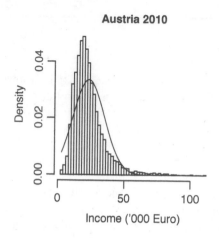
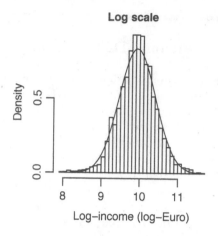

FIGURE 1.2

Histograms of the equivalised household income (eHI) of households in the
EU-SILC sample for Austria in 2010 on the linear and log scales.
The density of the normal distribution with the matching mean and standard
deviation is drawn by the solid line in each panel.

1.2.1   Poverty-Rate Curves

We may adopt the established definition of the poverty rate, with a reference
to the threshold of 60% of the median eHI, but then we have to carefully
qualify the resulting poverty rate by a link to this setting. But what if the
established threshold, denoted by τ, were a bit lower? Or higher? The poverty
rate is not difficult to evaluate for a single threshold; with minimum additional
programming effort we can evaluate it on a fine grid of thresholds τ and present
the threshold-linked poverty rates as a function of τ.

These poverty-rate functions are drawn in the left-hand panel of Figure
1.3 for Austria for each year 2004–2010. The estimated rates based on the
individuals are drawn by black and the rates based on the households by gray
colour. The curves are very difficult to distinguish because they have similar
shapes and the ranges of their values over $\tau \in (40, 80)\%$ are very wide. The
poverty rates for the households are higher than for individuals, uniformly so
for threshold τ above 55%. The curve for households in 2004 stands out for τ
in the range 30%–55%.

The plot is unsatisfactory because the relatively small inter-year differences cannot be discerned. The functions are distinguished more clearly by
contrasting them against one of them, or against another suitable function,
called the *basis*. For the contrast, we choose the *odds ratio*. Let F be a poverty-
rate function and B the basis, both with values in the range $(0, 1)$. Then their
odds ratio is defined as $F(1 - B)/\{B(1 - F)\}$; it is the ratio of the odds of
F and B, defined respectively as $F/(1 - F)$ and $B/(1 - B)$. The odds is an

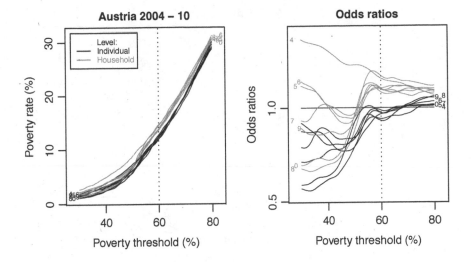

FIGURE 1.3
Estimated annual poverty-rate curves for Austria in 2004–2010 and their odds ratios.
The odds ratios (right-hand panel) are with respect to the individual-level rates in 2004 as the basis; individual (black lines) and household level (gray lines).

increasing function, so two poverty-rate functions retain their contrasts when related to the same basis.

The functions plotted in the right-hand panel of Figure 1.3 are the odds ratios of the poverty rates for a year and the individuals or households with the rate for the individuals in 2004 as the basis. The poverty rates have been declining since 2004 for small thresholds ($\tau < 50\%$), except for little or no change between 2005 and 2006 and a reversal in 2009, which was largely recouped in 2010. No clear trend can be discerned for higher thresholds. The resolution is very high for small thresholds τ for which the poverty rates are low. The resolution for the thresholds $\tau > 70\%$ is poor. It can be improved by drawing a version of the panel with the range of the vertical axis restricted, e.g., to $(0.95, 1.15)$, or the horizontal axis to a narrower range, such as $(50, 80)\%$. This is explored in Figure 1.4. The curves are not exact restrictions of the curves in Figure 1.3, because more smoothing is now applied. Details of how curves are smoothed are postponed till Section 8.4.

Markers for the years are added for individual-level curves at the left-hand margin, but a curve is still difficult to follow through the plotted range. This can be resolved by plotting the functions in even narrower ranges, but then some other detail is lost. As an alternative, the diagram can be supplemented by listing the order of the values for one or a few key thresholds. For example,

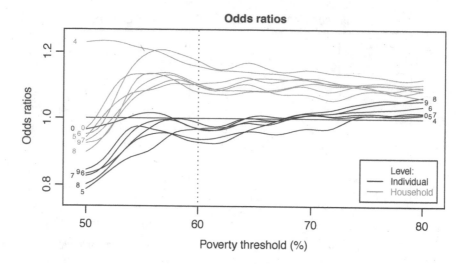

FIGURE 1.4
Poverty odds ratios in Austria in 2004–2010, plotted for a restricted range of thresholds.
The basis are the individual-level rates in 2004.

the estimated values at $\tau = 60\%$ and individual level are 0.929 (year 9), 0.938 (7), 0.968 (5), 0.969 (8), 0.970 (10), 0.986 (6) and 1.000 for the basis year 4. As a shorthand, these estimates can be presented as 9–7—5-8-10–6–4, where the length of the hyphen gives a coarse indication of the distance between the consecutive values for the listed years. Of course, a table of the estimates for a fine grid of values of τ would supersede such a shorthand, but such a table is much more difficult to digest than a diagram and takes up much more space. A solution is provided by software (R) in which any details can be extracted by expressions that are compact and easy to compose.

The poverty rate is an increasing function of τ with a steep gradient, so the qualification by the threshold (such as $\tau = 60\%$) is essential. Nevertheless, if most (or all) countries have poverty-rate functions of similar shape and the functions do not intersect (or are even parallel), then the poverty threshold for which they are compared is not important—we arrive at the same conclusion for any reasonable value of the threshold, even if different figures (values of the estimates) are reported.

Figure 1.5 presents the annual poverty-rate functions for the 15 countries that took part in EU-SILC in both 2004 and 2010. The countries' acronyms are listed in Appendix 1.A.7. The shapes of the functions are similar, but they intersect in many instances. Therefore, any comparison of the countries has to be qualified by the threshold τ. The diagram indicates that there is a modicum of change in the relative positions of the countries in the two years, but none of the changes are substantial. In 2004, Estonia (EE) and Italy (IT) had the

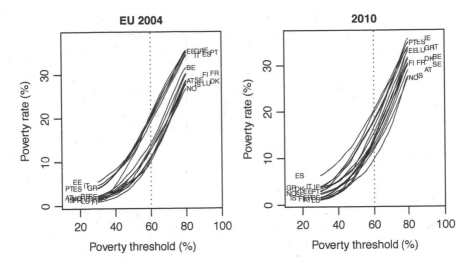

FIGURE 1.5
Annual individual-level poverty rates for the countries in EU-SILC in 2004
and 2010.

highest estimated poverty rates for thresholds close to 30% (5.4% and 4.6%,
respectively), but in 2010 Spain had the highest rate (6.1%), followed by Italy
(3.9%) Ireland (3.8%) and Greece (3.7%). Norway had the lowest estimated
poverty rate for $\tau = 80\%$ in both years (27.2% and 27.7%), followed by Iceland
in both 2004 and 2010 (28.2% in both years).

In 2004, the countries can be classified into two groups according to their
poverty rates with the threshold τ around 60%: Iceland, Norway, Denmark,
Finland, Sweden, Luxembourg, Austria, France and Belgium have poverty
rates in the range 10.0%–14.3% (listed in the ascending order), and Italy,
Greece, Spain, Estonia, Portugal and Ireland have rates in the range 19.1%–
20.9%. The two lists of countries might invite us to interpret the result as
a North-West to South-East division. The two groups of countries are well-
separated for τ in the range 45%–70%, and to a lesser extent even for smaller
τ, with the sole exception of Ireland, which has relatively low poverty rates
for small τ, but at $\tau \doteq 45\%$ joins the countries with higher poverty rates.

A weaker division appears for year 2010 around $\tau = 35\%$. We may refer to the poverty with respect to such a low threshold as extreme. At this
threshold, Luxembourg, Austria, Finland, Iceland, Belgium, France, Sweden
and Norway have estimated rates of (extreme) poverty in the range 1.4%–
3.0%, and Estonia, Portugal, Ireland, Greece, Italy and Spain have rates in
the range 3.9%–7.5%, but Denmark's estimated rate of 4.1% is in the latter
group.

Figure 1.6 presents the individual-level poverty rates in Figure 1.5 scaled
by dividing them by the rates for Austria. The diagram is reduced to the

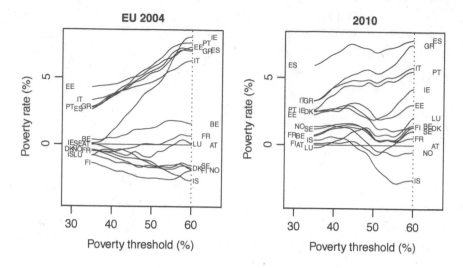

FIGURE 1.6
Annual individual-level poverty rates for the countries in EU-SILC in 2004
and 2010, related to Austria.

range 35%–60% to provide more detail. Each country's symbol is now clearly
distinguished. Similar diagrams can be drawn for other ranges of τ and using
another country as the standard. For wide ranges, plotting the differences
(contrasts) instead of ratios may be more suitable because the ratios converge
for τ close to 80%.

This analysis is incomplete because it leaves unaddressed the issue as to
whether the observed differences among the estimated curves can be attributed
to the differences among the (underlying) population-related curves, or are a
result of happenstance, and would not be replicated on other (hypothetical)
samples drawn by the same national sampling designs. We address this general
issue in Section 2.3.

1.3   Comparisons

The principal use of the poverty rate of a country in a year is to compare
its values for two different countries in the same year or for two years in the
same country; in brief, to compare two settings that differ in a single aspect
(factor), such as the country, year, or the like. Trusting the face validity of
the poverty rate as a summary, and ignoring for the moment its sampling
variation, we would use the comparison for assessing the poverty in the two

settings. We made such comparisons in Figures 1.3 and 1.4 over time (years) in one country and in Figures 1.5 and 1.6 for countries in a given year.

In this section, we explore some methods of comparing the income across two settings (populations), A and B, that do not require the intermediation of the poverty rate. We motivate the development by the following idea. Suppose we have the percentiles $X_A^{(m)}$ and $X_B^{(m)}$, $m = 1, \ldots, 99$, of the distribution of eHI in the respective settings A and B. Then setting B is associated with greater prosperity if for every m we would prefer $X_B^{(m)}$ in setting B over $X_A^{(m)}$ in setting A. This is different from a straightforward (numerical) comparison of $X_B^{(m)}$ and $X_A^{(m)}$. Without an adjustment of X_A (or X_B), such a comparison is ill-advised because of the different costs, needs and standards in the two settings, such as caused by inflation for comparing two years. We may omit some of the percentiles from these comparisons. For studying poverty, we may discard higher percentiles and focus on $m < P$ for $P = 50$, or even smaller. For studying inequality, we may omit some of the percentiles around 50, and switch the signs of the comparisons for either the higher or the lower percentiles.

The starting point for any of the outlined comparisons is the quantile-quantile (q-q) plot of $X_A^{(m)}$ against $X_B^{(m)}$. For eHI in Austria in 2004 and 2010, it is displayed in panel A of Figure 1.7, with the scales in thousands of Euro. The entire q-q curve lies above the identity line, drawn by dots; every estimated quantile is greater in 2010 than its counterpart in 2004. The dashed lines correspond to increases by 20, 25 and 30%. The lowest quantiles in 2010 are more than 30% greater than their counterparts in 2004; all the other quantiles are greater by between 20% and 30%. The plot on the multiplicative scale in panel B has the advantage that the dashed lines (percentage differences) are parallel, but the fine details are still difficult to discern in it.

Panel C displays the plot of the differences of the quantiles against the quantiles in 2004. The resolution of this plot is much higher than of the original q-q plot in panel A because the differences are in a relatively narrow range, (2000, 12 200), compared to (5200, 57 400) in panel A. Now we can see clearly that the lowest quantiles are indeed greater in 2010 by more than 30% and that the other quantiles are greater by between 20% and 30%. Although we refer to percentage changes, the vertical axis of the plot is linear in Euro. This is put right in panel D by plotting the ratio of the quantiles on the vertical axis. The values of the ratio for the lowest quantiles now clearly stand out. The estimates of the lowest quantiles have much greater ratios than the higher quantiles. We return to this analysis in Section 2.3.

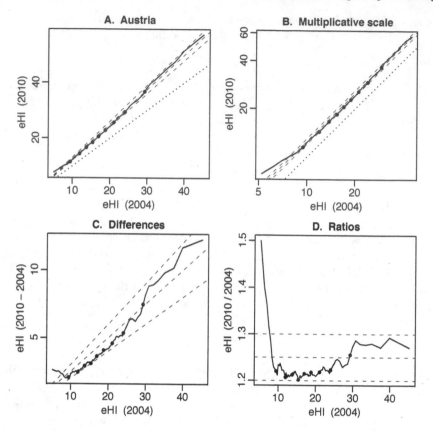

FIGURE 1.7

Quantile–quantile plot of the values of eHI in 2004 and 2010 in Austria.
Linear scale (panel A), multiplicative scale (B), with the differences plotted
against eHI in 2004 (C) and the ratios (D). The deciles are marked by knots.
All monetary values are in thousands of Euro.

1.4   Sampling Weights

The sampling weights of the observations can be motivated as their relative
importance with respect to one another. They are factors in most estimators
and related computations. We should therefore explore their distribution, ide-
ally in the population (values W_k, $k = 1, \ldots, N$) or, as second best, in the
sample (w_j, $j = 1, \ldots, n$), and maybe adjust the weights if their distribution
has some undesirable properties. Since we work with a collection of surveys,
we study several distributions. Figure 1.8 displays summaries of the distribu-

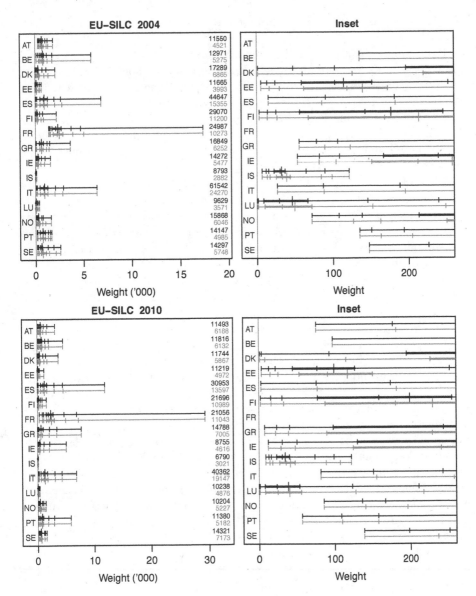

FIGURE 1.8

Distributions of the sampling weights in the datasets in EU-SILC in 2004 and 2010.

Each segment is marked by vertical ticks at the percentiles 0, 1, 5, 25, 50, 75, 95, 99, 100 of the sampling weights, and by a longer tick at their mean. The sample sizes are given at the right-hand margins of the left-hand panels for individuals (black) and households (gray). The insets on the right are restricted to the range $(0, 260)$ for the horizontal axes.

tions of the household-level weights for the 15 countries in EU-SILC in 2004 and 2010.

Each country is represented in the left-hand panel by a pair of horizontal segments, one for individuals (black) and one for households (gray), with its sample percentiles 0, 1, 5, 25, 50, 75, 95, 99 and 100 marked by vertical ticks. For better orientation, each segment is drawn thicker over the interquartile range. The weights in all EU-SILC samples are constant within households; each individual is associated with the same weight as his or her household, but for individual-level summaries the weight of a household is counted once for each of its members.

The percentiles and all the other summaries of the weights that we evaluate are with no weighting applied. The sample means are marked by longer ticks. The magnitudes of the weights differ across the surveys a great deal and some of the percentile ticks are difficult to discern for most countries. Iceland and Luxembourg, the two least populous countries, have very small weights that are in extremely narrow ranges in relation to the most populous countries, France in particular. Even the largest weight in the data for Luxembourg in 2004, 392.15, is several times smaller than the smallest weight for France, 1400.58.

The right-hand panel reproduces the quantiles with the scale for the weights reduced to the range $(0, 260)$. Luxembourg has not only the smallest minimum weight, but even its 5th percentile is very small, 1.45 for individuals. In fact, 239 individuals (2.5%) from 71 households (1.9%) have weights smaller than 1.0. Each of these observations represents less than one member of the respective population. Such an anomaly has arisen in the process of adjusting the weights for nonresponse and discrepancy with the available population totals. The anomaly is present for Luxembourg also in the sample for 2010. Very small weights occur also in the dataset for Denmark in 2004, but to a much lesser extent. A nine-member household in the 2004 sample has weight 1.55, and the next smallest weights are 2.07 and 2.48, for a six-member household each. No segments are drawn in the right-hand panels for Austria and France in 2004, nor for France in 2010, because even the smallest weights in the corresponding years are off the horizontal scale.

The dispersion of the weights can be regarded as a measure of departure from simple random sampling. However, the linear scale for comparing weights is not appropriate; for instance, the pair of weights 100 and 110 represents much smaller difference than, say, 20 and 30. We do not alter the properties of some common estimators by dividing the weights by a positive constant, such as their mean or median. The dispersion of the weights should be assessed after adjusting for their (average) size by an appropriate scaling.

As an alternative to scaling for this purpose, the log transformation of the weights has some merit. Scaling corresponds to the addition of a constant on the log scale, and so it has no impact on the dispersion of the log-weights. Figure 1.9 displays the plots of the means and standard deviations of the weights and of their log-transformed values. From the left-hand panel we merely learn

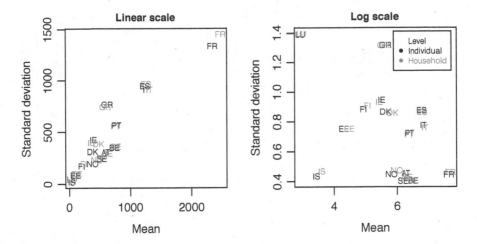

FIGURE 1.9
Means and dispersions of the weights and log-weights in national surveys in
EU-SILC in 2010.

that the national standard deviations of the weights (with no transformation
applied) tend to be greater for surveys with greater mean weights. Greece
may be regarded as an outlier in this respect. The right-hand panel, in which
the moments are evaluated for the log-transformed weights shows that the
weights in the sample for Luxembourg have the smallest log-mean, and yet,
together with Greece, by far the largest log-standard deviation. Iceland, the
other country with small population size, has the next smallest log-mean, but
one of the smallest log-standard deviations, together with several other countries, including France which has by far the largest mean (and log-mean) of
the weights.

Instead of the sampling weights we might study their reciprocals, the (adjusted) probabilities of inclusion in the sample, possibly after a suitable scaling. The log-transformed probabilities are the negatives of the log-weights, so
the dispersions of these two variables coincide. Neither scaling nor the choice
between probabilities and weights are an issue after the log transformation.

A more complete summary of the weights is presented in Figure 1.10. Each
country is represented by a segment delimited by the smallest and largest
scaled weight in the sample. Scaling is by dividing by the sample median of
the weights. The percentiles 1, 5, 25, 75, 95 and 99 are marked by ticks and
the two quartiles are connected by thicker lines. As a result of the scaling,
the median for each sample is equal to unity, and so the countries are easy to
compare by the extent to which their samples have extreme weights. Striking
are the extended right-hand tails, indicating extreme asymmetry of the weights
for every country, except, perhaps, for Sweden.

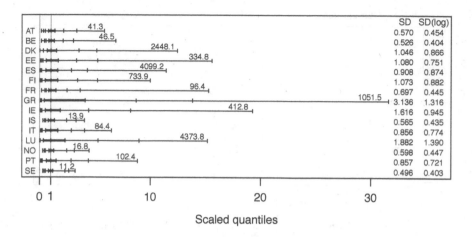

FIGURE 1.10
Scaled percentiles and standard deviations of the weights; EU-SILC in 2010.
The columns at the right-hand margin give the standard deviations of the
scaled weights (SD) and of the log-weights. The figure printed at the maximum
for each country is the ratio of the largest and smallest weight in the sample.

A commonly used summary of the sampling weights is the ratio of the
largest and smallest weight in the sample. Its value is printed in the diagram
for each country (e.g., 41.3 for Austria). The values as random quantities lack
robustness, because they depend on the units that appear in the sample rarely
(with small probabilities). The sample standard deviation of the scaled weights
(SD) or the standard deviation of the log-weights, SD(log), are preferable as
summaries of the dispersion of the weights. They are listed at the right-hand
margin of Figure 1.10. The three sets of summaries are not in good agree-
ment; for example, their values for Denmark and Estonia are in discord. The
scaled weights for Greece and Luxembourg have extremely high dispersions,
and those for Austria, Belgium, Iceland, Norway and Sweden have the least
dispersion.

A similar diagram can be drawn for the reciprocals of the weights (adjusted
sampling probabilities), in which the extent and extremeness of the small
weights (large probabilities) is easier to explore. The scaled quantiles of these
scaled 'probabilities' are on a much wider scale because the values of their
maxima are 356.4 for Spain and 261.5 for Luxembourg. In contrast, they are
only 3.6 for Iceland and 3.8 for Norway. Note that the log-standard deviations
for probabilities and weights coincide and they are not altered by scaling.

1.4.1   Trimming

Extremely large weights tend to inflate the sampling variances of most estimators. Trimming is commonly applied to reduce the dispersion of the weights. In its simplest form, an upper bound $w_{\max}$ is set and every sampling weight that exceeds it is reduced to $w_{\max}$. A more symmetric view of the problem suggests that we should also set a lower bound $w_{\min}$ and increase every weight that is smaller than $w_{\min}$ to this value. In some perspectives, this is problematic because we increase the influence of an observation; we are in general more comfortable with reducing it, appearing more cautious and conservative.

Trimming is discontinuous—some weights are changed but others are not. Two observations may have substantially different weights (on either side of the threshold $w_{\max}$) before trimming, and their weights are nearly equal afterwards. The dispersion of the weights can be reduced also by *shrinkage*. We set a focus w_*, such as the median of the weights or their geometric mean, and a shrinkage coefficient $b \in (0, 1)$, and transform the weights w_j to

$$w'_j = (1 - b)w_j + bw_* .$$

This transformation can be applied to all weights or only to those that exceed w_*. Shrinkage may be applied more effectively on the log scale, transforming the weights w_j to

$$w'_j = w_j^{1-b} w_*^b . \tag{1.1}$$

There are no established rules for setting the parameters of trimming or shrinkage. More radical application, with b closer to 1.0, induces greater changes of the weights. In general, it reduces the sampling variance of the estimators but increases the bias. The setting for which the mean squared error or a similar criterion is minimised is difficult to find when only the sample to be analysed is available. The issue can be explored by simulation studies with artificial populations. Such studies are easy to set up but they are inconclusive, because a vast variety of plausible features of the population would have to be considered, including the association (correlation) of the surveyed variables with the sampling weights. Moreover, the optimal weight adjustment may be specific to (classes of) estimators.

A criterion for trimming is based on the idea of controlling the influence of every observation. With one of them, we set a target for the maximum of the ratios $w_j^2 / (w_1^2 + \cdots + w_n^2)$ and apply the selected method of weight adjustment (shrinkage) with parameters for which this target is attained even for the observation with the largest weight. The dependence of such a fix on a single observation is a weak point in this approach. It is better to set a target for the dispersion of the scaled weights or of the log-weights. Suppose the standard deviation of the log-weights is S, and we would like to reduce it to $S^* < S$. This is accomplished by applying the shrinkage given by Equation (1.1) with $b = 1 - S^*/S$.

Suggested Reading

See Elliot (2008) and references therein for more about trimming. Valliant, Dever and Kreuter (2013) is an invaluable resource on the practical details of survey design and calculation and adjustment of sampling weights. Salverda, Nolan and Smeeding (2009) is a comprehensive source of reference to poverty and inequality.

1.A   Appendix. Programming Notes

This appendix documents the code used in the evaluations described in this chapter. The reader is welcome to adjust it to suit his or her own preferences, apply it to datasets other than those used in the chapter, and use it for other purposes. The code can be downloaded from the website quoted in the Preface. A comprehensive dictionary of the variables in EU-SILC, with their detailed descriptions, is obtained with the database.

1.A.1   Data Input

The data, extracted from the EU-SILC database in the form of country-by-year `Stata` datafiles, is not stored in the `R` workspace, but is read for every analysis. A function is compiled for data input and its reduction to the relevant cases and variables. It contains an option for input of data from `Excel` with comma separated values (`.csv`) and options for other data formats are easy to add.

```
EUread <- function(Dire, cou, Year, Ext, VRS, IH=1)
{
###   Data input -- individual or household level

## Dire    Data-file directory
## cou     Country (2-letter acronym)
## Year    Year of the survey
## Ext     Data-file extension
## VRS     The selected variables (at least two)
## IH      Individual level (1)  or  household level (2)

## Check the variables
VRS <- unique(c("hid", VRS))

## Read the dataset; formats "dta" and "csv" are implemented
if (Ext == "dta")
Dat <- read.dta(paste(Dire, cou, EUYear(Year), ".",Ext,sep=""))[, VRS]

else if (Ext == "csv")
```

```
Dat <- read.csv(paste(Dire, cou, EUYear(Year), ".", Ext, sep=""),
    header=T, sep=",")[, VRS]

## Add another data format here
else
    stop(paste("Do not know how to read format", Ext,"."))

## If a single variable selected
if (length(VRS) == 1)
    Dat <- matrix(Dat, dimnames=list(NULL, VRS))

## Reduction to household level
if (IH==2)  Dat <- as.matrix(Dat[!duplicated(Dat[,"hid"]), ])

## Listwise deletion
if (length(Dat) > 0)   Dat[apply(is.na(Dat), 1, sum)==0, ]
}  ## End of function EUread
```

The function has five mandatory and one optional argument. The latter, IH has a default value, 1. The values of Dire and Ext are the same in every application, so defaults for them could be specified and the arguments made optional. However, the function is itself called in other functions in which the corresponding arguments are optional, and the default is used in every such application. The output of the function is a matrix, unless the argument VRS is set so as to extract a single variable; then the result is a vector. If the argument VRS is set to an empty vector, the vector of household Id.s is returned; see the first expression of the function. The value of VRS can be specified in the following ways:

- the column numbers or names of the variables to be extracted;

- the column numbers to be dropped (with negative signs); or

- a logical vector of the same length as the number of columns in the original dataset.

Only the first option is practical in our setting. In general, the options cannot be mixed. Examples are given below. The code of the function EUread is stored in file EU1io.an1 and in the author's operating system, the function introduced to the R workspace by the expression

```
source("~/Splus/EUsilc12/EU1io.an1")
```

The presence of the function in the workspace can be checked by the expression ls(), which lists all the objects in the workspace. If there are too many of them, more elegant options are

```
ls(pattern="EUre")    ## You don't have to type the entire name
"EUread" %in% ls()
exists("EUread")
```

Details of any system-defined function `fnc` can be displayed by the expression
`help(fnc)`; `help` is itself a system-defined function. The expression `args(fnc)`
returns the arguments of the function `fnc`.

The functions `read.dta` and `read.csv` are defined in the R library
`foreign`; the library has to be declared once in every R session by the expression

```
library(foreign)
```

Other data formats can be included in `EUread` by copying the conditional
expression for reading a `csv` file, together with the condition `else if (Ext
== "csv")`, and making the appropriate changes. Note that some additional
arguments may have to be added, such as the way data items are separated
(comma or semicolon). In the author's environment, the data files extracted
from `Stata` data file have names `COyr.dta`, where `CO` is the acronym of the
country (e.g., `AT` for Austria; see Appendix 1.A.7) and `yr` is the year written
in the two-digit format (e.g., 07 for 2007). The argument `Year` of `EUread` can
be given in any reasonable form because it is converted to the appropriate
format by the function `EUYear`:

```
EUYear <- function(yea)
{
###   Formatting the year for the datafile name
##  yea   Year of the survey

if (nchar(yea)==1) yea <- paste("0", yea, sep="")
if (nchar(yea)==4) yea <- substring(yea, 3,4)

yea
} ## End of function EUYear
```

As an example, the dataset for Austria from year 2004 for the household Id.
(variable `hid`), eHI (`hx090`) and the sampling weights (`db090`) is extracted by
the expression

```
AT04i <- EUread(Dire="~/Splus/EUsilc12/DATA/", "AT", 4, "dta",
                VRS=c("hx090", "db090", "hid"))
```

The element `"hid"` would be added to in `VRS` even if it were omitted, and is
moved to the first column. In the author's file system, the datasets are stored
in the directory `~/Splus/EUsilc12/DATA/`. The name of the sought data file
is `AT04.dta`.

Some analyses are conducted at the household level. The corresponding
dataset can be obtained from `AT04i` by discarding all records with duplicated
values of `hid`. This is done by adapting the penultimate expression of the
function `EUread` and applying it on its own. As an alternative, `EUread` can be
applied with the value of the optional argument `IH` set to 2.

A data file extracted from `Excel` can be read using the function
`read.table` or `read.csv`. These two functions differ only in what they regard as the default for the argument `sep` which indicates how elements are

separated (by a space or a comma). If uncertain, even after using the function `help`, simply specify the value of the argument `sep`. The argument `dec` is relevant with data in which decimal comma is used, often accompanied with the semicolon separator, in which case `sep=";"` should be specified.

The result of applying some functions `read.xxx` is a `data.frame`. It is a structured list. Its elements are the values of variables. Each element of the list is a vector of the same length, with names of the vector the same across the elements; it has the appearance of a matrix. The elements can be a mix of categorical and continuous variables, the former numeric, character or logical, and the latter numeric. A data frame is converted to a matrix by the function `as.matrix`. However, the result is a numeric matrix only if all the elements (variables) can be coerced to be numeric. In the EU-SILC database, the country name is not numeric (e.g., `"AT"` for Austria), but we never need to use this variable because it is constant within each (national) dataset. When specified, the regions of a country, variable `"db040"`, are a categorical variable with values C01, C02, ..., so they can be coerced to be numeric after removing the first two characters. This can be accomplished by the function `substring`; for example, the third character is extracted by the expression

```
AT04[,"db040"] <- substring(AT04[,"db040", 3, 3])
```

for Austria, using a dataset `AT04` which contains the column named `"db040"`. See function `EUYear` for another application of `substring`. Special arrangements are required for Italy where the regions have values ITC, ITD, ITE, ITF and ITG. The problem is resolved by the expression

```
IT04[,"db040"] <- match(substring(IT04[,"db040", 3, 3]), LETTERS)
```

LETTERS is the (system-defined) vector of the (26) capital letters. The function `match` returns the position of the first match(es) of its arguments. The match for C in LETTERS is 3 because it is the third letter of the alphabet.

1.A.2 Estimating a Quantile

The function for estimating a quantile has three arguments: the variable we study, the sampling weights, and the probability for the quantile:

```
QuantW <- function(vec, wei, p=0.5)
{
###   Quantiles with weights  (used for median as the default)

## vec     Vector of values
## wei     Vector of weights
## p       The probability for the quantile
```

The weights `wei` are sorted according to `vec`, and the order number of the unit associated with the sought quantile is found:

```
## The sorted list of weights   --  cumulative totals
swe <-   cumsum(wei[sort.list(vec)])
```

```
##   The quantile of the cumulative weights
qxw <- p * swe[length(vec)]

##   The order of the quantile
ord <- sum(swe < qxw)
```

The values of vec for this and the preceding unit are then combined—the
quantile is found by linear interpolation.

```
##   The corresponding values of vec and wei
qvc <- c(min(vec), sort(vec))[ord+seq(2)]
qwe <- c(0, swe)[ord+seq(2)]

##   Their linear combination
sum(qvc * (qxw - rev(qwe))*c(-1,1) ) / (qwe[2]-qwe[1])
} ##  End of function QuantW
```

The function QuantW could be combined with data input, but this is not
practical when several quantiles of a variable are estimated. It is preferable to
input the relevant dataset, and then apply this function for each quantile (or
percentile) we seek. This is implemented in the function QuantC for individual-
or household-level estimation of a set of quantiles in two years.

```
QuantC <- function(Cou, YR12, pct=seq(5, 95)/100,
      dire="~/Splus/EUsilc12/DATA/", ext=".dta", ih=1)
{
###   Evaluation of a set of quantiles for a country and two years

##  Cou   The country name
##  YR12  The (two) years to be compared
##  pct   The probabilities for evaluating quantiles
##  dire  The data directory
##  ext   The file extension
##  ih    The level of the analysis (1 individual, 2 household)

##  From percentages to probabilities
while(max(pct) > 1) pct <- pct/100

##  Check on the data formats
if (length(ext) == 1) ext <- rep(ext, 2)

##  Initialisation for the data and the weights
dat <- list()
wei <- list()

##  Read the datasets and extract the weights
for (i in seq(2))
{
dat[[i]] <- EUread(dire, Cou, YR12[i], ext[i],
      c("hx040", "hx090", "db090"), IH=2)
wei[[i]] <- dat[[i]][, "db090"]
```

```
##  Adjustment for individual level
if (ih==1)  wei[[i]] <- wei[[i]] * dat[[i]][, "hx040"]
}


##  Initialisation
Qnts <- c()


##  Evaluation of each quantile
for (pt in pct)
Qnts <- rbind(Qnts,
   c(QuantW(dat[[1]][, "hx090"], dat[[1]][, "db090"], pt),
     QuantW(dat[[2]][, "hx090"], dat[[2]][, "db090"], pt) ) )

dimnames(Qnts) <- list(Dround(100*pct,1), 2000+as.numeric(YR12))
Qnts
}  ##  End of function  QuantC
```

The function has the same arguments as EUread, except for VRS (the set of variables), which in the application of EUread inside QuantC is specified inflexibly as household size, eHI and the sampling weights, c("hx040", "hx090", "db090"). The only additional argument is the vector of percentages (or probabilities) for which the percentiles are to be estimated. The output is a matrix of percentiles labelled by the percentages and years.

Although the estimates refer to individuals, we prefer to work with household-level data because it is a smaller dataset. Also, the record for an individual may be incomplete (or even missing), but the corresponding record for the household is sometimes complete, owing to imputation or reporting by proxy. Using complete individual-level data with both the weights w and values of eHI constant within households is equivalent to using household-level data with weights w multiplied by the household size (variable "hx040").

The function Dround rounds its first argument to the number of decimal places given by the second argument. The system-defined function round accomplishes this task, but we want to retain the trailing zeros as an indication of the precision of rounding in any display. Also, we want the values listed in a table to be vertically aligned by the decimal point. For example, 1.21996 would be rounded by round to three decimal places as 1.22, whereas we prefer 1.220, as done by Dround. See Section 1.A.4 for details.

1.A.3   Summarising the Weights

The weights are summarised by the function EUwei

```
EUwei <- function(Cou, yr, dire="~/Splus/EUsilc12/DATA/", ext="dta",
         dgt=3)
{
###   Summary of the sampling weights

##   Cou     Country
```

```
##   yr       Year
##   dire     The data directory
##   ext      The data file extension
##   dgt      The number of digits in rounding


##   The dataset (individuals)
dat <- EUread(dire, Cou, yr, ext, c("hid", "db090"))


##   Exclude observations with zero weights
dat <- dat[dat[, "db090"] > 0, ]


##   Households
weiH <- dat[!duplicated(dat[, "hid"]), "db090"]


##   The counts
CNT <- c(Individual=nrow(dat), Household=LeUni(dat[,"hid"]))


##   The quantiles
qua <- c(0,1,5,25,50,75,95,99,100)/100


##   The output-list
lapply(list(Range=rbind(Individual=quantile(dat[, "db090"], qua),
        Household=quantile(weiH, qua) ),
     Counts.Moments=rbind(CNT, c(EUeff(dat[, "db090"]), EUeff(weiH)),
        cbind(MeVa(dat[, "db090"]), MeVa(weiH) )),
     Log.scale=cbind(Individual=MeVa(log(dat[, "db090"])),
        Household=MeVa(log(weiH))) ), round, dgt)
}   ## End of function  EUwei
```

The function establishes the sample sizes and a set of percentiles of the weights at the two levels (individuals and households) and evaluates the effective sample sizes, ranges, means and standard deviations on the original (linear) and log scales. The function `EUeff` evaluates the effective sample size:

```
EUeff <- function(vec)
sum(vec)^2/sum(vec^2)
```

explained in Section 2.3, Equation (2.3).

The function `EUwei` returns a list; its elements are rounded. As an example, it is applied to Austria in 2004 by the following expression:

```
ATwei4 <- EUwei("AT", 4, dgt=2)
```

A more extensive example applies `EUwei` for a set of countries and years:

```
##   The countries in EU-SILC 2004
EU04a <- c("AT","BE","DK","EE","ES","FI","FR","GR","IE", "IS","IT",
        "LU","NO","PT","SE")


##   The weights summarised en masse
EUwei410 <- list("2004"=lapply(EU04a, EUwei, 4),
                "2010"=lapply(EU04a, EUwei, 10))
```

where `EU04a` is the vector of acronyms of the 15 countries with data in EU-SILC in 2004. It is advisable to give names to the elements of `EUwei410`, which are themselves lists:

```
for (yr in names(EUwei410))
names(EUwei410[[yr]]) <- EU04a
```

The function `EUwei` can be adapted in several ways. For example, the percentages or probabilities in `qua` could be declared as a new argument and the output (long or condensed) could be controlled by another (logical) argument.

1.A.4   Some Auxiliary Functions

This section introduces a few simple user-defined R functions that are useful for exploring large datasets and complex lists. These functions are used in some code in other appendices.

The function `LeUni` returns the number of unique values of a vector.

```
LeUni <- function(vec)
length(unique(vec))
```

The function `MeVa` evaluates the mean and standard deviation of a vector.

```
MeVa <- function(vec)
c(Mean=mean(vec), St.dev.=sd(vec))
```

The function `ExtrL` extracts a specified element of a list.

```
ExtrL <- function(lst, ele)
### Extract a single element  ele  of a list  lst
lst[[ele]]
```

Of course, we can extract such an element directly. However, when working with a list the elements of which are themselves lists, with the same format (e.g., a list in which every element is a list for a country), then we can extract country-level elements by the compact expression

```
lapply(lst, ExtrL, ele)
```

where `lst` is the 'grand' list and `ele` an index or a vector of indices (integers, logicals or names).

The functions `ExtrR` and `ExtrC` extract, respectively, rows and columns of a matrix.

```
ExtrR <- function(mat, rws)
## Extract row(s) rws of matrix mat
mat[rws, ]
```

```
ExtrC <- function(mat, cls)
## Extract column(s) cls of matrix mat
mat[, cls]
```

They are useful for reformatting lists of matrices. The function

```
ExtrE <- function(vec, ele)
##  Extract element(s) ele  of vector  vec
vec[ele]
```

is similar to `ExtrL`, but operates very differently. It can be used to extract a part of a vector (or matrix, which is regarded by R as a vector), but also a sublist of a list.

The function `Dround` is an adaptation of the system-defined function `round`. It returns the character version of its argument, rounded to the specified number of decimals, but with the trailing zeros added.

```
Dround <- function(vec, dgr)
{
###    Rounding with trailing zeros

##  vec    A vector of numerics
##  dgr    The number of digits

if (dgr <= 0) stop("Dround can be used only with decimals.")

##  The substrings on either side of the decimal point
vcr <- strsplit(as.character(round(vec, dgr)), split=".", fixed=T)

##  Deal with whole numbers
for (i in seq(length(vcr)))
if (length(vcr[[i]])<2) vcr[[i]] <- list(vcr[[i]],"")

##  The number of characters behind the decimal zero
ncr <- sapply(lapply(vcr, ExtrE, 2), nchar)

##  Deal with the complete rounding
ncr[is.na(ncr)] <- 0

##  The number of zeros to be added
ncr <- dgr - ncr

##  The sequences of zeros to be pasted
pst <- 0
for (i in seq(2, dgr+1))
pst <- c(pst, paste(pst[i-1],"0", sep=""))

pst[dgr] <- paste(".", pst[dgr], sep="")
pst <- c("", pst)

paste(round(vec, dgr), pst[ncr+1], sep="")
}  ##  End of function  Dround
```

For example, the result of `Dround(2.1996, 2)` is "2.20", not 2.2, which would be obtained by the expression `round(2.1996, 2)`. This is handy for copying and pasting matrices from R to LaTex. Note that the mode of the result is character, not numeric.

The user-defined functions introduced in this section are declared in file `EU1e1.an1`.

The function `lapply` operates on a list and a function. It applies the function on every element of the list and returns a list of the elementwise results. The function `sapply` differs from `lapply` only by returning a vector or matrix when it can reasonably be done, e.g., when the partial results from each element of the list have the same length (or the same dimensions), as is the case in the application in `Dround`.

1.A.5 Plotting a Set of Curves

A set of curves is plotted by the function `EUpov2C`. Its principal argument is a matrix or list of matrices of values to be plotted. Such an object is generated by a function, such as `QuantC`, so the estimation and graphics functions have to be coordinated.

```
EUpov2C <- function(res, xlm=c(),ylm=c(), IH=1, shr=1,offs=c(), ttl="")
{
###    Poverty curves for a set of countries

##  res    The result object (list)
##  xlm    The limits for the x axis
##  ylm    The limits for the y axis
##  IH     Individual (1) or household level (2)
##  shr    The extra gap in the right-hand panel
##  offs   The offset for the labels at the margins
##  ttl    The title
```

Using the function for the first time we advise to leave the optional arguments at their default values, and change them later either by experimenting or by deducing their roles from their description above (and below) or from the code. First the values for the specified level (individual, `IH=1`, or household `IH=2`) are extracted; the result should be a matrix. If it is not, the function fails.

```
##  Extract the relevant values
vls <- sapply(res, ExtrR, IH)

##  Check that the result is a matrix
if (!is.matrix(vls)) stop("The results list is not well structured.")
```

The values of the threshold are extracted from the row names of the matrix `vls`, and the names of the countries from the column names. The extents of the axes are set by default to the ranges of the values (thresholds and quantiles), but these defaults can be overruled by the user (arguments `xlm` and `ylm`).

```
##  The valuation points
pts <- as.numeric(rownames(vls))

##  The countries
cos <- colnames(vls)

##  The restriction for the horizontal scale
if (length(xlm) > 0)  pts <- pts[pts > xlm[1] & pts < xlm[2]]

##  Number of evaluation points
npt <- length(pts)

##  The vertical range
if (length(ylm) == 0)  ylm <- range(unlist(res))

##  Add a bit of extra space at the top and bottom
ylm <- ylm + c(-1,1) * diff(ylm)/25
```

The vector `offs` specifies the horizontal locations of the symbols for the countries; the next section gives more details. The values in `vls` are smoothed. Details of smoothing and of the function `KernSq` are given in respective Sections 8.4 and 8.A.1.

```
##  The offsets for labelling at the margins
if (length(offs) == 0)  offs <- c(2.8, 1.5, 0.2)

##  Smoothing the values of  vls
vls <- apply(vls, 2, KernSq, 1.5)
```

A plot can be drawn by a single expression if its contents are simple, such as one curve or a set of points. For a more complex plot, we draw first its empty shell which sets the extents of the axes, their labels and the title of the plot.

```
##  The empty shell of the plot
plot(range(pts)+5*shr*c(-1,1), ylm, type="n", xaxs="i", yaxs="i",
   xlab="Poverty threshold (%)", ylab="Poverty rate (%)", main=ttl,
   cex.lab=0.9, cex.axis=0.9, cex.main=0.9)
```

The three arguments `cex.xxx` control the font sizes of the items printed in the plot. Experimentation with their values is a better instructor than a comprehensive description, although all the answers can be found by consulting `help(plot)` and `help(plot.default)`.

The key expression of the function is the loop for drawing the curves.

```
##  The (smoothed) poverty curves
for (eu in cos)
lines(pts, vls[, eu], lwd=0.5)
```

The function is concluded by marking the countries' acronyms at the left- and right-hand margins, and drawing vertical dots at the conventional threshold $\tau = 60\%$.

```
##  The markers at the left-hand margin
srt <- sort.list(vls[1, ])
text(pts[1]-shr*EUloc(vls[1, ], offs), vls[1, srt], cos[srt],
     cex=0.5, adj=1)

##  The markers at the right-hand margin
srt <- sort.list(vls[npt, ])
text(pts[npt]+shr*EUloc(vls[npt, ], offs), vls[npt, srt], cos[srt],
     cex=0.5, adj=0)

##  The vertical dots at 60%
abline(v=60, lty=3, lwd=0.9)


vls
}  ##  End of function  EUpov2C
```

The function EUloc is explained in the next section. It spreads the symbols into the number of columns given by length(offs). The vertical coordinates of the symbols at the left-hand margin are in the first row and the coordinates at the right-hand margin are in the last row of vls.

1.A.6   Final Touch in a Diagram

We can label a set of curves in a plot by marking them at their limits, at the left- and right-hand margins, or directly on each curve. If there are many curves, we want to prevent the labels from being printed one over the other. We arrange this by printing them in columns. The function EUloc evaluates the column in which each label is to be printed.

```
EUloc <- function(vec, ofs)
{
###  Placing labels in several columns

##  vec    Vector of values
##  ofs    Offsets

nvc <- length(vec)
nfs <- length(ofs)

rep(ofs, ceiling(nvc/nfs))[seq(nvc)]
}  ##  End of function  EUloc
```

The first argument (vec) is the vector of the values at the (left or right) limit of the curve and the second is a vector of the distances from the limit to the columns. Thus, the number of columns is given by the length of ofs. The function returns a vector in which the values of ofs are recycled as many times as to be of the same length as vec.

Suppose vls is a $2 \times K$ matrix, and its rows contain the values of the K plotted variables at their left- and right-hand limits, equal to pts[1] and pts[2], respectively. Further, nms are the column-names of vls, to be used as labels, and shr is a positive scalar. Then the expressions

```
## For the left-hand margin
srt <- sort.list(vls[1, ])
text(pts[1]-shr*EUloc(vls[1, ], offs), vls[1, srt], nms[srt],
     cex=0.5, adj=1)

## For the right-hand margin
srt <- sort.list(vls[2, ])
text(pts[2]+shr*EUloc(vls[2, ], offs), vls[2, srt], nms[srt],
     cex=0.5, adj=0)
```

will add the labels at both margins, as in Figure 1.5. The vector srt is for permuting a row of vls into ascending order. The two pairs of expressions can be replaced by a single loop, using $(-1)^i$, $i = 1, 2$, for ± 1 as the factor of shr.

As an alternative, each curve can be marked at a given value of the horizontal axis. This is implemented in the function

```
EUpla <- function(mat, xva, frt, mrk, cx=c(1.7,0.6),clr="black")
{
### Marking a set of curves

## mat     Matrix
## xva     The x values
## frt     The markers (x-value No.s)
## mrk     The markers (names/labels)
## cx      cex's
## clr     The colour for marking

## The y coordinates of the markers
quaf <- c()
for (rg in seq(length(frt)))
quaf <- c(quaf, mat[frt[rg], rg])

## The x-values
pps <- xva[frt]

points(pps, quaf, pch=16, cex=cx[1], col="white")
text(pps, quaf, mrk, cex=cx[2], col=clr)
} ## End of function   EUpla
```

The first argument is a matrix in which each column contains the values of a curve for the values of x given by the second argument, xva. The third argument, frt, holds the indices of xva at which the labels for the curves, given in mrk, are to be placed. The function first evaluates the vectors pps and quaf which hold the coordinates of the marking spots. Then it prints white discs on these spots, to interrupt the curves, followed by printing on the same spots the labels in the colour specified by the argument clr. The size of the white spots and of the labels (characters) is given by the argument cx. The labels can be printed in any colour. Apart from black as the default, we use only shades of gray, gray1, ..., gray99, ordered from black to white. Just the one value, gray50, gets us a long way. Note that the order of the expressions is important. For example, a visible (e.g., black) symbol placed at a spot would be wiped out if a sufficiently large white disc were later placed there.

Suppose we have the matrix EUvals of $n = 100$ values (rows) each for $p = 15$ countries (columns), and the matrix is labelled by the values of x and the countries. Suppose we want to mark the curves to the left of the centre, that is, for values of x below the average, but not at the extreme left. To accomplish this, the function EUpla is applied by the expression

```
EUpla(EUvals, as.numeric(rownames(EUvals)), 20+2*seq(ncol(EUvals)),
     colnames(EUvals), cx=c(2.0, 0.6))
```

For the countries, the labels will be placed at the 22nd, 24th, ..., 50th elements of x, and the corresponding values of EUvals for countries 1, 2, ..., 15. The values of cx are set by trial and error. The default, (1.7, 0.6), is usually suitable for printing labels that are single characters or digits. In this example, each mark has two letters, so it takes up more space. That is why the first element of cx is set to a value greater than the default.

The book has many diagrams, each generated in R as a postscript file. A lot of repetitive typing can be spared and some uniformity of the style enforced by using a function for this process.

```
Figure <- function(name, expr, hgt=2.75, mfr=1, mrm=1, mgm=0.8,
     dire="~/papers/EUsilc12/Figures/Fig", wdh=5.5, lb=c(3,3,1))
{
### Generating a postscript file

## name    The name of the file, with prefix Fig
## expr    The expression that generates the diagram
## hgt     The height of the diagram  (inches)
## mfr     The mfrow argument of function par
## mrm     The mar argument of function par
## mgm     The mgp argument of function par
## dire    The directory for placing the postscript file
## wdh     The width of the diagram  (inches)
## lb      The lab argument of function par

## Close all graphics devices
graphics.off()

## Open the postscript file for the diagram
postscript(paste(dire, name, ".ps", sep=""), width=wdh, height=hgt,
           pointsize=11, horizontal=F)

## Deal with the defaults for some parameters of par
if (length(mfr)==1) mfr <- rep(mfr, 2)
```

```
if (length(mrm)==1) mrm <- mrm*c(4,4,2,1)

##  Set the R graphics parameters
par(mfrow=mfr, mar=mrm, mgp=mgm*c(3,1,0), lab=lb)

##  Drawing the diagram
res <- expr

##  Close the current graphics device
dev.off()

res
} ##  End of function  Figure
```

This function specifies the destination of the generated document by the argument name usually a word comprising four characters, the expression expr that generates the diagram, usually the application of a function, the height (hgt) of the diagram (the width is another optional argument, the value of which, 5.5, is used throughout), and other arguments. The values of hgt and wdh are in inches (1 inch = 2.54 cm), but in practice only the relative magnitudes of the values matter, because a diagram can be resized in LaTex or another text-processing software. Further arguments of Figure include the 'panelling' of the diagram (into c rows and r columns), mfr, with the default for a single panel, specification of the margins on the four sides of the plot (mfr), the setting of the margins for the axes (title, labels and lines), as a multiple of the vector c(3,1,0), and lb for controlling the annotation of the axes. With the default c(3,3,1), three values are marked on either axis, although other rules that aim to make these labels 'pretty' are in force and may overrule this. So, four values, or only two, are sometimes marked.

The expressions graphics.off() and dev.off() are for closing all the graphics devices and the current device. This is useful when several postscript files are generated in an R session. Also, a graphics file is not completed until the device used is closed, or a new one opened. An expression expr that contains several evaluations has to be delimited by braces, { and }. We may want to inspect the object returned by the expression expr (the result of its last evaluation). That is why the result is assigned to object res and the function is concluded with res. An example of applying Figure in this chapter is

```
AT410eR <- Figure("PovG",
    QuantG(AT410e, "Austria", qua=seq(9)*10, fcs=c(1.2, 1.25, 1.30),
        sca=1000),   5.4, c(2,2), 0.8, 0.6)
```

which generates the postscript file FigPovG.ps and deposits it in the author's subdirectory reserved for the figures. It is displayed in Figure 1.7. The matrix AT410eR is generated by the function QuantG. See Supplementary materials for details.

1.A.7   Countries in EU-SILC

The following countries have samples in EU-SILC in 2010.

Since 2004:
AT: Austria; BE: Belgium; DK: Denmark; EE: Estonia; ES: Spain;
FI: Finland; FR: France; GR: Greece; IE: Ireland; IS: Iceland; IT: Italy;
LU: Luxembourg; NO: Norway; PT: Portugal; SE: Sweden.
Since 2005:
CY: Cyprus; CZ: Czech Republic; DE: Germany; HU: Hungary;
LT: Lithuania; LV: Latvia; NL: the Netherlands; PL: Poland; SI: Slovenia;
SK: Slovakia; UK: the United Kingdom.

Since 2007:
BG: Bulgaria; RO: Romania.

Since 2009:
MT: Malta.
The elements of the list EU410a are the names of the countries in EU-SILC
in a particular year. For example, EU410a[[3]] or EU410a[["6"]] refer to the
same element, and its value is the vector of the labels of the countries in 2006:

```
"AT" "BE" "CY" "CZ" "DE" "DK" "EE" "ES" "FI" "FR" "GR" "HU" "IE" "IS"
"IT" "LT" "LU" "LV" "NL" "NO" "PL" "PT" "SE" "SI" "SK" "UK"
```

2

Statistical Background

This chapter gives a brief background to the statistical terms and concepts used throughout the book. Readers not interested in these details can skip the chapter, although they may have to treat some computer code dealt with later as a black box and their capacity to work with the code creatively may be impeded. The Appendix of Longford (2008) presents the statistical perspective and terminology used in this volume.

2.1   Replications. Fixed and Random

Central to a lot of statistical theory and relevant to much of the practice is the concept of replication. A statistical process, such as sampling, has a protocol— an instruction for its implementation that leaves no ambiguity about any of its steps and addresses all possible contingencies. A replication of such a process is defined as its independent repeat. Independence means the absence of any learning or using any information (experience) from any other application of the same process.

With the aid of the term 'replication', we can define the terms 'fixed' and 'random' as follows. A quantity, a set or some other object is said to be fixed if it attains (or is assigned) the same value in every replication. For example, a stratum in sampling is fixed because it stands for the same domain, with the same set of units (members), and units with unaltered attributes (values of variables). In contrast, a sample or its summary, such as the sample mean of a nonconstant variable X, is random when replicate samples contain different sets of subjects, with different values of X, and therefore have different values of the sample mean. The value of the summary is the same in every replication only is some degenerate situations which are rarely of any interest in practice.

2.2   Estimation. Sample Quantities

Recall that an inferential target is a population quantity about which we would like to make an inferential statement. An ideal statement reproduces the value of the inferential target when applied to any conceivable sample. A more realistic goal is an inferential statement that comes close to the (unknown) target in a specified way. This specification entails a criterion for quantifying the loss (harm, damage, or the like) when instead of a target θ we conclude with an estimate $\hat{\theta}$. A commonly used criterion is the squared error, $(\hat{\theta} - \theta)^2$, which we would like to minimise on average in a large number of replications of the sampling and evaluation processes. In the search for the ideal, we have two devices at our disposal: the sampling design, that is, the protocol for selecting a sample, and the choice of an estimator. Once we are committed to a sampling design, or the survey is in progress or already concluded, the choice of an estimator is the only device left. In practice, we can afford to conduct the survey only once, so we have only one 'shot' at the target, and cannot explore any replications of an estimator in a direct way.

Estimation is concerned with defining sample quantities that are good substitutes for population quantities as their targets. Suppose a binary variable X is defined in a population, such as a country, which has N members, so that the values of the variable, $X_1, X_2, \ldots, X_N$, are each equal to either zero (negative) or unity (positive). The rate of positives in the population, denoted by p, is defined in the obvious way. We define the sample rate as the proportion of positives in the sample. The corresponding *estimator* is defined as a computational procedure or a mathematical formula. Let $x_1, x_2, \ldots, x_n$ be the values of X for the subjects in the sample. The number of subjects, n, is the sample size. Note that subject 1 differs from member 1; in general, x_1 does not coincide with X_1. In any case, the ordering (indexing) of the members of the population, as well as of the subjects in the sample, is arbitrary and immaterial. The sample rate of positives is defined as

$$\hat{p} = \frac{x_1 + x_2 + \cdots + x_n}{n}. \tag{2.1}$$

In this formula, not only the values x_i, $i = 1, \ldots, n$, but sometimes even the sample size n are random—subject to the vagaries of the sampling process. In contrast, the values $X_1, X_2, \ldots, X_N$ are fixed because the population we study is considered at a particular moment, with no regard for any changes (births, deaths or migration) in the past or in the future.

The sampling design exerts an influence on the properties of $\hat{p}$. In many settings it would take little effort to design a survey in which $\hat{p}$ given by Equation (2.1) would have transparently poor properties. For example, the probability of inclusion in the survey may be associated with X. At an extreme, if members with $X = 0$ are excluded from the sample with certainty, then we might incorrectly conclude from $\hat{p} = 1$ that $X = 1$ for every member of the population.

The sampling design can be introduced in the expression for $\hat{p}$ as

$$\hat{p} = \frac{\sum\limits_k I_k X_k}{\sum\limits_k I_k}, \tag{2.2}$$

where I_k is the indicator of inclusion in the survey. That is, $I_k = 1$ if member k is a subject (in the sample), and $I_k = 0$ otherwise; I_k is a binary variable. This variable is random because its values in a replication are different for at least some units. The sampling design is defined (completely characterised) by the joint distribution of the random vector $\mathbf{I} = (I_1, I_2, \ldots, I_N)$. Independence of the elements (indicators) I_k corresponds to a special kind of design. For example, if the sample size is fixed (to $n > 0$), that is, when every sample that could be realised comprises exactly n subjects, $\text{var}(I_1 + I_2 + \cdots + I_N) = 0$. However, for independent $I_1, I_2, \ldots, I_N,$, this variance, equal to $\text{var}(I_1) + \cdots + \text{var}(I_N)$, is positive, unless every unit is included in the sample with probability zero or one. In practice, the indicators are correlated, but when the sample size n is large most of these correlations are small. With the notation used in Equation (2.2), both summations are over the population, $k = 1, \ldots, N$. Nevertheless, $\hat{p}$ is a sample quantity; it depends on the values of X only for the subjects that are in the sample.

2.2.1 Weighted Sample Median

In this section, we describe an estimator of the population median. A simple solution is the sample median, constructed by the method described in the definition of the population median in Section 1.1.1, but applied to the sample. When the subjects in the sample are associated with unequal weights this procedure has to be adjusted to reflect the uneven importance of the observations.

We sort the subjects in the ascending order of the studied variable. The corresponding subjects are represented by segments of lengths equal to their weights, and these segments are assembled into a single segment. The weighted sample median of the equivalised household income (eHI) is defined by the value of eHI for the subject whose segment contains the centre of the entire segment. An illustration is given in Figure 2.1, with a sample of size $n = 15$. The one-half of the total weight falls on the segment of subject 6 which has the eighth highest value of eHI, equal to 10 237 Euro. With a more detailed definition, we combine the values of eHI of this subject and the immediately preceding subject, with coefficients related to the lengths of the subsegments for the weight split by the half of the total weight. The segment for subject 6, of length 36.05, is split by the one-half of the total weight into subsegments of length 4.35 and 31.70. We use them as the coefficients in the combination

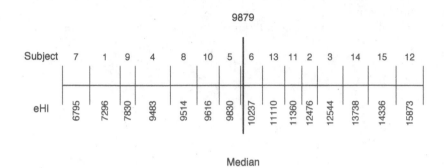

FIGURE 2.1
Weighted sample median.
An illustration with computer generated values.

of the values of eHI for subjects 5 and 6:

$$\frac{9830 \times 31.70 + 10\,237 \times 4.35}{36.05} = 9879.11 \text{ Euro},$$

only slightly greater than the value of eHI for subject 5.

2.3   Sampling Variation. Bootstrap

An estimate, such as the weighted sample median or the estimated poverty rate at a given threshold, should be accompanied by an assessment of the quality of the estimator that was applied. The assessment should reflect all the sources of uncertainty present in the process of drawing the sample and its evaluation (data manipulation). We assume first that drawing the sample is the sole source of uncertainty; only the sample varies across hypothetical replications of the survey, never the values of the variables of any members of the population. We have to describe the variety of values of the estimator that would have been obtained with replicate samples that could equally well have been realised. Since replication of a survey is not possible, we have to resort to methods that use no information other than the realised sample and the sampling design, and possibly some information about the population that is available from external sources.

Bootstrap (Efron and Tibshirani, 1993; Davison and Hinkley, 1997) is one such method. With a simple random sample of size n, we regard the realised sample provisionally as a population, and draw from it simple random samples

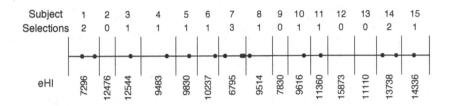

FIGURE 2.2
Bootstrap of a sample with unequal sampling probabilities.
An illustration with the same dataset as in Figure 2.1. The selections are
marked by black dots.

with replacement, each with the same sample size n. Replicate estimates are
obtained by evaluating the estimator on the replicate samples. The key theoretical result is that, for sufficiently large n and a sufficiently large number of
replications, the distribution of the replicate estimates approximates well the
sampling distribution of the originally considered estimator.

In practice, we have to deal with sampling designs that differ from simple
random. For sampling designs in which units are selected independently of one
another, we apply a resampling scheme in which the probabilities of inclusion
in a replicate sample are proportional to the sampling weights w_j. Thus, we
draw a segment composed of the subsegments of length proportional to w_j,
$j = 1, \ldots, n$, and select n_b points on the segment completely at random—the
selections are independent draws from the (continuous) uniform distribution
on $(0, w_+ = w_1 + w_2 + \cdots + w_n)$. An illustration is given in Figure 2.2 with
$n_b = n = 15$. Unlike in Figure 2.1, the order of the subjects on the segment
is immaterial. The values of the studied variable (eHI) play no role in the
selection of the replicate (bootstrap) sample. A subject may be selected into a
bootstrap sample more than once. The number of such selections in a sample
may by chance be out of proportion with the sampling weight of the subject.
In the illustration, subject 7 is selected three times, whereas subject 12, with
only slightly smaller sampling weight (44.69 vs. 44.66), is not selected at all.

The purpose of drawing a bootstrap sample is to obtain a sample that is
for all purposes like the original sample, except that it is easy to obtain many
of its replicates, and these replicates differ from one another in all aspects
(approximately) as much as replicates of the original sample, regarded as a
random entity, would. To satisfy these criteria, we have to adjust the size of

the bootstrap sample. Instead of n, it is set to

$$n_b = \frac{w_+^2}{w_1^2 + \cdots + w_n^2},\qquad\qquad(2.3)$$

called the *effective sample size*. See Potthoff, Woodbury and Manton (1992) for its derivation. It can be shown that $n_b \leq n$, with equality only when $w_1 = \ldots = w_n$. The units in a bootstrap sample are associated with equal weights. Application of the estimator to this sample yields a bootstrap estimate. We replicate the drawing of a bootstrap sample and evaluation of the estimator a large number of times. The expectation and sampling variance of the original estimator are approximated (estimated) by their sample versions evaluated on the bootstrap estimates.

We apply bootstrap to estimating the poverty rate in Austria in 2010, using the threshold of $\tau = 60\%$. The sample comprises 11 493 individuals in 6188 households. The effective numbers of individuals and households are 9155 and 5065, respectively. The estimate of the rate is $\hat{\theta} = 12.120\%$. The average of the $M = 1000$ replicate bootstrap estimates is $\hat{\theta}_b = 12.100\%$ and their standard deviation, the bootstrap standard error, is $\hat{\sigma} = 0.496\%$. We quote values to three decimal places, more than what their precision warrants, but with coarser rounding we could not compare the values of $\hat{\theta}$ and $\hat{\theta}_b$. The bootstrap estimates are summarised by the histogram in Figure 2.3. The approximate 95% confidence interval, delimited by $12.100 \pm 2 \times 0.496 = (11.108, 13.091)$, is marked by vertical dots. An alternative to this interval is defined by the 2.5 and 97.5 percentiles of the bootstrap estimates, equal to $(11.120, 13.056)$. In the former interval we rely on the normality of the bootstrap estimates, whereas in the latter we estimate extreme percentiles from a relatively small sample.

Is 1000 bootstrap samples enough? We answer this question indirectly, by assessing how much precision we get about the bias and standard error. (Note that the standard error is also estimated). The bias is estimated by the average of a random sample of size 1000, so its standard error is $\sigma/\sqrt{1000}$, where σ is the standard error of the estimator, estimated by 0.496%. Therefore $\hat{\sigma}/\sqrt{1000} \doteq 0.016\%$; compare it with the estimated bias $12.120 - 12.100 = 0.020\%$. It is plausible that the estimator is unbiased, but the biases of 0.045% and -0.010% are also plausible. We conclude that the bias of the estimator $\hat{\theta}$ is negligible; it is highly unlikely to attain a value, such as 0.10%, that might be regarded as important.

The squared standard error $\hat{\sigma}^2$ is associated with 999 degrees of freedom. That is sufficient for the normal approximation to its distribution. Its sampling variance is estimated by $2\hat{\sigma}^4/999 \doteq 0.000121$, and therefore the standard error by 0.0110. An approximate 95% confidence interval for the sampling variance is $(0.224, 0.268)$, and for the standard error $(0.473, 0.518)\%$. Greater precision can be attained by generating more replicates, but that is hardly necessary. See Appendix 2.A.1 for an implementation of the bootstrap in R.

Important assumptions in our implementation of the bootstrap are that

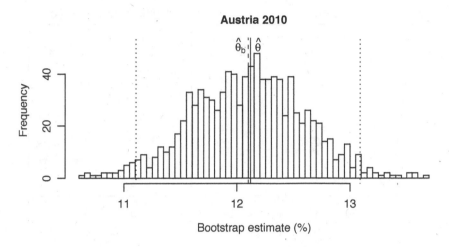

FIGURE 2.3

Bootstrap estimation of the poverty rate in Austria in 2010.
The estimate is marked by a solid vertical line, the average of the (1000)
bootstrap estimates by dashes, and the approximate 95% confidence interval
by dots.

the observational (sampling) units are mutually independent and that there
are many of them. The former condition is not satisfied because the sampling
design involves some clustering. If the clusters were identified in the database,
bootstrap could be applied to the clusters, selecting clusters as intact units,
although their small number threatens the other assumption. In fact, households are treated in the bootstrap as intact units, even when the analysis is
conducted at the individual level. If a household is selected, all its members are
included in the bootstrap sample. This is equivalent to analysing households
with their weights multiplied by the numbers of their members.

The bootstrap standard errors have three components of imprecision: estimation (using only a finite number of replicates), the theoretical error of the
bootstrap that would be present even in a congenial setting, and the error
due to the correlation of the observed units. However, we use the estimated
standard errors throughout only for orientation, and never for any formal
statistical procedures.

For an application of bootstrap, we return to the analysis in Section 1.3 to
assess to what extent the twists and turns of the curves displayed in Figure
1.7 are a phenomenon associated with the population or merely a result of the
vagaries of sampling. We generate bootstrap estimates of the quantiles for the
two years and draw the bootstrap version of the plot in panel D. A set of 100
bootstrap quantile-quantile (q-q) curves is drawn in Figure 2.4 by thin gray
lines, with the deciles marked by small black discs. The bootstrap estimates

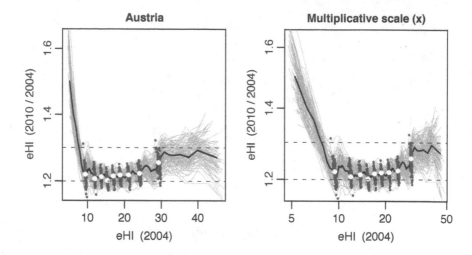

FIGURE 2.4

Bootstrap versions of the quantile–quantile curves for eHI in Austria in 2004 and 2010.

Percentiles of eHI for 2004 (in thousands of Euro) vs. ratios of percentiles for 2010 by 2004. The original estimates are connected by a black line and the deciles are marked by white discs.

of the percentile ratios are drawn by the black line, interrupted at the deciles by white discs.

Each line starts at the second (bootstrap) percentile and ends at the 98th. The left-hand panel has higher resolution for higher values and the right-hand panel, with the horizontal axis on the multiplicative scale, for smaller values of eHI. The horizontal coordinates of the left-hand limits vary a great deal because there is a considerable uncertainty about the second percentile. The uncertainty about the ratios of low percentiles is also considerable. We can conclude with confidence that the ratio is greater than 1.20 for percentiles 2–8, because all the corresponding bootstrap percentile ratios are greater. However, the ratio of the second percentiles could be anywhere in the range (1.40, 1.75). The plausible ranges of the ratios for the percentiles between 10 and 15 are much narrower, containing much of the interval (1.17, 1.30), then getting narrower further, to (1.19, 1.24) around the 60th percentile (around 19 000 Euro) and then widening again towards covering the entire range (1.20, 1.30) for the 90th percentile. Past the 90th percentile, the range of plausible values (or the confidence interval) is wider, about as wide as for the smallest percentiles. For example, we can conclude that the ratio of the 95th percentiles is very likely in the range (1.21, 1.34).

A 'cleaner' summary of the estimates of the ratios of the percentiles is given in Table 2.1. The table lists the estimated percentiles 2–9 and the

TABLE 2.1
Estimates of the ratios of percentiles of eHI for Austria, 2010 vs. 2004.

	Percentile								
	2	3	4	5	6	7	8	9	10
Estimate	1.515	1.404	1.356	1.310	1.280	1.250	1.233	1.226	1.218
St. error	0.085	0.058	0.043	0.037	0.032	0.027	0.027	0.028	0.026
	20	30	40	50	60	70	80	90	95
Estimate	1.212	1.214	1.202	1.212	1.214	1.218	1.222	1.250	1.276
St. error	0.019	0.014	0.015	0.014	0.014	0.014	0.016	0.025	0.031

estimated deciles. For percentiles $10-90$ that are not deciles, we can reliably approximate the estimates and standard errors from their counterparts for the deciles by linear interpolation. The figures for the 95th percentile are added to indicate the increasing uncertainty for the highest percentiles. The estimate for the 98th percentile is 1.278, with estimated standard error 0.056.

2.4 Horvitz–Thompson Estimator

Historically, a lot of emphasis in methods for survey analysis was placed on unbiased estimation of population quantities, such as the population total $Y_+ = Y_1 + Y_2 + \cdots + Y_N$. The Horvitz-Thompson estimator (HT; Horvitz and Thompson, 1952) is an example of an unbiased estimator.

Let π_k be the probability of inclusion of member k in the sample. It is the expectation of the inclusion indicator I_k; $\mathrm{E}(I_k) = \pi_k$. Any linear combination of the sample values $s = c_1 y_1 + c_2 y_2 + \cdots + c_n y_n$, that is,

$$s = \sum_{k=1}^{N} C_k I_k Y_k,$$

has the expectation

$$\mathrm{E}(s) = \sum_{k=1}^{N} C_k \pi_k Y_k,$$

which coincides with the population total Y_+ when $C_k = 1/\pi_k = W_k$, the sampling weight of member k. The sampling weights w_j for subjects $j = 1, \ldots, n$, as the coefficients in $s = w_1 y_1 + w_2 y_2 + \cdots + w_n y_n$, define the HT estimator. More generally, a linear population summary $S_D = D_1 Y_1 + D_2 Y_2 +$

$\cdots + D_N Y_N$ is estimated without bias by

$$s_{\mathrm{D}} = \sum_{k=1}^{N} D_k I_k W_k Y_k .$$

Thus, we add the weights w_j as a factor in the sample-version of this expression, $\sum_j^n d_j w_j y_j$, and no other work is needed.

Although HT estimator can be applied to any linear population summary, it is problematic when the weights W are in a wide range, because one or a few contributions $w_j y_j$ may exert undue influence on the sample total s. Denote by π_{kh} the probability that members k and h are both included in the sample, with the convention that $\pi_{kk} = \pi_k$. The sampling variance of HT estimator is

$$
\begin{aligned}
\mathrm{var}(s) &= \sum_{k=1}^{N} \sum_{h=1}^{N} (\pi_{kh} - \pi_k \pi_h) \, W_k W_h Y_k Y_h \\
&= \sum_{k=1}^{N} \sum_{h=1}^{N} \pi_{kh} \, W_k W_h Y_k Y_h - Y_+^2 .
\end{aligned}
\tag{2.4}
$$

When the events of inclusion of a member into the sample are independent, $\pi_{kh} = \pi_k \pi_h$ for $k \neq h$, this reduces to

$$\mathrm{var}(s) = \sum_{k=1}^{N} (W_k - 1) Y_k^2 ,$$

for which HT estimator can be applied again, as

$$\widehat{\mathrm{var}}(s) = \sum_{j=1}^{n} w_j (w_j - 1) y_j^2 .$$

Note that these expressions do not involve the 'corrected' terms $(y_j - \bar{y})^2$. In the sampling designs used in the European Union Statistics on Income and Living Conditions (EU-SILC), $\pi_{kh} \neq \pi_k \pi_h$. We do not want to estimate the double summation in Equation (2.4) because its estimator would contain too many terms; also the probabilities π_{kh} are usually not known or the errors in their estimation may accumulate in Equation (2.4). Some simpler expressions for $\mathrm{var}(s)$ and $\widehat{\mathrm{var}}(s)$ are obtained when the sample size n is fixed, but this assumption is not realistic. See Särndal, Swensson and Wretman (1992) for details and numerous extensions.

For estimating the poverty rate, we have to improvise further, because the median eHI is estimated, and so the assigned status, poor or not poor, is not the same across replications for a given member. By ignoring this issue, we obtain the standard error 0.591% for the poverty rate with the threshold $\tau = 60\%$ in Austria in 2010. This is higher than its bootstrap counterpart, by about 20%, contrary to what we anticipated. For detailed studies of this and related issues, see Deville (1999) and Osier (2009).

In practice, the weighted estimator

$$s = \frac{\sum_{k=1}^{N} I_k W_k Y_k}{\sum_{k=1}^{N} I_k W_k} \tag{2.5}$$

is often applied. It is also known as a ratio estimator. Each observation y_j is assigned its sampling weight and the estimator is 'standardised' by dividing by the sample total of the weights. The properties of this estimator are not easy to derive analytically because it is a ratio of two (correlated) random variables. The expectation of the denominator is the population size N. If its sampling variation can be ignored, then s is approximately unbiased and for the sampling variance we obtain an expression similar to the variance of HT estimator. We use estimators of the form of Equation (2.5) throughout the volume.

Horvitz-Thompson estimator is unbiased, but its variance is very large when the weights are widely dispersed. In modern survey analysis, biased estimators are often applied, when they are more efficient—when they have smaller mean squared errors (MSE). Such estimators require some additional information about the population, often in the form of a model (Särndal, Bengtsson and Wretman, 1992), and have the desirable properties only when certain conditions, however mild, are satisfied or are palatable. Arranging for such information and exploiting it effectively are at the core of modern survey design and analysis.

Estimation with criteria different from MSE, using asymmetric loss functions, is described in Longford (2013b).

2.5 Fragility of Unbiasedness and Efficiency

In the context of a dataset collected by a particular design and a target to be estimated, there may be several alternative estimators, and the analyst's choice is guided by an adopted criterion for the quality of an estimator. The default choice for the criterion is minimum MSE. This is often interpreted as the combined criterion of no bias and minimum variance. The argument that a bias could always be corrected for is in general not valid, because the bias is usually unknown and has to be estimated. Suppose estimator $\hat{\theta}$ is biased for θ and $\widehat{\Delta}$ is an unbiased (and efficient) estimator of the bias $\mathrm{E}(\hat{\theta}) - \theta$. Then $\hat{\theta} - \widehat{\Delta}$ is unbiased for θ, but its MSE, equal to its sampling variance $\mathrm{var}(\hat{\theta} - \widehat{\Delta})$ may exceed the variance of $\hat{\theta}$, and even the MSE of $\hat{\theta}$, equal to $\mathrm{var}(\hat{\theta}) + \{\mathrm{E}(\widehat{\Delta})\}^2$. By 'correcting' an estimator for its bias we may inflate its variance so much that the resulting estimator, though unbiased, is less efficient than the original biased estimator. Similarly, restricting our search to unbiased estimators is a poor strategy because we may reject a biased estimator that has smaller MSE than the minimum-variance unbiased estimator.

The properties of no bias and efficiency are retained by linear transformations. Suppose $\hat{\theta}$ is unbiased for θ. Then $a + b\hat{\theta}$ is unbiased for $a + b\theta$ for any fixed scalars a and b, trivially so for $b = 0$. Similarly, if $\hat{\theta}$ is efficient for θ, then $a + b\hat{\theta}$ is efficient for $a + b\theta$. However, nonlinear transformations spoil these properties. We refer to this feature as fragility—if we want the original claims made about an estimator to remain valid, we should treat it only by linear transformations, akin to the label *'Do not fold! Do not bend!'*, stuck on a posted envelope. By way of an example, suppose $\hat{\theta}$ has the distribution $\mathcal{N}(\theta, \sigma^2)$ and it is efficient for θ. Suppose σ^2 is known. Then $\hat{\theta}^2$ is not a good estimator of θ^2. Its MSE is

$$\mathrm{MSE}\left(\hat{\theta}^2\right) = \theta^2 + \sigma^2.$$

In this case, the bias can be corrected by subtracting σ^2 from the naive estimator $\hat{\theta}^2$. This estimator, $\hat{\theta}^2 - \sigma^2$, is clearly more efficient than $\hat{\theta}^2$. When σ^2 is not known and is estimated, the bias is corrected by $\hat{\theta}^2 - \hat{\sigma}^2$ with an unbiased estimator of σ^2, but the properties of this estimator are more difficult to establish because they depend on the properties of $\hat{\sigma}^2$.

A common example of incomplete understanding of the fragile nature of unbiasedness and efficiency is in the estimation of a sampling variance, for instance, in ordinary regression. Whilst we insist on unbiased estimation of the residual variance, by $\hat{\sigma}^2$ with denominator $n - p$, where n is the sample size and p the number of regression parameters, we ignore an estimator that is more efficient, with denominator $n - p + 2$. We are more comfortable with the standard error σ because it is defined on the same scale as the outcomes. But we estimate it naively, as $\sqrt{\hat{\sigma}^2}$, even though the resulting estimator is neither unbiased nor efficient. Longford (2013a) elaborates on this issue and proposes criteria other than minimum MSE for estimating σ^2 and its common transformations.

2.5.1   Lognormal Distribution

We illustrated in Figure 1.2 that the log-transformed income (log-income) has a distribution close to normal. Exponentiation and log transformation are frequently applied in studies of income. Models are fitted to log-income, and for more convenient interpretation we transform fitted quantities back to the original monetary scale. Note that good properties of the fitted values are undermined by such transformations.

A lognormal distribution is defined by the exponential transformation of a normally distributed variable. We denote it by $\mathcal{LN}(\mu, \sigma^2)$, where μ and σ^2 are the respective mean and variance of the original normal distribution. If the distribution of a variable X is $\mathcal{N}(\mu, \sigma^2)$, then $X - \mu$ has distribution $\mathcal{N}(0, \sigma^2)$. Therefore, the expectation and variance of $\exp(X)$ are $\exp(\mu)E(\sigma^2)$ and $\exp(2\mu)V(\sigma^2)$, where the functions E and V do not depend on μ. In Appendix 2.A.2, we derive the identities $E(\sigma^2) = \exp(\frac{1}{2}\sigma^2)$ and $V(\sigma^2) = \{\exp(\sigma^2) - 1\}\exp(\sigma^2)$.

Suggested reading

Atkinson and Marlier (2010) present extensive analyses of recent EU-SILC data. For the reader interested in Bayesian analysis, which implements a paradigm different from the frequentist (replication-based), we recommend Gelman *et al.* (2003). More details about the lognormal distribution can be found in Crow and Shimizu (1998).

2.A Appendix

2.A.1 Bootstrap

We describe the function for bootstrapping the estimator of the poverty rate. It can serve as a template for bootstrap estimation of other targets. The function involves data input, so its arguments are the country's acronym and year (mandatory) and the directory and file extension (optional). Further arguments are the threshold τ and the number of bootstrap replicates, both optional.

```
EUBoots <- function(Cou, yr, dire="~/Splus/EUsilc12/DATA/", ext="dta",
                    STD=60, nbr=100)
{
###    Bootstrap of the poverty rate estimator

## Cou     The country's name
## yr      The year
## dire    The data directory
## ext     The data file extension
## STD     The threshold
## nbr     The number of bootstrap replications
```

The threshold can be given as a percentage or a probability.

```
## Check on the threshold  --  has to be a scalar fraction
STD <- STD[1]
while (STD > 1.5) STD <- STD/100
```

The data input (here for households)

```
##  Data input  --  household level
dat <- EUread(dire,Cou,yr,ext, VRS=c("hx040","hx090","db090"), IH=2)
```

is followed by estimation of the median eHI and evaluation of the poverty rate estimator.

```
##  The weights for the individuals
weiP <- dat[, "db090"] * dat[, "hx040"]
```

```
##  The median equivalised income  --  individuals
MDi <- QuantW(dat[, "hx090"], weiP)
```

```
##  The poverty rate
PRTb <- sum((dat[, "hx090"] < STD * MDi) * weiP) / sum(weiP)
```

To the object PRTb, a scalar at this point, we attach the bootstrap replicate estimates.

```
##  Revert to the household-level weights
weiP <- dat[, "db090"]
```

```
##  Bootstrap replications
PRTb <- c(PRTb, replicate(nbr,
{
##  The bootstrap subjects (sample)
Sbj <- EUbts(weiP)
```

```
##  The household sizes used as the weights
weiB <- dat[Sbj, "hx040"]
```

```
##  The bootstrap median income (with equal weights)
MDc <- QuantW(dat[Sbj, "hx090"], weiB)
```

```
##  The bootstrap estimate
sum((dat[Sbj, "hx090"] < STD * MDc) * weiB)/ sum(weiB)
}))  ## End of the replicate expression, function, and  c(...)
}    ##  End of function  EUBoots
```

The (system-defined) function **replicate** executes its argument (an expression) the specified number of times, and collects the results of the final evaluation of the expression. In this application, the argument is a set of three expressions, enclosed in braces, concluded with a bootstrap replicate estimate of the poverty rate. The braces are essential; without them the second expression (weiB <- ...) would be interpreted as being outside the expression-argument of **replicate**, and a syntax error would result.

The function EUbts draws a bootstrap sample; its sole argument is the vector of sampling weights.

```
EUbts <- function(wei)
{
```

```
###    Selection of the bootstrap sample (weights)
##  wei    The sampling weights

##  The effective sample size   (randomly rounded)
n <- sum(wei)^2/sum(wei^2)
n <- floor(n) + (runif(1) > (n%%1))

##  Draw a sample
sample(seq(length(wei)), n, replace=T, prob=wei)
}  ##  End of function EUbts
```

The effective sample size, n, is determined first, and then a sample is drawn with replacement, with the probabilities proportional to the (sampling) weights. The result is a set of n indices. Note that the bootstrap selects households (with weights given by variable db090), but in the replicate estimate the household size is used as the weight, so that the estimate refers to individuals.

The following expression applies the function EUBoots to Austria in 2010. The result is a vector of 1001 values, comprising the realised value followed by 1000 bootstrap replicate estimates. The values are multiplied by 100 to be in percentages.

```
Time <- proc.time()

##  Application  (converted to percentages)
ATBoot10 <- EUBoots(Cou="AT", yr=10, nbr=1000) * 100

Time <- proc.time() - Time
```

The system-defined function proc.time returns the user (CPU), system and elapsed times, in seconds, since the beginning of the R session. The difference, Time, evaluated by the code is used like a stopwatch that measures the time taken by the execution. It recorded 4.8 sec. of CPU time. The bootstrap estimates are summarised by the function MeVa, which returns the sample mean and standard deviation; see Appendix 1.A.4. We have to drop the first element of EUBootsR from its argument:

```
ATBoot10R <- round(MeVa(ATBoot10[-1]), 3)
```

The result is

```
    Mean   St.dev
   12.099   0.496
```

The bias of the estimator is estimated by

```
ATBoot10R[1] - ATBoot10[1]
```

yielding −0.021. Note that it is subject to randomness; a repeated application yields a (slightly) different result.

2.A.2   Moments of the Lognormal Distribution

The expectation of the lognormal distribution $\mathcal{LN}(0, \sigma^2)$ is

$$\frac{1}{\sigma\sqrt{2\pi}} \int_{-\infty}^{+\infty} \exp(x) \exp\left(-\frac{x^2}{2\sigma^2}\right) dx$$

$$= \exp\left(\frac{1}{2}\sigma^2\right) \int_{-\infty}^{+\infty} \frac{1}{\sigma\sqrt{2\pi}} \exp\left\{-\frac{1}{2\sigma^2}(x-\sigma^2)^2\right\} dx, \qquad (2.6)$$

after completing the square for the arguments of the exponentials in the integrand. The integrand in the second line is the density of $\mathcal{N}(\sigma^2, \sigma^2)$, so its integral is equal to unity. Therefore, the expectation of $\mathcal{LN}(0, \sigma^2)$ is equal to $\exp(\frac{1}{2}\sigma^2)$.

For the variance, we use the formula $\text{var}(X) = \text{E}(X^2) - \{\text{E}(X)\}^2$. We have,

$$\text{E}(X^2) = \frac{1}{\sigma\sqrt{2\pi}} \int_{-\infty}^{+\infty} \exp(2x) \exp\left(-\frac{x^2}{2\sigma^2}\right) dx$$

$$= \exp(2\sigma^2) \int_{-\infty}^{+\infty} \frac{1}{\sigma\sqrt{2\pi}} \exp\left\{-\frac{1}{2\sigma^2}(x-2\sigma^2)^2\right\} dx,$$

and the latter integral is equal to unity because the integrand is the density of $\mathcal{N}(2\sigma^2, \sigma^2)$. Hence the variance of $\mathcal{LN}(0, \sigma^2)$ is

$$\text{var}(X) = \exp(2\sigma^2) - \exp(\sigma^2) = \exp(\sigma^2)\left\{\exp(\sigma^2) - 1\right\}.$$

The expectation and variance of the general lognormal distribution $\mathcal{LN}(\mu, \sigma^2)$ are $\exp(\mu + \frac{1}{2}\sigma^2)$ and $\exp(2\mu + \sigma^2)\{\exp(\sigma^2) - 1\}$, respectively.

The partial expectation of the lognormal distribution is used in Section 3.3.

$$\int_0^y u f(u; \sigma^2) \, du,$$

where f is the density of $\mathcal{LN}(0, \sigma^2)$, is derived similarly to Equation (2.6). After the transformation $x = \log(u)$, and by reorganising the arguments of the exponentials, we obtain

$$\frac{1}{\sigma\sqrt{2\pi}} \int_{-\infty}^{\log(y)} \exp(x) \exp\left(-\frac{x^2}{2\sigma^2}\right) dx$$

$$= \exp\left(\frac{1}{2}\sigma^2\right) \Phi\{\log(y); \sigma^2, \sigma^2\}.$$

3

Poverty Indices

This chapter deals with some common summaries of low income in a country that complement the poverty rate. First we discuss summaries that weigh the extent of shortage, then summaries that are concerned entirely with income inequality, and conclude with summaries motivated by the exploration of the weights in Section 1.4.

3.1 Poverty Index

We use the term 'index' for any univariate (single-number) summary of a variable. In the literature, 'indicator' is sometimes used for this purpose (e.g., Atkinson *et al.*, 2002), but we reserve the latter term for binary variables, with values 0/1 or False/True, which *indicate* positiveness of a variable, belonging to a category, and the like. A poverty index is any univariate summary of the extent of poverty, defined by the shortfall of equivalised household income (eHI) with respect to the adopted standard. A value of eHI below the standard indicates poverty. With a given standard S, the shortfall is defined as the difference $S-$ eHI if eHI $\leq S$ and as zero otherwise. The notation $(S-$eHI$)_+$, that is, the positive part of $S-$ eHI, can be used. Denote this variable by T, and its values on the realised sample by t_j, $j = 1,\ldots,n$.

A wide class of indices is defined by the formula

$$\frac{1}{n}\sum_{j=1}^{n}g(t_j) \tag{3.1}$$

where g is a nondecreasing function such that $g(0) = 0$. We refer to g as the *kernel*, to the value t_j as the *score* of unit (individual or household) j, and to $g(t_j)$ as the transformed score. For example, the poverty rate is an index based on the indicator of shortfall; $g(t) = 1$ for $t > 0$, and $g(t) = 0$ otherwise. The mean in Equation (3.1) is intended as an estimator of its population version

$$\frac{1}{N}\sum_{i=1}^{N}g(T_i),$$

that is, the expectation of $g(T)$, $\mathrm{E}\{g(T)\}$. Key properties of a useful kernel are *universality* and *additivity*. Universality means that any two units with the same transformed score $g(t)$ are regarded as identical for all purposes for which the index is designed and used. Additivity means that transformed scores $g(t_1)$ and $g(t_2)$ together amount for all purposes to the same quantity as $g(t_3)$ and $g(t_4)$ when $g(t_1) + g(t_2) = g(t_3) + g(t_4)$. That is, the transformed score is an appropriate currency for the purpose of the index (quantification of poverty). Without these properties, averaging (or taking expectations) of the transformed scores is not meaningful. We consider only kernels g that are nonnegative and such that $g(t) = 0$ corresponds to the absence of poverty. The scoring scheme defined by a kernel g is additive if and only if $ag(t)$ is additive for any positive constant a; the kernels form classes of equivalence. Any two kernels in the same class are such that one is a scalar multiple of the other. Note that the same factor a has to be used for all units.

For different purposes (perspectives, priorities, frameworks, or the like), different kernels may be additive. The use of a kernel g should be delineated accordingly. Deciding which function is appropriate (additive) for a particular perspective is a nontrivial task. It would seem obvious that we should discard the indicator function as a kernel, and therefore the poverty rate as an index, if we want to distinguish between a smaller and a greater shortfall. An obvious alternative is the identity kernel, $g(t) = t$, also called the linear kernel. Are the consequences of the shortfall of 1000 Euro the same as of the shortfall of 400 Euro and another of 600 Euro? If the answer to this and similar questions is affirmative, then the identity kernel is appropriate. We implement the identity kernel by truncating eHI at zero. Any negative value of eHI is redefined to zero, so that the maximum possible score is S, attained by an individual or household with negative or zero eHI. We show below on an example that this is sensible to do.

The linear poverty gap in a particular population (a country) is defined by the average of the shortfalls of its members, be it individuals or households. The members with zero score, for whom eHI $\geq S$, are counted in this average. The poverty gap is well-defined for any standard S given by a threshold τ, so we can define (linear) poverty-gap curves as functions of τ, similarly to poverty-rate curves in Section 1.2.1. Figure 3.1 displays the estimated poverty-gap curves for individuals in the 15 countries that participated in the European Union Statistics on Income and Living Conditions (EU-SILC) in 2004 and 2010. The vertical axes have the same extent and are on multiplicative scale; the resolution with it is much higher. The linear poverty-gap curves are based on the values of eHI truncated at zero. For Denmark in 2010, the poverty gap is evaluated also without truncation of eHI. The corresponding curve is drawn by dashes and marked DK_0. The curve is much flatter than its counterpart with truncation or the curves for all the other countries, some of which have no negative values of eHI in the sample. It seems therefore appropriate to truncate the negative values. The difference for Denmark is so large because the data contain many negative values of eHI, some of them quite substantial. In fact, 0.8% of the population is estimated to have negative income in Den-

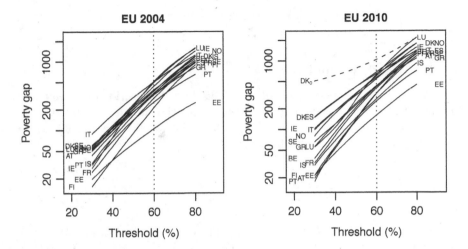

FIGURE 3.1
National poverty gaps, defined by the identity (linear) kernel, as functions of
the poverty threshold (individual level); 2004 and 2010.
The dashes indicate the mean poverty gap for Denmark in 2010 evaluated
with eHI not truncated at zero.

mark in 2010. In the data, 63 households, with 130 individuals in total, have
negative income. For nine households, eHI is smaller than $-100\,000$ Euro, and
the smallest (largest negative) value is nearly $-400\,000$ Euro. Only Spain has a
comparable frequency of negative values of eHI (national estimate 1.1%), but
the values are much less extreme than in Denmark; the smallest value of eHI
is $-42\,600$ Euro. The estimated rate of negative eHI in 2010 is smaller than
0.25% in all the other thirteen countries. In 2004, the highest estimated rate of
negative eHI was in Greece, 0.69%, followed by Estonia (0.45%) and Denmark
(0.42%). The lowest values of eHI in Greece were around $-60\,000$ Euro (two
households) and $-40\,000$ Euro in Denmark (also two households). The lowest
value in 2004, by a wide margin, was recorded in Belgium, $-167\,800$ Euro. See
Appendix 3.A.1 for the code on which this exploration is based.

Estonia had the lowest estimated poverty gap for $\tau > 40\%$ and Luxembourg and Norway had the highest poverty gaps for $\tau > 60\%$ in both years.
The curves intersect a lot for both years and the countries are ranked differently at the two extreme thresholds, $\tau = 30\%$ and $\tau = 80\%$. The ratio of the
estimated poverty gaps at 80% and 30% in 2004 is only 12.7 for Estonia and
13.9 for Italy, but 54.4 for Ireland and 57.6 for Finland. The poverty gap has
to be qualified by the threshold τ.

With a few exceptions, the poverty gaps at $\tau = 80\%$ are aligned according
to their average (or median) eHI; the Scandinavian countries with Luxemburg
and Ireland have the highest mean poverty gaps and Greece, Portugal and
Estonia the lowest. As a conclusion about the levels of poverty in the studied

countries, this is not credible. We resolve this apparent contradiction in Section 3.2.

3.1.1 Which Kernel?

In a particular perspective, small shortfalls have less serious consequences than what their magnitude might suggest. Thus, the shortfall of 1000 Euro is more serious than the aggregate of the shortfalls of 400 Euro and 600 Euro. This perspective is catered for by a convex kernel. A function g defined on an interval is said to be convex if

$$\frac{g(a) + g(b)}{2} \geq g\left(\frac{a + b}{2}\right)$$

for any two values a and b in the interval. The quadratic function, $g(t) = t^2$, is convex. Used as the kernel, the transformed score for the shortfall of 1000 Euro is 1 million, whereas for 400 and 600 Euro it adds up to only slightly over half a million. Maybe squaring the shortfall is too radical a transformation. We moderate the discord by considering a continuum of kernels,

$$g_r(t) = t^r$$

for $r \in (1, 2)$, although in principle any positive exponent r could be adopted.

Suppose we agree that the shortfall of 1000 Euro is 25% more serious than the total of the shortfalls of 400 and 600 Euro. So, we are looking for the exponent r for which

$$\frac{1000^r}{400^r + 600^r} = 1.25. \tag{3.2}$$

We could proceed by trial and error. For $r = 1.25$, chosen as a guess, we obtain the transformed scores of 5623.4 for 1000 Euro and 4758.4 for the aggregate of the shortfalls of 400 and 600 Euro. The ratio of these values is 1.18, suggesting that r should be increased slightly.

A more elegant solution is provided by the Newton (linearisation) method. Let t_A and t_B be two initial guesses of the solution of the equation $f(t) = 0$ for a continuous function f. Suppose $t_A < t_B$. If the function $f(t)$ were linear in the range (t_A, t_B), then the solution of the equation would satisfy the identity

$$\frac{f(t_A)}{t_A - t} = \frac{f(t_B)}{t_B - t}.$$

By solving it for t we obtain an (improved) approximation to the solution:

$$t_C = \frac{t_A f(t_B) - t_B f(t_A)}{f(t_B) - f(t_A)}. \tag{3.3}$$

We iterate this identity till convergence. That is, we apply it with t_B and t_C relabelled as t_A and t_B, respectively, obtaining an updated value of t_C, and

TABLE 3.1
Record of convergence of the Newton method.

Iteration	t_A	t_B	$f(t_A)$	$f(t_B)$	Conv. crit.
0	1.0000	2.0000	−0.2500	0.6731	0.0000
1	2.0000	1.2708	0.6731	−0.0518	0.1074
2	1.2708	1.3229	−0.0518	−0.0098	1.2086
3	1.3229	1.3350	−0.0098	0.0002	1.9095
4	1.3350	1.3348	0.0002	0.0000	3.6332
5	1.3348	1.3348	0.0000	0.0000	6.0815
6	1.3348	1.3348	0.0000	0.0000	10.2390

Note: Solution of Equation (3.2) by iterations of Equation (3.3).

keep doing so until both $|t_B - t_A|$ and $f(t_B)$ become sufficiently small. This is implemented in R function EUnewton. See Appendix 3.A.2 for details.

A complete record of the iterations is listed in Table 3.1. We start with $t_A = 1.0$ and $t_B = 2.0$; we do not have to be more sophisticated with the initial setting. The next approximation is $t_C = 1.271$, with $f(t_C) = -0.052$, much closer to zero than $f(t_A) = -0.250$ or $f(t_B) = 0.673$, and one or two more iterations would suffice for almost any purpose. We set the required precision to eight decimal places, an apparent overkill, but even this precision is achieved after only six iterations. The solution is $x^* = 1.335$; we may settle for $\frac{4}{3}$. In the table, the decimal digits after the second are superfluous for any practical purpose, but we want to demonstrate the fast and unambiguous convergence.

Figure 3.2 illustrates the progression of the approximations. For a pair of provisional solutions, we join their points $\{t, f(t)\}$ by a straight line, and the new approximation is defined by the intersection of this line with the horizontal axis. The diagram confirms that the adjustments (corrections) after the third iteration are minute. If we started with approximations $t_A = 1.0$ and $t_B = 1.25$, we would have saved only one iteration. We can be quite indiscriminate about the choice of t_A and t_B, because the function f is 'well behaved', without any sudden changes of curvature. Over the plotted range, f does not differ substantially from a linear function. If it were linear, the exact solution would be found by a single application of Equation (3.3).

In a more profound exercise of choosing a kernel g, we would construct several examples and ask one or several experts (a commission) to compare the aggregate severities of poverty in pairs or collections of hypothetical households, adjust them so that they would be judged as (approximately) equal by the agreed criterion, and thus hone in on a suitable exponent r, or a narrow range of its plausible values. Of course, we may find that none of the power transformations lead to additivity. Also, we may hedge our bets and work with a range of exponents and address any inconsistencies in the results by

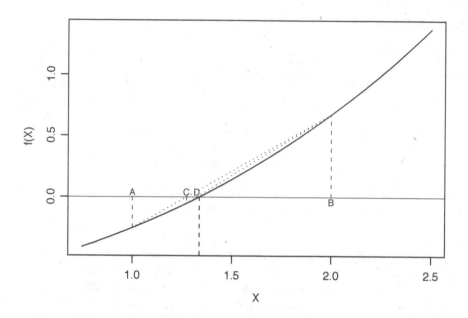

FIGURE 3.2
Newton method.
Illustration for solving Equation (3.2).

declaring an impasse. Working with poverty-gap curves instead of the poverty gap for a single threshold τ is another application of this general principle.

The class of kernels (transformations) can be extended by linear combinations. For exponents $r_1 > 0$ and $r_2 > 0$ and a factor $b > 0$, the transformation $t^{r_1} + bt^{r_2}$ is also a kernel. Finding the parameters r_1, r_2 and b is a somewhat more involved problem, requiring much more input (at least three examples of sets of shortfalls adjudged to be equivalent), but if an exact analytical solution fails or we give up on it, setting the parameters by trial and error can be just as effective. However, the rewards for such generality are meager. Figure 3.3 displays the poverty gap curves for the 15 countries in 2010 for exponents $r = 1.0, 1.1, \ldots, 1.5$ and shows that their relative sizes and shapes depend on r very weakly. Concern about the exact value to which r should be set is not well-justified.

For small thresholds, Denmark has the highest poverty gap for all exponents in the diagram. In Figure 1.5, we noted that its poverty rate for $\tau = 30\%$ is more in line with some Mediterranean countries than the more prosperous countries of northern and western Europe. These estimates are clearly skewed by the relatively high frequency of zero and negative income in Denmark in 2010. For higher exponent r the influence of these values is even stronger. The numerous zero values may be an artefact of how income data is collected.

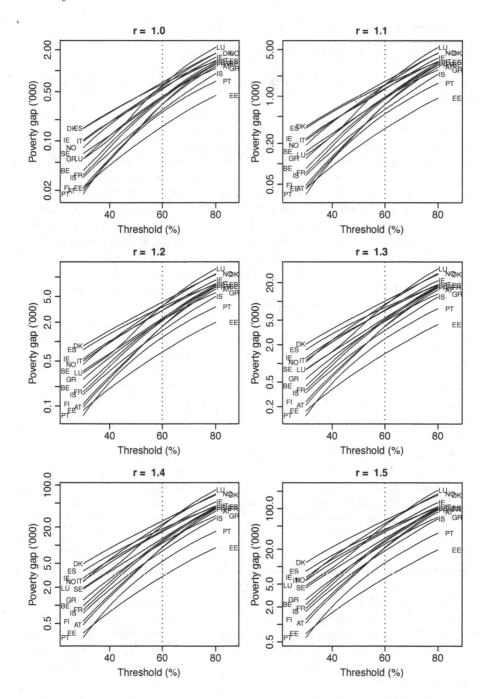

FIGURE 3.3
National poverty-gap curves for a range of exponents r; year 2010.

3.2   Relative and Log-Poverty Gaps

The poverty gap regards the shortfall of one Euro in one setting on par with one Euro in another setting. It is not well-suited for comparing the extent of poverty in a country in one year with the extent a few years later. An adjustment has to be made for the inflation (or change of the standard) and we would have to enter into a difficult discussion as to which inflation figures to apply. A comparison of the extent of poverty in two countries is similarly problematic, even if the same currency is used (after a conversion, if applicable). Suppose the standard (60% of median eHI) in country A is 14 000 Euro and in country B it is 4000. The shortfall of 2000 Euro in A cannot be regarded as severe as the same shortfall in B, where it amounts to one-half of the standard. Summarising (averaging) the shortfall on the linear scale seems not to be appropriate—the linear kernel is not universal for the two domains. That is, the kernel used may be additive in both countries, but it is not additive in their union. This provides an explanation for why some countries with high poverty rates have relatively low poverty gaps.

The relative shortfall is defined as the shortfall divided (scaled) by the standard S. Thus, the (absolute) shortfall of 2000 Euro on the standard of 14 000 Euro corresponds to the relative shortfall of 14.3%, and the same shortfall on 4000 Euro to 50%. The relative poverty gap is defined as the mean relative shortfall. It can be evaluated by dividing the poverty gap by the standard. Its kernel is $g(t) = t/S$. The 'scaled' version of the kernel t^r is t^r/S^r. These kernels were originally proposed by Foster, Greer and Thorbecke (1984); see also Foster, Greer and Thorbecke (2010).

Figure 3.4 plots the estimated relative poverty gaps for the same context (countries and years) as in Figure 3.1. As anticipated, the countries with the lowest standards 'progress' in the comparisons and have higher relative poverty gaps than countries with the highest standards. Estonia's progression is extreme, from the lowest linear (absolute) poverty gap in Figure 3.1 to the highest relative poverty gap for all thresholds in 2004. We obtain the same division of the countries as for the poverty rates in Figure 1.5 (Mediterranean and Estonia vs. West and North), with Ireland straddling them.

In 2010, Spain has uniformly the highest relative poverty gap, by a wide margin. The division of the two sets of countries is much less clear in 2010, and is present only for thresholds $\tau > 60\%$. For the lowest thresholds, Denmark has a higher estimated rate than Greece and Estonia, but the differences are very small. The large number of households with negative eHI is a contributing factor for Denmark, even after truncation. Without truncation, its estimated rate would be exceptionally high for small thresholds. The corresponding curve is drawn in the right-hand panel by dashes.

The relative poverty gap increases with τ for all countries, as does the absolute poverty gap. As τ is reduced, the constituency of households that are

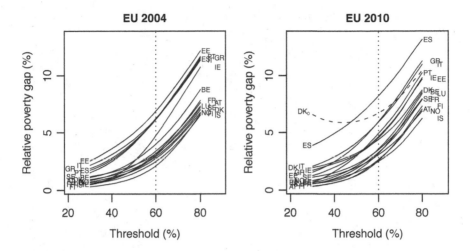

FIGURE 3.4

National relative poverty-gap curves in 2004 and 2010.
The curve for Denmark (2010) evaluated without truncation of eHI at zero is drawn by dashes.

classified as poor is reduced and their shortfalls are also reduced. However, the standard in the denominator, a percentage of the median eHI, is also reduced, and so an anomaly, whereby the relative poverty gap would be greater for a smaller threshold could occur. In fact, it would occur for Denmark if the negative values of eHI were not truncated. The decision to truncate is guided by the requirement of additivity.

Figure 3.5 compares the absolute and relative poverty gaps for Austria in years 2004–2010. The curves in panels at the top are difficult to distinguish, but their scaled versions, drawn in the bottom panels using the curves for 2004 as the basis, magnify the differences. With the exception of 2006 and 2009, the absolute poverty gaps decrease for small thresholds τ, whereas for the largest thresholds they increase. The relative poverty gaps are more closely aligned with the pattern of decreasing over time for all thresholds τ.

After scaling, the absolute and relative poverty-gap curves have similar shapes and associations. Both confirm that the poverty gap was reduced between 2004 and 2010 to about half for the lowest threshold $\tau = 30\%$. Note that scaling does not alter the multiplicative (percentage) comparisons, so long as the same basis (divisor) is used for the two years. The absolute poverty gap increased by a few percent for the highest threshold, $\tau = 80\%$. In contrast, the relative poverty gap decreased by a few percent over the six years. Higher values of eHI, and of its annual median, are the cause of this discrepancy. The advantages of scaling are obvious, but the values of the (absolute and relative) poverty gap are discarded by its application. The diagram does not

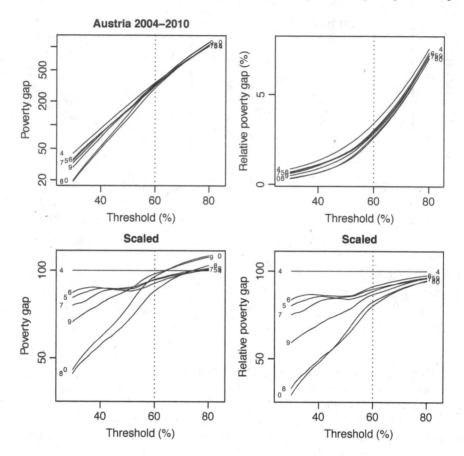

FIGURE 3.5
Absolute and relative poverty-gap curves for Austria in 2004–2010.
The panels at the bottom are the scaled versions of the panels above them,
with Austria in 2004 used as the basis.

help us to decide which index (scale) is better suited for the assessment of
poverty. The choice should be based solely on how the index is constructed,
and universality and additivity should be the principal criteria.

The relative poverty gap is unit-free. That is, if we convert the currency
used to another currency, or adjust it multiplicatively (e.g., for inflation), the
value of the relative poverty gap is not changed. Scaling has a similar property,
although it relates two currencies, one for the numerator (e.g., Austria in 2008)
and one for the denominator (Austria in 2004, the basis).

The relative version of the poverty index with exponent $r > 1$ is formed
by scaling by S^r, so that the index remains unit-free. Figure 3.3 suggests that
the choice of r is not important, say, in the range $(1.0, 1.5)$, but the relative
poverty-gap curves obtained for different values of r are not related in the

same way, because the operations of power and averaging do not commute. For instance, $(20^{1.3} + 30^{1.3})/2 \neq \{(20 + 30)/2\}^{1.3}$.

The log-shortfall is defined as $l = -\log(\text{eHI}/S)$ if $\text{eHI} < S$ and as zero otherwise. A household with no income would have $l_j = +\infty$ and would cause the evaluations of the average $\bar{l}$ to fail. This is prevented by truncating eHI at zero and adding a token Euro to each household's income. Then the largest possible value of l is $\log(S)$. Denote the value of l for subject j by l_j. For a small shortfall, when eHI is smaller than S by only a narrow margin, $\log(\text{eHI}/S) \doteq 1 - \text{eHI}/S$, which is the relative shortfall, denoted by u_j. When $u_j > 0$, $l_j = -\log(1 - u_j)$, but this identity holds also when $u_j = 0$. The log-poverty gap is defined as the average of the log-shortfalls. It is related to the geometric mean of the complements $1 - u_j$:

$$\bar{l} = -\frac{1}{n}\sum_{j=1}^{n}\log(1 - u_j) = -\log\left\{\sqrt[n]{\prod_{j=1}^{n}(1 - u_j)}\right\}.$$

The sampling weights are incorporated in the expression for $\bar{l}$ by replacing the arithmetic mean with the weighted mean. Since the arithmetic mean always exceeds the geometric mean, $\exp(\bar{l}) \geq 1 - \bar{u}$, with equality only when u_j are constant. Subjects with eHI in excess of the threshold do not contribute to $\bar{l}$. If all values of eHI are nonnegative, then the values of $\exp(-l_j)$ are in the range $(0, 1)$. The national log-poverty gaps for years 2004 and 2010 are plotted in Figure 3.6. They bear some resemblance to their counterparts in Figure 3.4. The curves for 2004 have very similar shapes except for Spain, Portugal and Ireland, which have much steeper curves than the rest. In 2010, Spain has uniformly the highest estimated log-poverty gap, followed by Denmark. Most of the curves have very similar shapes and only the curve for Portugal intersects several other curves; at $\tau = 30\%$, its log-poverty gap is the smallest, but at $\tau = 80\%$ it is the seventh smallest. In relation to the other countries, Portugal has in both years a much lower log-poverty gap than its relative poverty gap on the linear scale.

The log-poverty gap is scale free. It does not have a version that would correspond to the power kernel $g(t) = t^r$. With eHI and the standard S raised to power r, we obtain the log-shortfall $-\log(\text{eHI}^r/S^r) = rl$, in the same class of equivalence as the log-shortfall l. The log-poverty gaps derived from the power kernels are all equivalent.

3.3 Lorenz Curve and Gini Coefficient

The poverty rate and other poverty indices defined so far can be interpreted as measures of inequality (of income), because they are all related to a standard that lacks face validity as the borderline between poverty and prosperity. In a

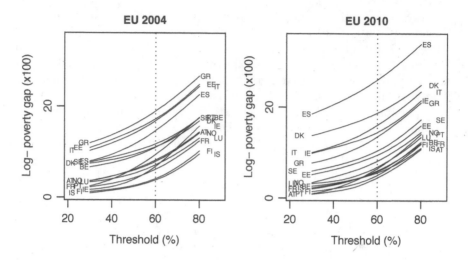

FIGURE 3.6

Mean log-poverty gaps for the countries in EU-SILC in 2004 and 2010. The values of the log-poverty gap are multiplied by 100.

country with extremely dispersed values of eHI, the poverty rate, as defined in Section 1.1, with the threshold $\tau = 60\%$ would be close to 50%. It would be equal to 50% only in the implausible case when none of the values of eHI are in the range delimited by 30th and 50th percentile of eHI.

A sample or population quantity s, derived from a single variable X, is called a measure of dispersion if its value for the linearly transformed variable $a + bX$ is equal to $|b|s$. The poverty gap is a measure of dispersion. Would any other measure of dispersion applied to eHI be suitable as an index of income inequality? The answer is negative because we are not interested in the distribution of eHI among those living in prosperity, such as households with eHI far greater than the national median.

Any difference of quantiles, such as the interquartile range, $Q(0.75) - Q(0.25)$, is also a measure of dispersion. We could define a poverty index as $Q(0.50) - Q(0.10)$, or similar, so that it would depend on large values of eHI only through the median $Q(0.50)$. Adaptations of this idea include averages of the differences $Q(0.50) - Q(p)$ for a selected set of probabilities p, all of them smaller than 0.50. Instead of differences, ratios can be considered; for example, the ratio of quintiles is defined as the ratio of the 80th and 20th percentiles of eHI. The ratio of the quartiles (percentiles 75 and 25) could be used instead. Ratios of more extreme percentiles, such as 90th and 10th, may be problematic when estimated from relatively small samples. Symmetric ratios of percentiles closer to the median, such as the ratio of the tertiles (percentiles 66.7 and 33.3), are not useful because they do not capture some important features

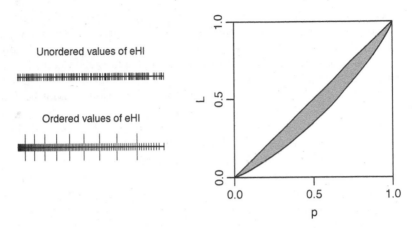

FIGURE 3.7

Lorenz curve for a set of 100 values.

The long vertical ticks on the left-hand side mark the values of the Lorenz curve at $p = 0.1, 0.2, \ldots, 0.9$.

related to poverty. However, it is meaningful to consider functions

$$R(p) = \frac{Q(1-p)}{Q(p)} \tag{3.4}$$

for p in a plausible range, such as $(0.70, 0.85)$.

The Lorenz curve (Lorenz, 1905) is based on similar ideas. We sort the values of eHI in the ascending order, as $X^{(1)} \leq X^{(2)} \leq \ldots \leq X^{(N)}$, form their cumulative totals $Y^{(1)} = X^{(1)}, Y^{(2)} = X^{(1)} + X^{(2)}, \ldots, Y^{(N)} = X^{(1)} + X^{(2)} + \cdots + X^{(N)}$, and for each probability p establish the value $Y^{(pN)}$, with an appropriate arrangement when pN is not an integer. The Lorenz function (or curve) at a probability p is defined as the ratio $L(p) = Y^{(pN)}/Y^{(N)}$, that is, the fraction of the national total of eHI earnt by the poorest $100p\%$ of the units (individuals or households). Apart from the illustration that follows, all the examples are for the estimated Lorenz curve for the individuals, associating each individual with his or her household's value of eHI.

Figure 3.7 illustrates the Lorenz curve. The values of eHI for a set of 100 households, generated by a random process, are marked by short vertical ticks at the top and are reproduced after being sorted in the ascending order underneath. The long ticks mark every 10th household, and the corresponding relative distances from the left-hand limit are the values of the Lorenz curve at the deciles, $p = 0.1, 0.2, \ldots, 0.9$.

The Lorenz curve is plotted in the right-hand part of the diagram together with the identity line. Obviously, $L(p) \leq p$ in general, and equality occurs only when every household has the same value of eHI. At the other extreme, the Lorenz curve is equal to zero for all $p \in (0, 1)$ when all income is earnt by a single household, an equally unrealistic scenario. Lorenz curves have a partial ordering. If the curve for one population (country in a year) is uniformly smaller than or equal to the curve for another, $L_1(p) \leq L_2(p)$ for all $p \in (0, 1)$, with inequality for at least some p, then the income in the former is said to be more unequal. When the curves intersect, the populations cannot be compared by this criterion.

The Gini coefficient (Gini, 1912) is defined as the area between the Lorenz curve and the identity line, scaled so that its values would be between zero and 100, to have the appearance of a percentage. The area on which the Gini coefficient is based is indicated in Figure 3.7 by shading. In fact, the Gini coefficient is defined as the 200-multiple of the area. In practice, the coefficient is estimated by its sample version, using numerical integration. See Appendix 3.A.4 for details. The Gini coefficient in the example in Figure 3.7 is equal to 0.2166. The Gini coefficients for two populations are compared straightforwardly.

Figure 3.8 displays the Lorenz curves for the countries in EU-SILC in 2004 and 2010. In the insets in the right-hand panels, the curves for the countries are easier to distinguish. The curves have very similar shapes; they intersect in a few instances, but at very acute angles. This suggests that comparisons by their Gini coefficients are appropriate. The quadratic function p^2 is drawn by dots for future reference.

The diagram shows that the Lorenz curves have not changed substantially over the six years. The prosperous countries of northern and western Europe have uniformly lower Gini coefficients than the countries in the Mediterranean and Estonia. In 2004, Estonia and Portugal stand out even from the other three Mediterranean countries and Ireland, and in 2010 the differences among the countries are reduced.

Figure 3.9 compares the estimates of the Gini coefficients more directly. It confirms the clear divide between the Scandinavian and west European countries (excluding Ireland) as one group and the Mediterranean countries (with Ireland and Estonia added to them) as the other. France is in-between the two groups. The Gini coefficient increased over the six years most in Spain and decreased most in Estonia and Portugal.

An unsatisfactory feature of Figure 3.8 is that differences that seem to be of substantive importance are difficult to discern in the diagram. Although the countries can be identified in the accompanying insets, a lot of detail is not presented with clarity. We indicated in the diagram that the quadratic function p^2 differs from the Lorenz curves by small quantities throughout. Therefore, by plotting the functions $p^2 - L(p)$ we can improve the resolution substantially and distinguish the national Lorenz curves much better. This is done in Figure 3.10. The countries are listed (from left to right) in the descending order of their values of $L(0.5)$, that is, ascending in $p^2 - L(p)$ for $p = 0.5$. From the position of their labels we can infer a departure from the shape of the curve for one country from the others. For example, Spain has

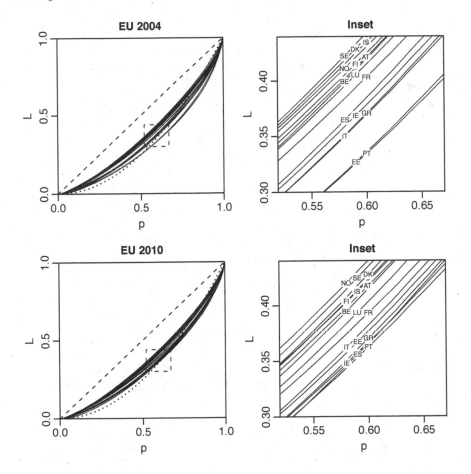

FIGURE 3.8
Lorenz curves for the countries in EU-SILC in 2004 and 2010.

the highest estimated value of the curve at $p = 0.5$ in 2010, but for p around
0.8 the values for Portugal, Ireland and Greece are higher. Note that higher
location corresponds to smaller value of the Lorenz curve, and is therefore
associated with higher values of the Gini coefficient. The resolution of the
diagram can be improved further by plotting the differences of the Lorenz
curves against the curve for one of the countries; that is, by using the Lorenz
curve for a country (and year) as the basis instead of the function p^2.

The Lorenz curve and the Gini coefficient can be distorted by a single
extreme (very high) value; a single person (or household) may introduce a
substantial change in both summaries. This problem can be explored by infiltrating the sample for a country and year by a single exceptional observation.
Figure 3.11 presents the Lorenz curves for a set of such samples for Austria

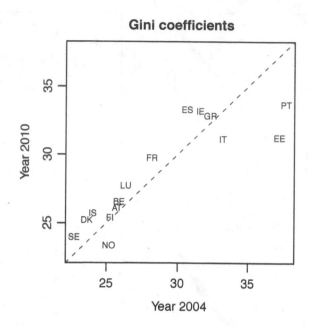

FIGURE 3.9

Gini coefficients for the countries in EU-SILC in 2004 and 2010.

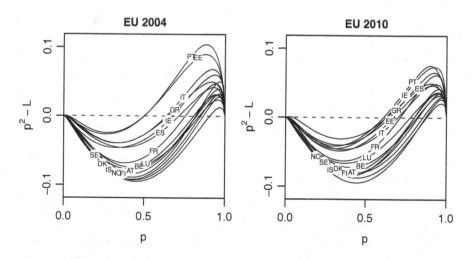

FIGURE 3.10

Transformed Lorenz curves; EU-SILC in 2004 and 2010.

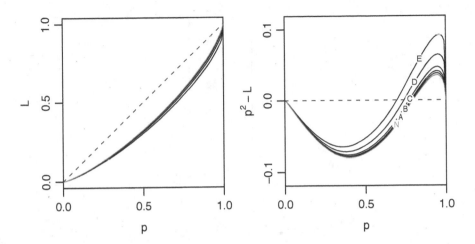

FIGURE 3.11
Lorenz curves for Austria in 2010, with data infiltrated by an extreme observation.
The Lorenz curves are: N: with no infiltration, A−E: with inflitration by an
observation with eHI equal to 0.5, 1, 2, 5 and 10 million Euro, respectively,
and weight 1339.

in 2010. The curve with no infiltration is drawn by gray colour and marked
by N, and the curves marked A − E are the (transformed) Lorenz curves evaluated after infiltrating the sample with a household with eHI equal to 0.5, 1,
2, 5 and 10 million Euro. The highest value of eHI in the dataset proper is
205 000 Euro. We set the weight of the infiltrated observation to 1339. The
value was chosen as the average of the products of the sampling weights and
household sizes, so that it would represent the typical collective weight of the
members of a household.

Curve A can be distinguished from curve N only in the vicinity of $p = 1.0$;
its Gini coefficient, 26.34, is greater than for curve N by 0.24. The other
curves have Gini coefficients 26.59, 27.09, 28.54 and 30.08 for the respective
inflitrations B − E, increasing with the infiltrating amount of eHI; relate these
figures with the Gini coefficients in Figure 3.9.

The influence of the highest values of eHI is undesirable because they
are related to poverty only peripherally. We can eliminate their influence by
evaluating the Lorenz curve and the Gini coefficients for the subset of the
subjects (or the subpopulation) with eHI below a certain limit, such as a
given multiple of the national median. As an alternative, the values of eHI
can be truncated at such a limit.

The lack of robustness discovered by infiltration can be put into perspective
by relating it to the sampling variation of the Lorenz curve and the Gini
coefficient. The transformation of the Lorenz curve has no impact on the

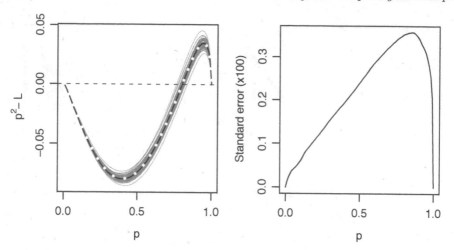

FIGURE 3.12

Bootstrap replicates of the transformed Lorenz curve for Austria in 2010 and the estimated standard errors.

The replicate Lorenz curves are drawn by thin gray lines and the estimated Lorenz curve by solid black line. The average of the replicates is indicated by white dots.

standard error. We applied 100 replications of the bootstrap for Austria in 2010. The results are presented in Figure 3.12 by the collection of the replicate Lorenz curves in the left-hand panel and the estimated standard error of the Lorenz curve, as a function of the probability p, in the right-hand panel. Of course, there is no sampling variation at $p = 0$ and $p = 1$. However, the standard error (the dispersion of the replicates) is not symmetric around 0.5. It increases roughly linearly for p up to 0.87, where the standard error is 0.0035, and then decreases precipitously to zero at $p = 1$. The transformed Lorenz curve with infiltration by a household with eHI equal to one million Euro (case B in Figure 3.11) exceeds its 'proper' counterpart by 0.0049 at $p = 0.87$ and by 0.0062 at $p = 0.99$, respectively, by 1.4- to 1.8-multiple of the maximum standard error.

The bootstrap estimate of the standard error of the Gini coefficient is 0.39. The estimate of the coefficient is 26.10, and its bootstrap version is 26.08. Infiltration of the sample by a household with one million Euro results in the increase of the Gini coefficient by 3.98, so the influence of the single outlier is far from trivial. The estimate of the standard error of the Gini coefficient for Austria in 2004 is 0.45. Assuming that the samples for 2004 and 2010 are independent, the difference of the Gini coefficients for the two years, estimated by 0.34, has standard error $\sqrt{0.39^2 + 0.45^2} = 0.60$. Hence, change of the Gini coefficient in the range $(-0.86, 1.54)$ is plausible.

The Gini coefficient is a summary of the Lorenz curve. Other, simpler, summaries can be devised. For example, the quintile share ratio (Langel and Tillé, 2011) is defined as $\{1 - L(0.80)\}/L(0.20)$, that is, the ratio of the total income by the 20% of (individuals or households) with the highest values of eHI and the 20% of those with the lowest values. The choice of the 20% is somewhat arbitrary, and the ratio curve

$$\frac{1 - L(1 - p)}{L(p)}$$

for p in a range that includes 0.20, such as $(0.10, 0.25)$, may provide much more detail. Note the parallels of this ratio with Equation (3.4).

Estimation of the Lorenz curve and Gini coefficient is addressed by Gastwirth (1972) and Moskowitz *et al.* (2007), the latter in a context different from poverty assessment.

The Lorenz curve and Gini coefficient are only tenuously connected with the mainstream statistical concepts, such as the normal distribution. Their versions are well-defined for any nonnegative distribution, but outside the confines of studying income inequality they are applied infrequently. We can partly make up for this separation by evaluating the Lorenz curve and Gini coefficient for a range of lognormal distributions and relate the values of the Gini coefficient to the variance of the lognormal distribution.

We can compare the Lorenz curve for a country and year (say, Austria in 2004) with its counterpart for the best fit to eHI by a lognormal distribution. The fit is obtained by matching the data-based value of the Gini coefficient, 25.76, with its counterpart for a suitable lognormal distribution. For the latter, we only have to find the matching variance, because the Lorenz curve does not depend on the log-scale mean μ in $\mathcal{LN}(\mu, \sigma^2)$. We first derive an expression for the values of the Lorenz curve for the lognormal distribution (with $\mu = 0$). Its value at probability p is the solution y of the equation

$$\int_{-\infty}^{y} u f(u; \sigma^2) \, \mathrm{d}u = p \int_{-\infty}^{+\infty} u f(u; \sigma^2) \, \mathrm{d}u, \tag{3.5}$$

where $f(u; \sigma^2)$ is the density of $\mathcal{LN}(0, \sigma^2)$. The equation is interpreted as seeking the value y such that the total income up to y is $100p\%$ of the total of all income. This interpretation is literal after both sides of the equation are multiplied by the population size. We evaluate the integrals by relating the integrands to the normal density. First, the integral on the right-hand side is equal to the expectation of the lognormal distribution, that is, $\exp(\frac{1}{2}\sigma^2)$. The integrand on the left-hand side is expressed as the $\exp(\frac{1}{2}\sigma^2)$-multiple of the density of $\mathcal{N}(\sigma^2, \sigma^2)$. Details are given in Appendix 2.A.2. Equation (3.5) is equivalent to

$$\exp\left(\frac{1}{2}\sigma^2\right) \Phi\{\log(y); \sigma^2, \sigma^2\} = p \exp\left(\frac{1}{2}\sigma^2\right),$$

and its solution for y is

$$y_p = \exp\left\{\sigma^2 + \sigma \Phi^{-1}(p)\right\}.$$

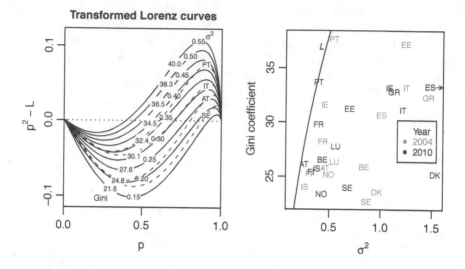

FIGURE 3.13
Lorenz curves and Gini coefficients for lognormal distributions.

It is evaluated on the quantiles of the standard normal distribution for a fine grid of probabilities, such as $0.01, 0.02, \ldots, 0.99$, and the Gini coefficient is approximated by their numerical integration. The transformed Lorenz curves for a set of lognormal distributions and the Gini coefficients of the lognormal distributions, as a function of the log-scale variance σ^2, are drawn in Figure 3.13.

The Lorenz curves for four countries are drawn by dashes and marked by their acronyms at the left-hand margin. None of these curves is approximated well by a curve for a lognormal distribution. The right-hand panel plots the function that relates the lognormal variance σ^2 to the Gini coefficient and marks the corresponding values for the countries in 2004 (gray) and 2010 (black). Except for Austria and Portugal, both in 2010, the countries' values of the Gini coefficient are smaller than the Gini coefficient based on the fitted log-scale variance $\hat{\sigma}^2$. In many instances, the difference is very large. The fitted variance for Spain in 2010, 2.08, is off the scale for the horizontal axis. The next largest variance in 2010 is for Denmark, equal to 1.53. Both countries have many households with negative values of eHI, but they are truncated at zero in the evaluations. In conclusion, the distributions of eHI in most countries are distinctly different from lognormal and the lognormal fit for them should not be used for estimating the Lorenz curve or the Gini coefficient.

3.4 Scaled Quantiles

Alternative measures of inequality compare the (extreme) quantiles with the median. We may plot the scaled quantile function $Q_X(p)/Q_X(\frac{1}{2})$, or mark a selection of scaled quantiles on a horizontal line. The latter option is particularly attractive because the summaries can be presented in a compact diagram, as illustrated in Figure 3.14. A country is represented by a horizontal segment for each year 2004–2010, on which the vertical ticks mark the scaled percentiles 2, 5, 10, 25, 75, 90, 95 and 98. The country with least inequality would have its scaled quantiles closest to the scaled median, equal to 1.0. The wider the country's segment, and the further its scaled quantiles are from 1.0, the greater the inequality. The linear scale of the horizontal axis causes the scaled quantiles above the median to be exaggerated. On the multiplicative scale, the low quantiles would be widely spread, and would distort the image because of a few small values of the second scaled percentile (e.g., for Spain in 2009 and 2010). Separate panels could be drawn for the percentiles below 50 and above, corresponding to the focus on respective low and high values of eHI. Also, a different set of percentiles can be used, but they should be the same for all the countries in a diagram.

The diagram shows that Portugal has the greatest relative concentration of wealth—its 95th and 98th scaled percentiles are the highest, although they have been decreasing since 2005 or 2006. Iceland and Ireland have single years (2009 and 2010, respectively) in which the estimated scaled 98th percentiles are much greater than in the other years. Denmark, Sweden and Norway have the least concentrated relative wealth.

Spain has by far the smallest estimates of the second scaled percentiles in 2009 and 2010, and Denmark has much smaller estimates in these two years than earlier (recall that there are many negative values of eHI). Ireland and Sweden have a year each (2010 and 2006, respectively) with much smaller estimated second scaled percentiles than in the other years.

Estonia, Spain, Greece and Italy have the smallest estimates of the second scaled percentile, although the estimates for one or two years each are quite small also for Denmark (2009 and 2010), Ireland (2010) and Sweden (2006). Austria, Belgium, Finland and Luxembourg have the highest estimated second scaled percentiles. Extremely low relative income is very rare in these countries, although the details of how income is defined and varying data quality may be factors that contribute to this impression.

The quantiles are estimated by their sample versions. Longford (2012a) describes a method that improves on this approach, especially for high quantiles.

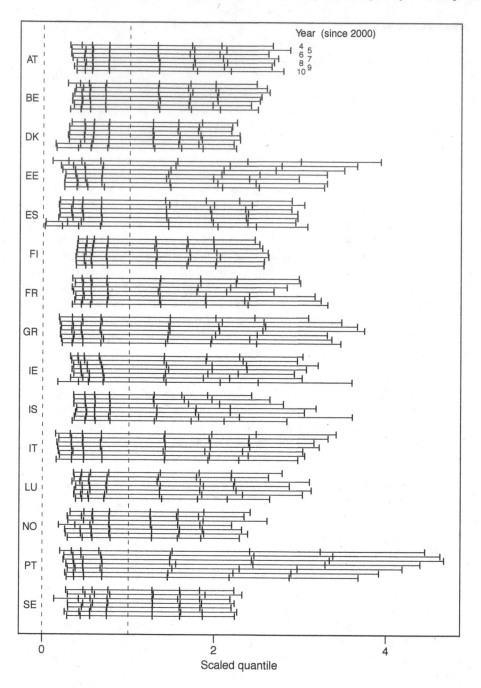

FIGURE 3.14
Scaled quantiles for the countries in EU-SILC in 2004–2010.

3.4.1 Permutation Test

The assessments made based on Figure 3.14 involve a judgement as to the statistical significance of the observed differences. We can estimate the associated standard errors by bootstrap. In this section, we describe an alternative that is specific to comparing a small number of independent samples.

Suppose we want to compare the second scaled percentiles of eHI for Denmark in 2008 and 2009. For the realised samples, we have the (estimated) difference of $0.295 - 0.141 = 0.154$, based on samples of sizes $n_{2008} = 5778$ and $n_{2009} = 5866$ households. We take the $n_{2008} + n_{2009} = 11\,644$ households and assign them to pseudo-years 2008 and 2009 completely at random, subject to the sample sizes in the realised survey. By construction, the difference of the second scaled percentiles evaluated on these two pseudo-samples is entirely due to chance, corresponding to the null hypothesis. We replicate the assignment to pseudo-years many times. The distribution of the replicate differences based on these pseudo-samples corresponds to the null-distribution of the differences of the second scaled percentiles. If the realised value is exceptional among the replicate pseudo-values, the difference is statistically significant. Note that when a household is assigned to a pseudo-sample, its sampling weight and composition (number of its members) is carried forward. The sampling weights play no role in the assignment of the households to the years; each household has the probability $n_{2008}/(n_{2008} + n_{2009})$ of being assigned to pseudo-year 2008.

In the application with 1000 replications, we obtained 18 pseudo-samples that had 2008–2009 differences greater than the sample difference, $\widehat{\Delta} = 0.154$, and 18 pseudo-samples that had differences of the scaled second percentiles smaller than $-\widehat{\Delta}$. Therefore, the (estimated hypothesis-test related) p value for the difference of the scaled second percentiles is 0.036. Nominally, this indicates statistical significance. High percentiles, scaled or without scaling, are estimated with far greater sampling variation, and so even the extreme observed differences should not be regarded as important.

Note that in the example of the two years' samples from Denmark we have broken an important rule by declaring the hypothesis after data inspection. Therefore the verdict (the p value) can only be treated as an illustration of the test.

3.5 Income Inequality. Kernels, Scores and Scaling

The term 'inequality' is associated with variation. No variation, when every household has the same value of the relevant variable, such as eHI, represents the extreme of absolute equality, and greater variation is naturally interpreted as more inequality. Hence, measures of dispersion of eHI are suitable indices of income inequality. We have to distinguish between population and sample versions of these indices. With the population versions, we are not concerned that the values for one or a few units may unduly influence the value of the index. Such units have exceptionally large values of eHI, and these we may want to be reflected in the value of the index with the appropriate strength. In contrast, a unit with an exceptional value in the realised sample may have a strong influence, but such units may be absent from some replications of the survey, and therefore the estimator of the index has very large sampling variance. This suggests that the index should be estimated more robustly. An alternative is to define indices that are easier to estimate, while remaining good characteristics of dispersion.

A drawback of the variance of eHI is that it is confounded with inflation or the currency used. This is resolved by evaluating (and estimating) the variance on the log scale. As an aside, the influence of exceptionally large values is greatly reduced by the log transformation, although very small values may become influential instead.

Indices of inequality can be constructed from means. The well-known arithmetic mean of N quantities $y_1, y_2, \ldots, y_N$ is defined as

$$\mu = \frac{1}{N} \sum_{j=1}^{N} y_j$$

Its generalisation is the mean of the powers,

$$\mu^{(q)} = \frac{1}{N} \sum_{j=1}^{N} y_j^q \,,$$

where the exponent q is a positive scalar. Note that $\mu^{(1)} = \mu$. More generally, we define a nonnegative kernel function g and set

$$\mu_g = \frac{1}{N} \sum_{j=1}^{N} g(y_j) \,.$$

Only increasing functions of g are of any practical use. Without proof we state the result that when g is a convex function, then $\mu_g \geq g(\mu)$, and equality holds only when all the values y_j are identical. This is a special case of Jensen's inequality (Needham, 1993). Moreover, we can measure the dispersion of y_j indirectly by μ_g for a suitable function g. We can transform μ_g to the scale of the same currency as eHI as $g^{-1}(\mu_g)$ and standardise it as $\alpha_g = g^{-1}(\mu_g)/\mu$; here g^{-1} stands for the inverse of g. For identical values of y_j, $\alpha_g = 1$; otherwise $\alpha_g > 1$.

The Atkinson index (Atkinson, 1970) is based on these ideas applied to the power functions $g(y) = y^q$, $q > 1$. These functions are a sufficiently wide class for most purposes. The exponent q controls how much relative weight is assigned to the largest values of y. We could turn our focus to small income by using the reciprocal of eHI, $1/y$, but this is problematic when some households declare zero or negative income. Such values can be truncated, but this entails some arbitrariness regarding the values on which we intend to focus.

The term 'kernel' was used also in connection with assessment of poverty in Section 3.1. This is no accident, for the two types of kernels have similar roles. For poverty indices we use the shortfall and for inequality we use eHI. These variables, defined in the studied population, are referred to as *scores*. A common characterisation of the poverty, income inequality and some other indices is that they are scaled averages of scores transformed by a kernel. This class is sufficiently flexible for most purposes, but can be further expanded by their linear combinations with positive coefficients.

Suggested Reading

Alternative approaches to estimation of the sampling variances for poverty indices are developed by Kovačević and Binder (1997) and Osier (2009). For more background on methods for finding the roots and extremes of a smooth function we refer to Gill, Murray and Wright (1981) and Lange (1999). More background on permutation tests can be found in Good (2005). The R contributed package `laeken` (Alfons and Templ, 2013) can be used for evaluating several indices related to poverty and inequality. Alfons *et al.* (2011) describe a method of generating (from an EU-SILC sample) an artificial population of a country on which simulations could be applied. Dagum (1997) derived a decomposition of the Gini coefficient to its components within and between subpopulations, akin to the decomposition of the overall variance in analysis of variance. Förster (2005) studies the differences among the member states of EU on several indices. Delhausse, Lüttgens and Perelman (1993) study these indices in Belgium.

3.A Appendix

3.A.1 Negative Values of eHI

This section documents the function `EUneg` for the basic description and exploration of the dataset for a country and a year. The focus of the exploration is on negative values of eHI. Although the filename of the dataset is the sole input into the function, it is preferable to organise the input so that only the country's acronym and the year (e.g., 9 for 2009) have to be given.

```
EUneg <- function(Cou, Yea, dire="~/Splus/EUsilc12/DATA/", ext="dta")
{
###   Summary of the negative values of eHI in a dataset

##  Cou   The country (two-leter acronym)
```

```
## Yea    The year     (digit, or 10)
## dire   The data directory
## ext    The extension of the dataset  (from Stata)
```

The arguments `dire` and `ext` are optional because the given defaults are used in all applications in the author's operating system; another analyst may organise his or her workspace differently. The country and the year are mandatory arguments, and are placed first in the list of arguments. The function can be executed by a short expression, such as

```
AT04e <- EUneg("AT", 4)
```

for Austria in 2004. The dataset is introduced into the R data directory by the function `EUread`, see Appendix 1.A.1:

```
## Reading the data  --  individual level
dat <- EUread(dire, Cou, Yea, ext, c("hid", "hx090", "db090"))
```

The total of the sampling weights will be required later.

```
Swe <- sum(dat[, "db090"])
```

The number of the individuals is established as the first dimension of the data and the number of households as the number of unique household Id's. Since this operation is executed several times (on different datasets), a function is declared for this purpose:

```
NiNh <- function(dt)
c(Indi=nrow(dt), Hhlds=sum(LeUni(dt[, "hid"])))
```

It is declared within the function `EUneg`, and so it does not appear in the user's permanent workspace. Its simple application is the next expression in `EUneg`:

```
## The key counts
CNT <- NiNh(dat)
```

The records with zero and negative values are extracted and counted

```
dt0 <- dat[dat[, "hx090"] == 0, ]
dat <- dat[dat[, "hx090"] < 0, ]
```

```
## Form the table of counts
CNT <- rbind(All=CNT, Zero.eHI=NiNh(dt0), Neg.eHI=NiNh(dat))
```

A vector is formed with the negative values of eHI for households, with the household size as its label:

```
dbt <- tapply(dat[, "hx090"], dat[, "hid"], mean)
names(dbt) <- table(dat[, "hid"])
```

The function `EUneg` returns a list that contains the counts `CNT`, the actual negative values and the estimates of the percentage of individuals with negative and zero values of eHI.

```
list(Counts=CNT, Cases=sort(dbt),
    Weight=round(100*c(Negatives=sum(dat[, "db090"]),
                        Zeros=sum(dt0[, "db090"]))/Swe, 2))
}  ##  End of function  EUneg
```

The code below is for application of the function **EUneg** to all 15 countries that are in EU-SILC in 2004 and 2010. The results are stored in the list **EU410n** for both years.

```
##  Start the stopwatch
Time <- proc.time()

##  Application en masse
EU410n <- list("2004"=lapply(EU04a, EUneg, 4),
               "2010"=lapply(EU04a, EUneg, 10))

##  Labelling (country)
for (yr in names(EU410n))
names(EU04n[[yr]]) <- EU04a

##  How long it took
Time <- proc.time() - Time
```

The function **EUneg** is declared and the code for its application is written in file **EU3ch.an1**. It is introduced into the R workspace and the list **EU410n** is evaluated by the expression

```
source("~/Splus/EUsilc12/EU3ch.an1")
```

on the author's computer. This is much more elegant than cutting and pasting, and scrolling and editing on the command line are very convenient for applying **source** to other files in the same directory and with similar names. The output is quite long. For orientation, we reproduce only the element of the list **EU410n[[2]]** for Sweden, that is, **EU410n[[2]][["SE"]]**:

```
$Counts
         Indi Hhlds
All      14321  7173
Zero.eHI    10     6
Neg.eHI     32    16

$Cases
         2          2          4          2          1          1          1
-73346.95  -15507.65  -14032.83  -8094.85  -6663.09  -3695.51  -3272.88

         4          2          2          2          2          2          1
 -2220.41   -1918.45   -1530.54   -945.03   -434.53   -221.30   -164.80

         3          1
```

```
 -160.94   -121.48

$Weights
Negatives    Zeros
     0.23     0.09
```

Thus, in the dataset of 14 321 individuals in 7173 households, 32 individuals from 16 households have negative values of eHI. The negative values are listed as a vector with its names set to the household size. The concluding element, (0.23, 0.09), gives the estimated percentages of individuals with negative and zero values of eHI.

The vector Time records the length of time the evaluations took:

```
Time
   user  system elapsed
 10.665   0.837  13.485
```

The function ExtrL, introduced in Appendix 1.A.4, enables us to extract parts of the information in the list EU410n. For example, the estimated percentages of individuals with negative eHI are extracted by the expression

```
rbind("2004"=sapply(EU410n[[1]], ExtrL, "Weights"),
      "2010"=sapply(EU410n[[2]], ExtrL, "Weights"))
```

The result is the matrix

```
            AT   BE   DK   EE   ES   FI   FR   GR   IE   IS   IT   LU
Negatives 0.00 0.29 0.42 0.45 0.10 0.02 0.08 0.69 0.03 0.08 0.33 0.19
Zeros     0.01 0.06 0.06 0.35 0.42 0.01 0.01 0.33 0.02 0.00 0.35 0.00
Negatives 0.00 0.12 0.84 0.19 1.08 0.02 0.08 0.23 0.14 0.10 0.24 0.19
Zeros     0.00 0.03 0.00 0.07 0.50 0.03 0.01 0.39 0.45 0.00 0.35 0.00

            NO PT   SE
Negatives 0.07  0 0.24
Zeros     0.05  0 0.12
Negatives 0.08  0 0.23
Zeros     0.02  0 0.09
```

The code can be adapted for other elements of data exploration, such as to count the numbers of missing values, exploring the distribution of the sampling weights, and the like. Recall that the function Dround would convert the numeric zeros 0 to character strings '0.00', better suited for a tabular display.

3.A.2   Newton Method in R

The Newton method is implemented by the user-defined function EUnewton as

```
EUnewton <- function(fnc, xx, maxit=20, tol=8, plt=F, long=T, dgt=5)
```

```
{
### Newton method for solving the equation  f(x) = 0

## fnc      The objective function
## xx       The (vector of two) initial solutions
## maxit    Maximum number of iterations
## tol      The precision (integer; number of decimal places)
## plt      Whether to draw a graph
## long     Whether to have a detailed output
## dgt      The number of digits in rounding
```

It suffices to specify the mandatory arguments fnc, a function, and xx, a vector of length (at least) two. The values of the objective function at the initial values xx are calculated first, together with the initialisation of the current number of iterations (starting at zero), and a scalar for assessing convergence (dist). The vector Conver (changed later to a matrix) records the results at the intermediate iterations.

```
## The first evaluation
fxx <- fnc(xx)

## Initialisation for the iterations
dist <- 0
iter <- 0

## The convergence monitor
Conver <- c(xx, fxx, dist)
```

The iterations are controlled by the conditional loop (while), in which the current iteration number is increased by one, the new pair of solutions is evaluated, together with the corresponding values of the objective function fnc, and the scalar dist, which arbitrates whether the iterations should be stopped, is updated.

```
## The iterations
while (iter < maxit & dist < tol)
{
## The iteration counter updated
iter <- iter + 1

## The new pair of (provisional) solutions
xx <- c(xx[2], sum(rev(xx) * fxx * c(1,-1)) / sum(fxx * c(1,-1)))

## The values of the objective function
fxx <- fnc(xx)

## The convergence criterion
dist <- -log(abs(xx[2]-xx[1]) + abs(fxx[2]) , base=10)

## Updating the convergence monitor
Conver <- rbind(Conver, c(xx, fxx, dist))
```

```
}  ##  End of the iteration loop  (while)
```

The function contains further expressions for plotting a diagram (as in Figure 3.2) and for organising the output. The function is declared in file EU3fn.an2. An expression for its application is in file EU3ev.an2.

```
##  The objective function
EUfnr <- function(r)
1000^r/(600^r + 400^r) - 1.25


Time <- proc.time()


##  The application
EUnewtR <- EUnewton(EUfnr, c(1,2), 10, 8, long=T, plt=T, dgt=4)

##  The application from a better pair of initial solutions
EUnewtS <- EUnewton(EUfnr, c(1,1.25), 10, 8, long=T, plt=T, dgt=4)


Time <- proc.time() - Time
```

The object Time has the same role as in the previous section.

The postscript file of Figure 3.2 is drawn using the code in file EU3gr.an2 and is named FigNwt1.ps.

```
EUnewtR <- Figure("Nwt1",
      EUnewton(EUfnr, c(1,2), 10, 8, long=T, plt=T, dgt=4),
      hgt=3.75, mrm=c(3.6, 3.6, 1, 1), mgm=0.75)
```

The objective function EUfnr has to be defined in such a way that it could be applied to a vector of values of its argument r. The results are in objects EUnewtR and EUnewtS. EUnewtR is reproduced in Table 3.1.

3.A.3   More on Poverty Indices

The function EUpovS evaluates the poverty gap for specified exponents and their absolute and relative versions. It is used in other functions in which the relevant data is read and preprocessed. The arguments of EUpovS are the vectors of values of eHI and the sampling weights, the thresholds, the exponent in the poverty gap and whether the absolute or relative poverty gap is to be evaluated:

```
EUpovS <- function(Dat, Wei, std, Quo, Rel=F)
{
###   The poverty gap for a set of thresholds

##  Dat    The values of eHI
##  Wei    The sampling weights
##  std    The thresholds (fractions of the median)
##  Quo    The exponent in the poverty gap
##  Rel    Whether to divide (scale) by the standard (relative pov.gap)
```

The median of eHI is estimated first, and then the value of the poverty gap is estimated for each value of the threshold.

```
##  The median equivalised income
MD <- QuantW(Dat, Wei)

##  The gaps, initialised
RT <- c()

##  The gap for each threshold
for (st in std)
{
dfi <- st*MD - Dat

##  Who has a shortfall
dfi[dfi > 0] <- 0

##  The weighted totals
RT <- c(RT, sum(Wei * dfi^Quo) )
} ##  End of the loop over the thresholds (st)

##  The weighted average
RT <- RT / sum(Wei)
```

Finally, the relative poverty gap is calculated (if applicable) and the resulting vector is labelled.

```
##  Whether to evaluate the relative poverty gap
if (Rel)  RT <- RT / (std*MD)^Quo

##  Labelling
names(RT) <- 100*std

RT
} ##  End of function  EUpovS
```

The function EUpovL for evaluating the log-poverty gap is adapted from EUpovS by changing the key expression in the loop over the thresholds, in which the log-poverty gap is evaluated:

```
##  The gap for each threshold
for (st in std)
{
dfi <- (1+Dat)/(st*MD)

whi <- Dat < 0
RT <- c(RT, sum(Wei[whi] * log(dfi[whi])))
}
```

The function returns the value of -RT/sum(Wei). The arguments Quo and Rel are dropped in EUpovL; the log-poverty gap does not have a scaled version, and the exponent is redundant in it.

The function EUpovS is used in the functions EUpovA and EUpovB. In the latter two functions, the appropriate dataset is extracted from the database, the values of eHI are truncated if applicable, and the absolute or relative poverty gaps (linear, with exponent $r = 1$, as the default) are evaluated for individuals and households. The result is a $2 \times H$ matrix, where H is the number of thresholds at which the poverty gap is evaluated. The column names of this matrix are the thresholds.

```
EUpovA <- function(Cou, Yea, dire="~/Splus/EUsilc12/DATA/", ext="dta",
               STD=60, TRC=T, quo=1)
{
###    Absolute poverty gap for a set of thresholds

## Cou     The country  (two-leter acronym)
## Yea     The year     (digit)
## dire    The data directory
## ext     The extension of the dataset
## STD     The threshold(s)
## TRC     Whether to truncate eHI at zero
## quo     The quotient  (power) of the poverty gap

## Check on the thresholds  --  have to be fractions
while (max(STD) > 1.5) STD <- STD/100

## Read household-level data
dat <- EUread(dire, Cou, Yea, ext, VRS=c("hx040","hx090","db090"),IH=2)

## Truncation at zero
if (TRC)  dat[dat[, "hx090"] < 0, "hx090"] <- 0

## The absolute poverty gap -- individual-level
RTi <- EUpovS(dat[, "hx090"], dat[, "db090"] * dat[, "hx040"], STD,
             quo, Rel=F)

## The absolute poverty gap -- household-level
RTh <- EUpovS(dat[, "hx090"], dat[, "db090"], STD, quo, Rel=F)

rbind(Individual=RTi, Household=RTh)
}  ## End of function  EUpovA
```

The following expression evaluates the linear poverty gap with truncation (at zero) for the set of countries in EU04a.

```
EU410g <- list("2004"=lapply(EU04a, EUpovA, 4, STD=seq(30, 80, 0.5)),
               "2010"=lapply(EU04a, EUpovA,10, STD=seq(30, 80, 0.5)))
```

The list EU410g has two elements, each one a list of length(EU04a), that is, 15 elements. The latter elements are 2×101 matrices; 101 is the length of the sequence 30, 30.5, 31, ..., 80. A much less elegant alternative to using the function `lapply` is a loop over the elements of EU04a, after initialising EU410g to a list comprising two empty lists.

For Denmark, the poverty gap is evaluated also without truncation:

```
DK10g <- EUpovA("DK", 10, STD=seq(30, 80, 0.5), ext="dta", TRC=F)
```

The function EUpovB differs from EUpovA only by the addition of the argument Rel=T in both applications of EUpovS (the default value of Rel is F). The functions EUpovA and EUpovB could be replaced by a single function in which there is an additional logical argument, say RL, which specifies whether absolute or relative poverty gaps are to be evaluated. In the body of this function, the argument Rel of EUpovS would be set to RL in both applications.

An element of the list EU410g is the principal input to the function EUpovG for plotting a set of poverty-gap curves. The function has several other arguments that control aspects of the plot to be drawn and whether individual- or household-level estimates are to be plotted.

```
EUpovG <- function(res, xlm=c(), ylm=c(), IH=1, shr=1, offs=c(),
                   ttl="", key=c("Relative", "(%)"), cx=0.5, cxs=0.9)
{
###  Plotting the poverty-gap curves

##  res    The results  (list)
##  xlm    The horizontal scale (optional)
##  ylm    The vertical scale (optional)
##  IH     Individual (1) or houshold (2) values
##  shr    Unit space for the diagram
##  offs   The offsets for the labels at the margins
##  ttl    Title for the diagram
##  key    Keywords on the y-axis
##  cx     The cex argument in  text
##  cxs    The value of the cex.... arguments in  plot
```

The values of the thresholds are extracted from the column names. The code would work even if these values differed across the elements of res (countries).

```
##  The thresholds
std <- as.numeric(unique(unlist(lapply(res, colnames))))

##  The extents of the axes (provided or default)
if (length(xlm) < 2)  xlm <- range(std)
if (length(ylm) < 2)  ylm <- 100*range(unlist(res))
```

The vectors xlm and ylm are the extents of the horizontal and vertical axes of the plot to be drawn. If they are not provided, the default is set so as to cover the entire range of the values.

```
##  Empty shell of the plot
plot(xlm+c(-1,1)*shr*9, ylm, type="n", main=ttl,
  xlab="Standard (%)",ylab=paste(key[1]," poverty gap ",key[2],sep=""),
```

```
        cex.lab=cxs, cex.axis=cxs, cex.main=cxs)
```

```
##  The poverty-gap curves
for (eu in res)
lines(as.numeric(colnames(eu)), 100*eu[IH, ], lwd=0.5)
```

These expressions are the core of the function. They result in drawing the poverty-gap curves for the countries in a plot with specified ranges of the axes x and y. The values of cex.lab, cex.axis and cex.main are controlled by the argument cxs because the default, 0.9, is suitable for a diagram with a single panel or two panels side by side, but a larger value, 1.1, is needed for diagrams with a matrix of panels, such as 2×2.

Next, the coordinates of the limits of the curves are established. They will be used for labelling at the left- and right-hand margins.

```
##  The values relevant for labelling, on the left
pts <- as.numeric(sapply(lapply(res, colnames), ExtrE, 1))
vls <- sapply(lapply(res, ExtrR, IH), ExtrE, 1)
```

```
##  The values relevant for labelling, on the right
pts <- cbind(pts, as.numeric(sapply(lapply(lapply(res, colnames), rev),
      ExtrE, 1)))
vls <- cbind(vls, sapply(lapply(lapply(res, ExtrR, IH), rev), ExtrE,1))
```

```
##  Convert to percentages
vls <- 100*vls
```

The vectors pts and vls have two columns each that hold the x- and y-coordinates of the left- and right-hand limits of the curves.

The offset is the gap between the limit of the curve and the label. It is specified by the argument offs. Its length indicates how many columns are to be used for the labels. There is a default, with three columns.

```
##  The offsets for labelling at the margins
if (length(offs) == 0)   offs <- c(2.8, 1.6, 0.4)
```

The two sets of labels are drawn by the following expression

```
##  The labels of the countries (or years, in which case 10 --> 0)
rsn <- names(res)
rsn[rns==10] <- 0
```

```
##  Drawing the labels
for (j in seq(2))
{
##  Sort out the horizontal placement
srt <- sort.list(vls[, j])
text(pts[, j]+(-1)^j*shr*EUloc(vls[, j], offs), vls[srt, j], rsn[srt],
      cex=cx, adj=2-j)
}
```

In the loop (j), j=1 is for placing the labels at the left-hand margin, next to the left-hand limit of each curve. The argument `shr` is for additional control over the placement of the labels (so that they would not overlap, but would not be too far apart either). The argument `adj` of the function `text` is for justification; `adj=1` is for right-justified text—all the text is placed to the left of the specified coordinates, and `adj=0` is for left-justified text. The default is `adj=0.5`; text is centered around the coordinates. The purpose of the vector `srt` is for the labels to be sorted in rows, from left to right.

3.A.4   Lorenz Curve and Gini Coefficient

We use the function `EUneg` (Appendix 3.A.1) as the template for evaluating the Lorenz curve and the Gini coefficient. In the function `EULrzA`, we replace the evaluations that follow data input with the expressions for truncating eHI at zero and calculating the Lorenz curve and the Gini coefficient.

```
##  Truncate negative values of eHI
if (TRC) dat[ dat[, "hx090"] < 0, "hx090"] <- 0
```

```
##  Application
EULrzW(dat[, "hx090"], dat[, "db090"] * dat[, "hx040"], plt=F)
}  ## End of function   EULrzA
```

The concluding expression is also the last in the function `EULrzA`, so its value is not assigned within the function. The additional argument `TRC` indicates whether the values of eHI should be truncated at zero.

The execution for all 15 countries in EU04a for year 2010 is by the expression

```
EULz10a <- lapply(EU04a, EULrzA, Yea=10)
```

The values of the Gini coefficient for the countries are extracted by the expression

```
round(sapply(EULz10a, ExtrL, 2)*100, 2)
```

obtaining the vector

```
   AT    BE    DK    EE    ES    FI    FR    GR    IE    IS    IT    LU
26.10 26.51 25.22 31.23 33.29 25.40 29.76 32.83 33.19 25.72 31.12 27.73
   NO    PT    SE
23.39 33.65 23.98
```

The function `EULrzW` used in `EULrzA` has the following arguments:

```
EULrzW <- function (dat, wei = 1, plt = T, dgt = 2)
{
##  Lorenz curve and Gini coefficient

##  dat  The values of income (eHI)
##  wei  The sampling weights (default 1 -- equal weights)
##  plt  The plotting option (logical; default T  -- to plot)
##  dgt  The number of decimal places in the graph
```

After the initial settings and checks, the vectors dat and wei are permuted to the ascending order of dat:

```
## Data sorted in ascending order
wei <- wei[sort.list(dat)]
dat <- sort(dat)
```

The cumulative totals of the weights are formed to identify the records at the percentiles 1, 2, ..., 99; npt was earlier set to 100, but could be defined as an additional argument of EULrzW.

```
## The total of weights per evaluation point
Swe <- sum(wei)/npt
```

```
## The cumulative totals of weights
cwe <- cumsum(wei)
```

```
## Initialisation
qnt <- c()
```

```
## The record at the percentile
for (i in seq(npt-1))
qnt <- c(qnt, sum(cwe < i*Swe))
```

The percentiles are evaluated by linearly combining the values of eHI at the percentiles of the cumulative weights in cwe.

```
## The linear combinations of the borderline values
dfr <- cwe[qnt+1] - cwe[qnt]
dfA <- seq(npt-1)*Swe - cwe[qnt]
```

```
## The percentiles of the values, with interpolation
qntV <- (dat[qnt+1]*dfA + dat[qnt]*(dfr - dfA)) / dfr
```

The values of the Lorenz curve are first evaluated at the unit to the left and to the right or the borderline, and then the pairs of values are combined.

```
## The values of the Lorenz curve
Lrz <- matrix(cumsum(wei*dat)[c(qnt, qnt+1)], ncol=2)
```

```
## Adjustments for the partials by interpolation
Lrz <- (Lrz[, 1]*(dfr - dfA) + Lrz[, 2]*dfA) / dfr / sum(wei*dat)
```

The Gini coefficient is obtained as the average of the values of the Lorenz curve, appropriately transformed.

```
## The Gini coefficient -- numerical integration
Gni <- 1 - 2 * sum(Lrz)/npt - 1/npt
```

To see how the latter expression is derived, consider first the integral under the Lorenz curve. Each value l_p of the Lorenz curve (Lrz) contributes to the approximated integral by δl_p, where δ is the distance of two consecutive values of p, equal to `1/npt`, 0.01 in our implementation. The points at the limits, $p = 0$ and $p = 1$, contribute with only half, that is, zero and $\frac{1}{2}\delta$, respectively. The Gini coefficient is approximated as the complement to unity of the double of this integral.

Finally, the values of the Lorenz curve are labelled, with zero and unity added to either limit.

```
##   Add the beginning and the end
Lrz <- c(0, Lrz, 1)

##   Labelling
names(Lrz) <- seq(0, 1, length=npt+1)
```

The result is a list comprising the values of the Lorenz curve and of the Gini coefficient:

```
list(Lorenz=Lrz, Gini=Gni)
}  ##  End of function  EULrzW
```

Before the concluding expression, the option to plot the Lorenz curve is executed. The frame of a plot, $(0, 1) \times (0, 1)$, with the axis labels and its diagonal ($y = x$) is drawn, followed by the Lorenz curve and shading of the area between the Lorenz curve and the diagonal. Finally, the value of the Gini coefficient is printed above the diagonal at a 45 degree angle. The value of the coefficient is printed to `dgt` decimal places, with zeros in the concluding decimal places if necessary. Details of the function `Dround` are in Appendix 1.A.4.

```
if (plt)
{
##   The identity line
plot(c(0,1), c(0,1), type="l", xlab="p", ylab="L",
     cex.lab=0.9, cex.axis=0.9)

##   The Lorenz curve
lines(seq(0, npt)/npt, Lrz)

##   Shading
polygon(seq(0, npt)/npt, Lrz, col="gray65")

##   The Gini coefficient
text(0.46, 0.54, paste("Gini coeff. ", Dround(Gni, dgt)), srt=45,
     cex=0.85)
}  ##  End of the plotting option
```

Another function, `EULrzF`, is compiled for plotting a set of Lorenz curves.

```
EULrzF <- function(lst, zm=c(), ttl="", qua=T, cxs=0.9)
{
###   Graphics for a set of Lorenz curves

##  lst   The list of results from   EULrzA
```

```
##   zm      Zooming in
##   ttl     The title
##   qua     Whether to draw the quadratic function (for reference)
##   cxs     The value of the cex.... arguments in the plot
```

The axes are drawn first, together with the identity line (dashes), followed by the countries' Lorenz curves.

```
plot(c(0,1), c(0,1), type="l", xaxs="i", yaxs="i", lty=2, main=ttl,
     xlab="p", ylab="L", cex.lab=cxs, cex.axis=cxs, cex.main=cxs)
```

```
##   A curve for each country (or year)
for (lt in lst)
lines(as.numeric(names(lt[[1]])), lt[[1]], lwd=0.5)
```

The arguments xaxs and yaxs are set to "i", the code for limits for both x and y axes to be specified exactly, without additional width of 4% shared out at the two limits. The argument cxs has the same role as in function EUpovG in Appendix 3.A.3.

As an option, the quadratic function $f(x) = x^2$ is drawn by dots.

```
if (qua)
{
pt0 <- seq(0,100)/100
lines(pt0, pt0^2, lty=3, lwd=1.25)
}
```

As another option, the plot is redrawn with the axis-ranges specified in the list zm

```
##   Zooming in
if (length(zm) > 0)
{
##   Drawing a rectangle (that is the inset in the next panel)
polygon(c(zm[[1]], rev(zm[[1]])), rep(zm[[2]], rep(2,2)),
        lty=2, lwd=0.75)
...
```

followed by starting a new plot and a loop for plotting the Lorenz curves. The default value, zm=c(), is for drawing no inset. If this is overruled, then it is essential to set the argument of mfrow in par or of mrf in Figure to c(1,2), c(2,2) or similar, to ensure that the two panels appear in the same diagram. The value of cxs should also be reviewed.

Figure 3.8 is generated by the expression

```
EULzF410 <- Figure("Lrz2", 100*rbind("2004"=EULrzF(EULz04a,
     zm=list(c(0.52, 0.67), c(0.30, 0.44)), ttl="EU 2004", cxs=1.1),
     "2010"=EULrzF(EULz10a, zm=list(c(0.52, 0.67), c(0.30, 0.44)),
     ttl="EU 2010", cxs=1.1) ),
          5.6, c(2,2), c(3.6, 3.6, 2, 0.8), 0.65)
```

The result in EULzF410 is the 2×15 matrix of the estimated Gini coefficients for the countries in 2004 and 2010.

3.A.5 Scaled Quantiles

The scaled quantiles are evaluated by the function EUSQua, which has a code similar to QuantC, see Appendix 1.A.2, adapted for the evaluation for a single year and individual level. The only statement added is to apply the scaling (division by the estimated median):

```
MDi <- MDi / QuantW(dat[, "hx090"], wei)
```

As an alternative to collecting the results in annual lists, a single list can be formed, with the annual lists as its elements.

```
EU410q <- list()

for (yr in as.character(seq(4,10)))
{
EU410q[[yr]] <- sapply(EU04a, EUSQua, yr)
colnames(EU410q[[yr]]) <- EU04a
}
```

If the 'years' yr were not converted to characters (words), then the first three elements of EU410q would be empty and the list for 2004 would be the fourth element. With the implemented arrangement, the first element of EU410q has name "4" and the length of EU410q is 7.

The scaled quantiles are plotted by the function EUquaF. The list of the annual lists of scaled quantiles, with the labelling produced by EUSQua, is its only argument. The result of applying EUquaF on EU410q, and submitting it to Figure, is Figure 3.14.

```
Figure("Qua1", EUquaF(EU410q), 8.2, 1, c(3.3,0.05, 0.3, 0.05),
       0.5, lb=c(2,2,1))
```

In the diagram, each country is reserved a horizontal strip in which the scaled quantiles are marked on the annual horizontal segments. Each year is associated with a particular location within this strip, stored in vector loc. The countries' labels are extracted from the submitted list.

```
EUquaF <- function(res)
{
### Scaled quantiles for the countries and years
## res   The results list (annual), obtained by function  EUSQua

## The relevant years
yrs <- names(res)
nyr <- length(yrs)

## The location of the line for the year
loc <- seq(0,1, length=nyr+3)[-c(1,nyr+2, nyr+3)]
names(loc) <- yrs

## The horizontal axis range
xrng <- range(unlist(res))
```

```
##   The countries
cos <- sort(unique(as.vector(sapply(res, colnames))))
nco <- length(cos)
```

After drawing the empty shell of the plot there are separate expressions for drawing the horizontal segments and vertical ticks (short segments).

```
plot(xrng+c(-0.1,0), -0.5-c(nco+0.75, 0), type="n", yaxs="i", yaxt="n",
   xlab="Scaled quantile", ylab="", cex.lab=0.9, cex.axis=0.9)

##   The tick-length (half)
shr <- 1/2/nyr

##   Loop over the years
for (yr in yrs)
{
##   The horizontal lines
segments(res[[yr]][1, ], -seq(nco) - loc[yr],
        res[[yr]][nrow(res[[yr]]), ], -seq(nco) - loc[yr], lwd=0.5)

##   The vertical ticks
segments(t(res[[yr]]), -rep(seq(nco), nyr) - loc[yr] - shr,
        t(res[[yr]]), -rep(seq(nco), nyr) - loc[yr] + shr, lwd=1.2)
}  ##   End of the loop over the years
```

The argument yaxt controls the printing of the y-axis label. With the value "n", the ticks and labels are omitted. They would not appear in the diagram anyway because the left-hand gap in the argument mar of the graphics function par, controlled by the argument mrm of the function Figure, is set to 0.05, just enough to show the line that delimits the plotting region. The countries' labels are added at the left-hand margin and the years are indicated in the strip for the first country.

```
##   The vertical dashes at zero and unity
abline(v=c(0,1), lty=2, lwd=0.5)

##   The countries' markers
text(-0.058, -seq(nco)-0.5, cos, cex=0.8, adj=1)

##   Location of the text in the plot
hrz <- rep(c(-1,1), ceiling(nyr/2))
hrz <- hrz[seq(nyr)] * shr/1.5

##   Text in the plot (the years)
text(2.96+hrz[1], -1+3*shr, "Year (since 2000)", cex=0.8, adj=0)
text(3.03+hrz, -1-loc[yrs], yrs, cex=0.6)
```

You do not have to use this function 'as is', but can make alterations in it to suit your taste and conventions.

3.A.6   Permutation Test

The permutation test for comparing the scaled percentiles in two settings (countries or years) is implemented in the function `EUperm`.

```
EUperm <- function(co2, yr2, dire="~/Splus/EUsilc12/DATA/", ext="dta",
        qua=2, nrp=1000)
{
###     Permutation test

##  co2    The countries to be compared
##  yr2    The years to be compared
##  dire   The directory for the datasets
##  ext    The extensions of the datasets
##  qua    The quantile/percentile
##  nrp    The number of pseudo-replicates

if (length(co2) + length(yr2) < 3)
      stop("Either two distinct countries,
              or two distinct years have to be given.")

##  Duplicate if necessary
if (length(co2)==1) co2 <- rep(co2, 2)
if (length(yr2)==1) yr2 <- rep(yr2, 2)
if (length(ext)==1) ext <- rep(ext, 2)

##  Work with a single value
qua <- abs(qua[1])

##  From percentage to fraction
while (qua > 1) qua <- qua/100
```

The two datasets for the test may be for two different countries (the same year) or for two different years (the same country), and even the datafile extensions could be different. Any of these three factors that are fixed can have the corresponding argument of length 1. The test is performed for a single percentile declared in qua.

The two datasets are read and stacked vertically, with the indicator for the dataset added as the last column (GRP):

```
##  Initialisation for the datasets
dat <- c()
n2 <- c()

##  Data input
for (i in seq(2))
{
##  Attach the indicator of the group (i)
dty <- cbind(EUread(dire, co2[i], yr2[i], ext[i],
      VRS=c("hx040", "hx090", "db090"), IH=2), GRP=i)
```

```
n2 <- c(n2, nrow(dty))

##  Stack the datasets vertically
dat <- rbind(dat, dty)
}  ##  End of the loop for data input  (i)
```

The quantile is evaluated on the realised datasets:

```
##  Initialisation
MDi <- c()

##  Evaluation for the realised samples
for (i in seq(2))
{
##  Who belongs to dataset  i  and their weights
whi <- dat[, "GRP"]==i
wei <- dat[whi, "db090"] * dat[whi, "hx040"]

##  The scaled quantile
MDi <- c(MDi, QuantW(dat[whi, "hx090"], wei, qua) /
        QuantW(dat[whi, "hx090"], wei) )
}  ##  End of the loop over the datasets (i)
```

The pairs of pseudo-estimates are then attached to MDi as additional columns.

```
##  Pseudo-estimation
MDi <- cbind(MDi, replicate(nrp,
{
##  Who is in the first sample
whi <- seq(sum(n2)) %in% sample.int(sum(n2), n2[1])

mdj <- c()

for (i in seq(2))
{
wei <- dat[whi, "db090"] * dat[whi, "hx040"]

##  Evaluation of the pseudo-sample
mdj <- c(mdj, QuantW(dat[whi, "hx090"], wei, qua) /
        QuantW(dat[whi, "hx090"], wei) )

##  Switch to the other group
whi <- !whi
}
mdj
}  ##  End of the expression for replication
))  ##  End of the  replicate  expression and cbind
```

When the vector whi is formed at the beginning of the long expression-argument of **replicate**, it indicates the first setting. After its execution in the loop over the settings for $i = 1$ it is switched to indicate the second setting.

The operation `%in%` establishes whether the elements of the first argument are matched in the second argument.

The function is concluded by calculating the differences of the pseudo-quantiles and generating their summaries.

```
## The post-processing
dfs <- apply(MDi, 2, diff)

list(MDi, Tails=c(mean(-abs(dfs[1]) > dfs[-1]),
     mean(abs(dfs[1]) < dfs[-1])), rbind(Households=n2,
     Individuals=tapply(dat[, "hx040"], dat[, "GRP"], sum)) )
} ## End of function  EUperm
```

The element `Tails` evaluates the fraction of the pseudo-estimates that are more extreme than the realised estimate (first element of `dfs`). The scaled fifth percentiles of eHI in Denmark in 2008 and 2009 are compared by the permutation test with 1000 replicates using the expression

```
DKprm89 <- EUperm("DK", c(8,9), qua=5, nrp=1000)
```

The key result is in the second element of the result-list,

```
$Tails
[1]  0.032  0.037
```

It indicates that 3.2% of the replicates are more extreme than the realised value (-0.0448) in the left-hand tail and 3.7% in the right-hand tail. The estimated p value is 0.069.

4

Mixtures of Distributions

4.1 Introduction

The classes of distributions, such as the normal, gamma, Pareto and the like, which are known by their names and have some attractive properties, form a very small part of the space of all distributions, or even of continuous distributions. A greater variety of distributions is obtained by transformations of these common (named) distributions, and it can be further extended by other operations, such as convolutions. But none of them offer as much flexibility as mixtures (Everitt and Hand, 1981; Titterington, Smith and Makov, 1985).

Suppose D_k, $k = 1, \ldots, K$, are continuous distributions with respective densities f_k, and let p_k be a set of probabilities $(0 < p_k < 1)$ that add up to unity; $p_1 + p_2 + \cdots + p_K = 1$. Such a set or vector of nonnegative scalars is called a composition. The mixture of the distributions with these probabilities is defined by the density

$$f(x) = p_1 f_1(x) + p_2 f_2(x) + \cdots + p_K f_K(x) \tag{4.1}$$

for $x \in (-\infty, +\infty)$. We say that the mixture has K components. The distributions D_k may belong to the same class, such as the normal, $\mathcal{N}(\mu, \sigma^2)$, and can even have some parameters in common, but they have to be distinct, otherwise the summation in Equation (4.1) is reduced to fewer terms. Mixtures of discrete distributions are defined similarly, with probabilities in place of the densities in Equation (4.1). The number of components may be infinite.

A draw from a mixture distribution is generated by the following two steps. First we make a draw from the multinomial distribution with probabilities $p_1, \ldots, p_K$. Suppose value h is drawn. Then we make a draw from the distribution D_h. In a random sample of large size from such a mixture distribution there are approximately $100 p_k \%$ units from distribution D_k. In many applications, no reference is made to this two-stage process, and a mixture model is posited merely to approximate a distribution, never assuming that the components may represent some meaningful subpopulations or that the data has been generated by such a process. On the other hand, the components may be associated with clusters, distinctly separated subpopulations, or may reveal some other features of the population. The distribution to which unit i belongs is denoted by k_i. When only the values of the observations are available we may not be able to recover the index k_i. The recovery is even more difficult when we do not know the distributions D_k, nor their number K.

Mixtures can comprise one or several distributions of both types, continuous and categorical, although of practical importance among such hybrids are only semicontinuous distributions, which are mixtures of a degenerate (zero-variance) and a continuous distribution. For example, consumption of a commodity or income from a specified source may be modelled by a continuous distribution for those with positive consumption or income, and the constant zero represents those who abstain from consumption or derive no income from the source. In this case, the assignment of a subject to its component (subpopulation) is trivial. That is not the case for mixtures of a nondegenerate and a degenerate discrete distribution, such as a Poisson distribution with extra zeros (Lambert, 1992). In this chapter, we deal only with finite mixtures of continuous distributions and focus on mixtures of normal distributions.

Figure 4.1 presents a few examples of mixtures of normal distributions. In each panel, the scaled densities $p_k f_k(x)$, the additive terms in Equation (4.1), are drawn by thin lines and the density of the mixture, their total, by a thick line. Panels A, B and C display densities of mixtures of two components. The components in panel B have the same expectation; $\mu_1 = \mu_2 = 0$. The density of such a mixture is symmetric, but cannot be matched by a single normal density. The density in panel C is bimodal. A three-component mixture of normal distributions can have one, two or three modes; panel D displays an example of the latter.

A mixture component is formally defined as the pair (p_k, D_k). Instead of a distribution D_k, we can refer to a density, f_k, a set of probabilities for discrete distributions, or to the corresponding distribution function F_k. In many contexts, all that matters is the scaled density $p_k f_k$ or scaled distribution function $p_k F_k$, so the pair (p_k, D_k) can be represented by this scaled density or distribution function.

Usually we do not know which unit originates from which distribution. In this case, the assignment of an observation x_i to one of the components is a standard problem. The assignment cannot be made with certainty when x_i is a plausible value of several components. Suppose the probabilities p_k and densities f_k are known. Then according to the Bayes theorem the probability that x_i belongs to component h is

$$\mathrm{P}(k_i = h \mid x_i) = \frac{p_h f_h(x_i)}{p_1 f_1(x_i) + \cdots + p_K f_K(x_i)}. \tag{4.2}$$

Denote this probability by $r_{i,h}$. Usually neither the probabilities p_k nor the distributions D_k are known, although the latter are often assumed to have a particular form. The choice is frequently guided by expediency (simplicity of the model-fitting algorithm), not any profound theory. For example, the distributions D_k may all be normal. The number of components K is usually not known either.

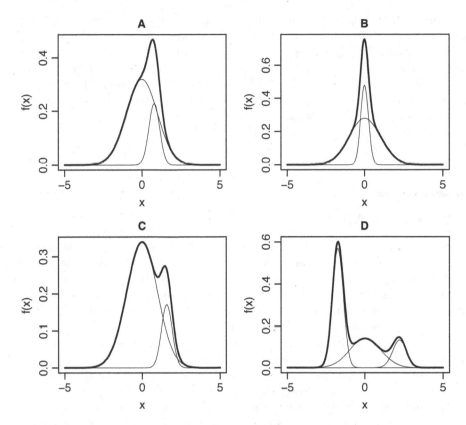

FIGURE 4.1
Densities of mixtures of normal distributions.

It is advantageous to borrow some terminology from linear algebra. We refer to the distributions D_k, or their classes, as the *basis* of the mixture, using an analogy with the basis of a linear space. The probabilities p_k can be regarded as the coordinates, although they are restricted to compositions. Mixtures generated by a finite basis are called finite. (The summation in Equation (4.1) could have infinitely many terms, letting $K \to \infty$.) A virtue of mixtures is that the space of distributions spanned by the basis of normal distributions is extremely wide. In fact, any continuous distribution can be approximated with arbitrary precision by a mixture of normals, with sufficiently many components K (Marron and Wand, 1992). Some other bases also have this property. For example, a histogram can be regarded as an approximation by a finite mixture of uniform distributions. The approximation can be made arbitrarily close with a sufficiently large sample and fine division of its support (the range of its values). The basis can be interpreted as a collection of models, one for each component. In our applications, each model is a random sample from a lognormal distribution. In general, these models may be much more complex, and may even differ in their complexity and type.

A permutation (reordering) of the components does not change the mixture. For example, the mixture model given by components (A_1, A_2, A_3), where $A_k = (p_k, D_k)$, is identical to the model (A_2, A_3, A_1). Therefore, we have to be careful about how we label the components to make unambiguous references to particular components. The problem does not arise when the bases of the components do not overlap.

4.2   Fitting Mixtures

In a typical problem, the number of components K is not given, and neither are the probabilities or parameters of the component (basis) distributions. In such a setting, fitting a mixture model, that is, estimating all the parameters involved in the components, including the probabilities p_k, by maximum likelihood, is an application of the EM algorithm (Dempster, Laird and Rubin, 1977; McLachlan and Peel, 2000).

The original motivation for the EM algorithm is the problem of missing values in a dataset. Its premise is that if no values were missing maximum likelihood estimation could be accomplished by a simple algorithm called the *complete-data algorithm*. Proceeding naively, we might estimate the missing values and apply this algorithm to the dataset completed with these estimates. This approach is deficient when some nonlinear transformations of the missing values are used in the complete-data algorithm. If a quantity θ is efficiently estimated by $\hat{\theta}$, then a nonlinear transformation $g(\theta)$ may be estimated by $g(\hat{\theta})$ inefficiently (Section 2.5).

The dataset that was planned to be recorded is called *complete* and the realised (recorded) dataset is called *incomplete*. The items of data that were not recorded are called *missing*. So, the complete data is a union (an amalgam) of the incomplete and missing data. We could deal with the complete dataset easily, by applying the complete-data algorithm, if the complete data were available. We can apply the algorithm to a *completed* dataset, but the result is problematic because no distinction is made in the algorithm between genuinely observed and imputed (estimated or constructed) values. In effect, we pretend to have more data than was collected.

In the EM algorithm, this general problem is resolved by estimating not the missing values directly, but their summaries for which the complete-data log-likelihood is a linear function. These summaries, known as linear sufficient statistics (for the missing values), may depend on some parameters that have to be estimated, and therefore estimation of the missing-data summaries and of the model parameters has to be iterated. The EM algorithm comprises iterations of pairs of steps. In the E (expectation) step, the summaries of the missing data are estimated by their conditional expectations given the data and the provisional values of the parameter estimates involved in the components' distributions. In the M (maximisation) step, the log-likelihood is maximised, with these estimated summaries substituted for their unknown values. In many settings, the algorithm converges slowly, but an iteration is performed so quickly that hundreds or even thousands of iterations can be executed in a short time. The real virtue of the algorithm is its easy implementation, not computational speed, although there are a variety of improvements. For a comprehensive review, see Meng and van Dyk (1997).

The EM algorithm can be applied even when we have not planned to collect the dataset that we declare as complete. Simply, the missing data is an analyst's construct, not a defect of the realised dataset. For mixtures, the missing data are the assignments of the observations to the components, the indicators $J_i(h)$, which are equal to unity if observation i belongs to component h, that is, when $k_i = h$, and are equal to zero otherwise. The complete-data log-likelihood is

$$L_{\mathrm{CMP}} = \sum_{k=1}^{K} \sum_{i=1}^{n_k} \log \left\{ f_k(x_{i,k}) \right\} , \qquad (4.3)$$

where f_k is the density of component k and the labelling of the units is changed so that $x_{1,k}, \ldots, x_{n_k,k}$ are the values of the (n_k) units in component $k = 1, \ldots, K$. If we knew which observation belongs to which component, the (complete-data) analysis would entail estimating the parameters of each component density f_k based on its n_k observations. This is a simple task with several classes of distributions, and the normal in particular. The marginal probabilities p_k would also be estimated straightforwardly, as $\hat{p}_k = n_k/n$, where $n = n_1 + \cdots + n_K$.

The incomplete-data log-likelihood is

$$L_{\mathrm{INC}} = \sum_{i=1}^{n} \log \left\{ f(x_i) \right\} , \qquad (4.4)$$

where $f(x)$ is the density of an observation with value x, given by Equation (4.1). Here the observations are labelled $i = 1, \ldots, n$. Direct maximisation of L_{INC} would be challenging because the derivative of each term in Equation (4.4) is an unwieldy expression that involves the ratios of densities f_k/f.

In the application of the EM algorithm to mixture models, we set first the initial (provisional) values of all the model parameters: the parameters involved in the densities f_k, $k = 1, \ldots, K$, and the values of the probabilities p_k. For example, when a component k has normal distribution, the initial values have to be set for the expectation μ_k and variance σ_k^2. Denote them by $\hat{\mu}_k^{(0)}$ and $\hat{\sigma}_k^{2,(0)}$ and let $\hat{p}_k^{(0)}$ be the initial value for p_k.

The E step in iteration t evaluates the expectations of the indicators $J_i(k)$ given the current (provisional) values of all the parameter estimates, set initially or estimated in the M step of the previous iteration. These expectations are the (estimated) probabilities $\hat{r}_{i,k}$. They are obtained by Equation (4.2), with the (provisional) estimates substituted for the probabilities p_k and the parameters involved in the densities f_k, $k = 1, \ldots, K$. With a set of (estimated) probabilities $\hat{r}_{i,h}^{(t)}$, the expectations and variances of the normal component distributions are estimated in the M step of iteration t by

$$\hat{\mu}_h^{(t)} = \frac{1}{n\hat{p}_h^{(t-1)}} \sum_{i=1}^n \hat{r}_{i,h}^{(t)} x_i$$

$$\hat{\sigma}_h^{2(t)} = \frac{1}{n\hat{p}_h^{(t-1)}} \sum_{i=1}^n \hat{r}_{i,h}^{(t)} \left(x_i - \hat{\mu}_h^{(t)} \right)^2,$$

where $\hat{p}_h^{(t-1)} = (\hat{r}_{1,h}^{(t-1)} + \cdots + \hat{r}_{n,h}^{(t-1)})/n$ is the estimate of the marginal probability from the M step in iteration $t-1$. Further details, including conventions for setting the initial estimates and for terminating the iterations, are given in Appendix 4.A.1.

The observations are assumed to be independent. The sampling weights are incorporated by evaluating the log-likelihood, a sum over the observations, with the sampling weights scaled so that their total would be equal to the effective sample size, defined in Section 2.3. Scaling of the weights has no effect on the estimates. The degree of freedom lost due to estimating a (component) mean is ignored in all evaluations, to conform with estimation by maximum likelihood. All the samples in the applications we consider are very large, although some components have small probabilities and their model-related sample sizes np_k are quite small.

4.3   Examples

We fit mixtures of lognormal distributions to the values of the equivalised household income (eHI) for households in the countries in the European Union Statistics on Income and Living Conditions (EU-SILC) in 2010. We give details for Austria, and then summarise the results for the other countries. We choose the lognormal distributions as the basis of the mixture, because a single lognormal distribution offers a much better fit than a single normal or another familiar distribution. The fit with a single lognormal distribution provides a benchmark to which we can relate the fit with a mixture of two (or more) lognormals. With equal sampling weights, we obtain the value of the incomplete-data log-likelihood L_{INC} given by Equation (4.4) for every model fitted. An established way of comparing model fits is by reference to the likelihood ratio statistic (LR), equal to the -2-multiple of the difference of the log-likelihoods of the alternative models. (One model has to be a special case—a *submodel*—of the other.) Information criteria (Akaike, 1973; Schwartz, 1978; Claeskens and Hjort, 2008; and others) are generally regarded as superior to LR. We regard such model comparisons as inappropriate because their outcomes are strongly affected by the sample size. We compare the fitted densities instead.

The data for Austria in the cross-sectional part of EU-SILC for 2010 comprises 6188 households. The weighted mean of $\log(1 + \mathrm{eHI})$ for the households is $\hat{\mu} = 9.912$ and the weighted standard deviation is $\hat{\sigma} = 0.509$. The weights make little difference; without them, the mean and standard deviation have respective values 9.932 and 0.510. The weighted sample mean of eHI (on the linear scale) is 22872.99 Euro with standard deviation 12991.73 Euro. The mean differs substantially from $\exp(\hat{\mu}) = 20162.51$, but is much closer to $\exp(\hat{\mu} + \frac{1}{2}\hat{\sigma}^2) = 22949.10$.

Table 4.1 displays the solutions for two to five components. All the mixture models are fitted using the sampling weights. The order (labelling) of the fitted components depends on the initial solution, and is essentially arbitrary. For a given labelling, we use the notation $k(K)$ for component k of the fit with K components. The components in the table are sorted in the descending order of the estimated probability $\hat{p}_k$. For $K = 2$, component 1(2) is dominant, with $\hat{p}_1 = 0.94$. The minority component 2(2), with $\hat{p}_2 = 0.06$, has extremely large standard deviation, $\hat{\sigma}_2 = 1.14$. As $\exp(1.14) \doteq 3.1$, $\hat{\mu}_2 \pm \hat{\sigma}_2$ corresponds to over three times greater or smaller value of eHI. It seems that component 2(2) fits mainly the tails of the distribution and, since $\hat{\mu}_2 < \hat{\mu}_1$, it has a greater weight on the small values of eHI. We can confirm this by plotting the fitted probability $\hat{r}_{i,2}$ as a function of log-eHI; see below.

The information in Table 4.1 is presented graphically in Figure 4.2. Each component $k(K)$, $2 \leq K \leq 5$, is represented by a horizontal segment extending from $\hat{\mu}_k - q\hat{\sigma}_k$ to $\hat{\mu}_k + q\hat{\sigma}_k$. The factor q is an argument of the R function by which the diagram is drawn. It is set to 0.5. For greater values of q either the horizontal axis is too wide or its scale does not cover all the segments. The size of the knot at $\hat{\mu}_k$ is related to the probability $\hat{p}_k$. The fitted marginal probabilities, converted to percentages, are listed at the right-hand margin.

The left-hand panel of Figure 4.3 reproduces the histogram of the values of log-eHI for Austria in 2010. Added in the panel is the fit by a single lognormal distribution (dashes), to show that it is unsatisfactory, especially at the mode of the distribution. The fit by the two-component mixture (thick solid line) is superior; the scaled densities of its components, $\hat{p}_1\hat{f}_1$ and $\hat{p}_2\hat{f}_2$, are plotted by thin solid lines and indicated by the math symbols. The scaled density for the majority component 1(2) is difficult to distinguish from the fit because the scaled density of minority component 2(2) is very small throughout.

The right-hand panel confirms that the fitted probabilities $\hat{r}_{i,2}$ are deterministically related to the values of log-eHI. Exceptionally large and small values x of log-eHI are assigned to component 2(2) with near certainty, because for them $\hat{p}_2\hat{f}_2(x) \gg \hat{p}_1\hat{f}_1(x)$. Values x close to the national log-mean, around 9.9, are assigned to component 1(2) with near certainty; there, $\hat{p}_1\hat{f}_1(x) \gg \hat{p}_2\hat{f}_2(x)$. There are two relatively narrow ranges of values of log-eHI for which $\hat{r}_{i,2}$ is in the range $(0.1, 0.9)$, where the assignment is with consider-

TABLE 4.1
Mixture model fits to log-eHI for Austrian households in 2010.

	Components					Deviance (Iterations)
$K = 2$						
$\hat{p}$	0.939	0.061				7116.41
$\hat{\mu}$	9.922	9.757				(124)
$\hat{\sigma}$	0.436	1.141				
$K = 3$						
$\hat{p}$	0.791	0.200	0.008			7089.98
$\hat{\mu}$	9.908	9.985	8.484			(410)
$\hat{\sigma}$	0.401	0.703	1.344			
$K = 4$						
$\hat{p}$	0.677	0.286	0.028	0.009		7033.40
$\hat{\mu}$	9.927	9.984	9.237	8.600		(841)
$\hat{\sigma}$	0.361	0.656	0.048	1.338		
$K = 5$						
$\hat{p}$	0.413	0.405	0.146	0.031	0.005	7029.90
$\hat{\mu}$	10.013	9.887	9.895	9.236	8.114	(782)
$\hat{\sigma}$	0.458	0.302	0.771	0.053	1.394	

able uncertainty. These ranges, (8.336, 9.127) and (10.772, 11.565), are delineated by short horizontal segments. They contain 455 (213+242) households, 7.35% of the sample.

The densities fitted with up to five lognormal components are plotted in Figure 4.4. The marginal densities are plotted by thick solid lines and the (scaled) component densities by thin lines of the indicated type. There is an appreciable difference in the fits for the two- and three-component mixtures. The differences between the four- and five-component solutions are imperceptible, even though the components of the two fits differ substantially. Both solutions have a secondary mode at 9.24. The fitted means of components 3(4) and 4(5), both with very small standard deviations (0.05), are close to this value; see Table 4.1. The fits of the four- and five-component models are very close at the principal mode 9.92. Note that the unequal weights of the households are not reflected in the backdrop histogram, and so it is somewhat distorted. On the basis of these fitted densities we choose $K = 4$.

4.3.1   Exploration of the Fitted Probabilities

The values of the fitted probabilities $\hat{r}_{i,k} \doteq 1$ and $\hat{r}_{i,k} \doteq 0$ correspond to the near certainty that unit i belongs, or does not belong, to component k.

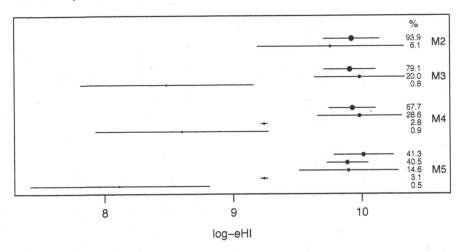

FIGURE 4.2
Mixture model fits to log-eHI for Austrian households in 2010.

Values closer to 0.5 correspond to greater uncertainty. In the exploration that follows, we regard the estimates $\hat{r}_{i,k}$ as if they were the genuine probabilities, even though they depend on $\hat{p}_k$, $\hat{\mu}_k$ and $\hat{\sigma}_k$, and are therefore subject to uncertainty even for a fixed value x.

A simple summary of the probabilities in the two-component fit is a table of rounded probabilities $\hat{r}_{i,1}$, or their complements $\hat{r}_{i,2}$. It is more practical to take the leading decimal digit instead of rounding; this corresponds to rounding downwards (function `floor` in R). We obtain the counts in the top part of Table 4.2. Thus, $5679 + 54 = 5733$ households (92.6% out of 6188) are classified with fair certainty as to whether they belong to one component or the other. Of these, $4890 + 43 = 4933$ (79.7%) have values of $\hat{r}_{i,1}$ smaller than 0.05 or greater than 0.95, but only 23 have values of $\hat{r}_{i,1}$ outside the range $(0.01, 0.99)$, all of them smaller than 0.01. The largest fitted probability $\hat{r}_{i,1}$ is 0.976, but for 1749 households $\hat{r}_{i,1} > 0.975$.

For the three-component fit we have to tabulate pairs of probabilities, so the summary is a two-way table, displayed in the bottom part of Table 4.2 for the respective components 1(3) and 2(3) in the rows and columns. Most of the households are assigned to the diagonal or just above it; these households have very small probabilities of being assigned to (the minority) component 3(3). For example, 67 households in the cell $(4,5)$ have $\hat{r}_{i,1} \in (0.4, 0.5)$ and $\hat{r}_{i,2} \in (0.5, 0.6)$, so $\hat{r}_{i,3} = 1 - \hat{r}_{i,1} - \hat{r}_{i,2} < 0.1$. Each of the ten households in cell $(0,0)$ has a substantial probability of being assigned to component 3(3). For them, $\hat{r}_{i,1} + \hat{r}_{i,2} < 0.2$, so $\hat{r}_{i,3} > 0.8$. The 41 households in cells $(0,1)$, $(0,2)$, ..., $(0,8)$, have nontrivial probabilities of being assigned to component 3.

The assignment probabilities can be presented by a *ternary plot*. It is

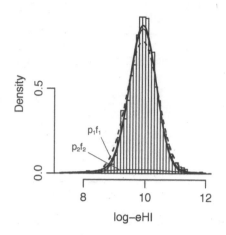 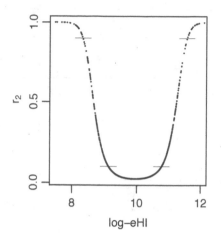

FIGURE 4.3
Log-eHI and fitted probabilities $\hat{r}_{i,2}$.

an equilateral triangle, with vertices corresponding to the components. Each household i is represented in it by a dot. Its distances from the vertices are proportional to the probabilities of not being assigned to the components, that is, $1 - \hat{r}_{i,k}$, $k = 1, 2, 3$. In the triangle with vertices at $(-1, 0)$, $(1, 0)$ and $(0, \sqrt{3})$ for the respective components 1, 2 and 3, household i is represented by the point with coordinates $(1 - \hat{r}_{i,3} - 2\hat{r}_{i,1}, \hat{r}_{i,3}\sqrt{3})$. Points close to a vertex correspond to near certainty of belonging to the component represented by the vertex. Points on or close to a side correspond to near certainty of not belonging to the component represented by the opposite vertex.

The ternary plot for the three-component fit for Austria in 2010 is displayed in the left-hand panel of Figure 4.5. A small amount of random noise is added in both horizontal and vertical directions to prevent massive overprinting of the points and to better indicate their density. The deviations are generated as a random sample from a centred normal distribution with a small standard deviation (set to 0.006 in the diagram).

The plot shows that only a few households are assigned with a high level of certainty to a component, namely to 2(3), but one of the three components can be ruled out for most households—their points are located on the sides of the triangle. Further, the component that can be ruled out is either 1(3), for a relatively few households with their points on or near side (2, 3), or 3(3), for the many households whose points are on or near side (1, 2). The nine households with $\hat{r}_{i,3} > 0.95$ are in fact those with the lowest values of eHI, between 67 Euro (one household) and 1500 Euro. Further 30 households have $\hat{r}_{i,1} < 0.01$—their points are also located on the side (2, 3). They comprise 15 households with values of eHI up to 3472 Euro, the lowest except for the nine households assigned to component 3(3) with near certainty, and 15 households

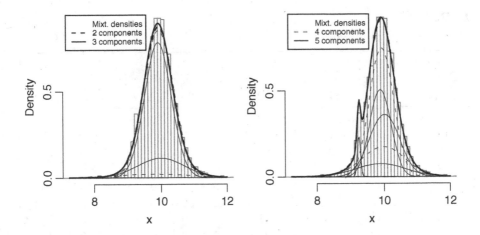

FIGURE 4.4

Fitted densities with two to five mixture components; Austria in 2010.
Some scaled densities with large standard deviations appear as flat lines indistinguishable from the constant zero.

with the highest values of eHI, exceeding 113 500 Euro. The latter households
have $\hat{r}_{i,2} > 0.95$. None of the households is assigned to component 1(3) with a
high level of certainty; the highest fitted probability $\hat{r}_{i,1}$ is slightly above 0.87,
but $\hat{r}_{i,1} > 0.870$ for 1323 households (21%) with eHI in the range (17 250,
21 844) and $\hat{r}_{i,1} > 0.850$ for 3101 households (50%) in the range (14 316,
26 316) Euro. The plot can be described as two curves, an arc from vertex
3 heading toward vertex 1, and a near-horizontal line from vertex 2 toward
vertex 1 (but stopping way short of it). In fact, the households in the arc are
in the ascending order of eHI, and this order continues along the side (1,2) in
the direction of vertex 2.

We can summarise these observations as follows. Component 3(3) stands
out and a few (only the most exceptional) households are assigned to it with
near certainty. In contrast, the distinction between components 1(3) and 2(3)
is much more difficult to draw, and very few households are assigned to either
of them with a high level of certainty. A household with a nontrivial probability
of belonging to 3(3) almost certainly does not belong to 1(3).

The fitted probabilities with four or more mixture components are in general more difficult to describe and some improvisation is called for. With the
studied dataset, one fitted component, 4(4), has marginal probability smaller
than 0.01 and only 69 households have probabilities of assignment to this
component greater than 0.1. We could therefore construct the ternary plot for
the remaining households and the other three components, and describe the
singled-out households separately. They are exceptional; the component has
a very large standard deviation and its main role is to account for extreme

TABLE 4.2

Fitted assignment probabilities $\hat{r}_{i,k}$ for models with two and three components.

Two components										
Digit	0	1	2	3	4	5	6	7	8	9
Count	54	11	12	23	23	29	42	78	237	5679
Three components										
0	10	1	4	1	2	5	10	15	3	42
1	0	0	0	0	0	0	0	16	32	0
2	0	0	0	0	0	0	16	36	0	0
3	0	0	0	0	0	13	50	0	0	0
4	0	0	0	0	18	67	0	0	0	0
5	0	0	0	14	126	0	0	0	0	0
6	0	0	30	279	0	0	0	0	0	0
7	0	47	853	0	0	0	0	0	0	0
8	0	4498	0	0	0	0	0	0	0	0
9	0	0	0	0	0	0	0	0	0	0

values (outliers). The right-hand panel of Figure 4.5 presents this ternary plot, based on the conditional probabilities $\hat{r}_{i,k}/(1 - \hat{r}_{i,4})$, $k = 1, 2, 3$. Components 1(4) and 2(4) account for more than 96% of the households. Component 3(4) has very small standard deviation (0.048). The density of eHI at its mode, 10 270 Euro, is high but after scaling by the marginal probability (0.028) it is so small that even a household with the modal value of eHI is assigned to it only with probability 0.52. The fitted probability $\hat{p}_{i,3}$ exceeds 0.1 for values of eHI in the range (9193, 11 291) Euro.

Thus, with a more complex model we have greater uncertainty about the assignment. This is not a problem, because the main goal of modelling is a good fit of the marginal distribution. Arguably, three components are sufficient for this, although we are on the safe side by chosing four components. The fifth component is clearly not necessary. One might wish to forego the uncertainty and define the assignment of each household i to the component with the highest fitted probability $\hat{r}_{i,k}$. The resulting assignment may be in disagreement with the marginal probabilities. It would inappropriately treat a household assigned to a component with probability 0.4 (when the other probabilities are, say, 0.30, 0.25 and 0.05) on par with a household with probability 0.95. See Basford and McLachlan (1985) for a study of this and related allocation rules. An alternative way of assigning is by random draws from the multinomial distributions with the estimated compositions $(\hat{r}_{i,1}, \hat{r}_{i,2}, \hat{r}_{i,3}, \hat{r}_{i,4})$. With such an assignment, we may commit more 'errors' of misplacing individual households, but will obtain sets of households that represent the components much better.

Figure 4.6 displays the plots of the fitted probabilities for four pairs of

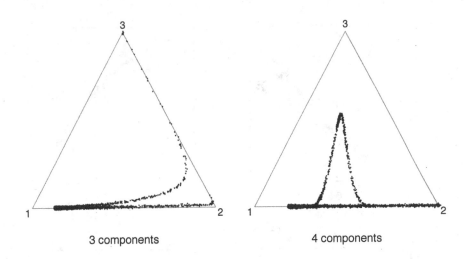

3 components   4 components

FIGURE 4.5
Ternary plots of the fitted assignment probabilities for the three- and four-component mixture model fits; Austria in 2010.

components. Each pair is formed by a component from the model with three and one with four components. The fitted probabilities $\hat{r}_{i,1}^{(3)}$ and $\hat{r}_{i,1}^{(4)}$ are related by a monotone function in panel A, except for the apparent intervention of the small-variance component 3(4). Panel B shows that large probability $\hat{r}_{i,1}^{(3)}$ is incompatible with large $\hat{r}_{i,2}^{(4)}$, since $\hat{r}_{i,1}^{(3)} + \hat{r}_{i,2}^{(4)} \doteq 1$ for many units i, but it is a precondition of nontrivial probability $\hat{r}_{i,3}^{(4)}$ (panel C). Being in component 3(3), with large variance, almost rules out being in component 3(4), with small variance (panel D).

We may be tempted to ascribe some meaning to each component and interpret it as a distinct subpopulation. This is not always appropriate because the main function of the components is to approximate (reconstruct) the marginal distribution. Our conclusion about the components is highly contingent on the basis we choose. With a different basis for the mixture, comprising some distributions other than lognormal, a different set of components may be obtained. With the basis of lognormal distributions, the first condition for an appropriate interpretation is that eHI in each subpopulation is lognormally distributed. Next, we have to resolve observations with uncertain assignment, which lie in ranges where more than one scaled density is substantial. And finally, we have to address sampling variation. That means not only estimation errors for all the parameters involved, but also the labelling (numbering) of the components and the possibility that a different number of components may fit best to a replicate dataset, in which case a different set of parameters is estimated.

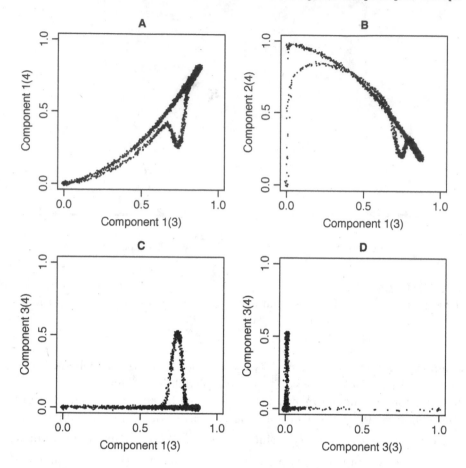

FIGURE 4.6
Assignment probabilities for alternative mixture models; Austria in 2010.

An exercise left for the reader is to fit a mixture of normal distributions (without the log transformation) and relate the fitted components to those in Table 4.1. No simple correspondence will be found.

4.3.2 Results for Several Countries

The results for the 15 countries studied in the previous chapters, or for all countries in EU-SILC in 2010, would take up several pages in tabular form, and they would be difficult to digest. A more compact summary of the results is presented in Figure 4.7 using the same layout as Figure 4.2. All but two countries, Denmark and Norway, are represented by their four-component model fits.

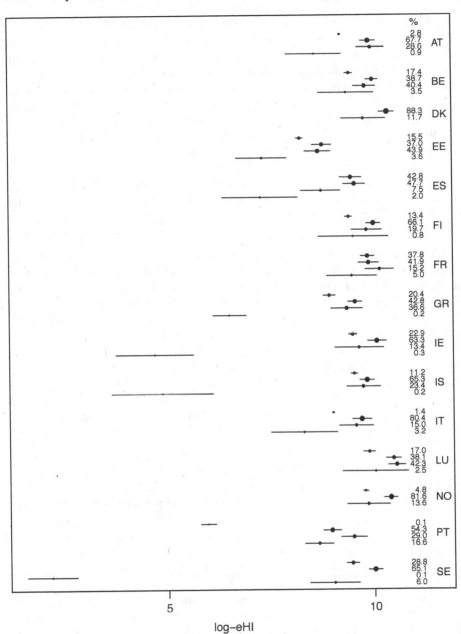

FIGURE 4.7

Fits of the four-component mixture model to the countries in EU-SILC since 2004; households in 2010.

Two components are fitted for Denmark and three for Norway. The components are in the ascending order of the fitted standard deviation.

For Denmark, a singularity occurs in the iterations of the EM algorithm, whereby the probability of a component converges to zero. The algorithm would crash (reach an underflow), but this is prevented by stopping the iterations when a fitted probability becomes smaller than 0.0001. If this occurs, the model with one fewer component should be fitted, because convergence may not have been reached for the components that have nontrivial fitted probabilities. A very small fitted standard deviation (smaller than 0.0001 in our implementation) indicates a set of (nearly) equal values. The mixture model should be refitted with these observations removed. For Denmark, singularity occurs not only when fitting four components, but also with three components. For Norway, the problem occurs only with four components. In the diagram, the two-component fit is displayed for Denmark and the three-component fit for Norway. The components of a country are sorted in the ascending order of the fitted variances.

Several countries have a component with very small marginal probability which has a small fitted mean and large standard deviation; Ireland and Iceland are extreme cases. The minority components for Greece, Italy and Portugal also have small fitted means but their standard deviations are not large. Several countries have a component each with extremely small fitted standard deviation but not exceptional mean: Austria, Belgium, Estonia, Finland, Ireland, Iceland, Italy and Norway. For each of these countries, the component fits a small bump in the distribution of eHI. For Belgium, the second and third components have similar fitted probabilities (0.40 and 0.39), so the issue of appropriate labelling arises. Similar 'plurality' ($\hat{p}_1$ and $\hat{p}_2$ of similar magnitude) occurs also for Spain, France, Greece and Luxembourg.

A component with a small standard deviation may indicate a frequently reported income, such as a rounded figure. However, with several sources of the income, several earners in a household and different household compositions, such a 'blip' in the distribution is likely to melt away and should not influence the mixture analysis. The relevance of such an hypothesis is best tested by fitting the mixture models to the data from 2009. This is left as an exercise for the reader. The robustness of the mixture model fits can also be explored by infiltrating the data with one or a few observations with exceptional values, as done in Section 3.3. In a more involved exercise, the dataset is infiltrated by a sample of values drawn from a distribution. Of interest is whether these new observations are detected as a component or they melt away among some of the original components or the fit becomes a set of components that cannot be straightforwardly related to the original components.

From the analysed data, we excluded all households with negative or zero values of eHI, because the model fitting algorithm would converge to identifying them as a separate component with zero expectation and very small standard deviation. This would lead to a singularity and a failure of the algorithm. Without these households, singularity was encountered for four components only for Denmark and Norway. For five-component models it occurs for several countries. In most cases, the fit comprises an easily identifiable pair of components formed by splitting a component of the four-component solution.

4.4   Improper Component

The estimated probabilities of assignment can be viewed as the result of a contest of the components for each observation. The components compete for x_i by means of their fitted scaled densities $\hat{p}_k \hat{f}_k(x_i)$, $k = 1, \ldots, K$. Higher value of one, with the other values held fixed, results in greater gain, as assessed by the fitted probability $\hat{r}_{i,k}$. Each (normal) component has its mode at its expectation, so it might appear to be most competitive for observations in its vicinity. However, a component with a large variance (flat density) loses out in the contest even for the observations with values near its mode to components with more concentrated densities.

If we accept that one (or a few) components should have the role of accounting for exceptional observations, why not define a component specifically for this purpose? Such a component should have a small density for the unexceptional values of X and be a clear winner for exceptional values. The constant satisfies this condition, but it is not a density because its integral is not defined (is infinite). Nevertheless, we can define such a component, called *improper*, because all the formulae used, and Equation (4.2) in particular, are well-interpreted with it. The EM algorithm is easy to adapt for an improper component. The component involves no parameters; the value of the constant has to be set. There are alternatives to the constant improper density, but they are of little practical use. For example, we could take the normal density, which has a mode and thin tails, and turn it upside down, so that it would have a specified antimode (a point of minimum density) and tails converging to a positive value or diverging to $+\infty$. We obtain the improper density

$$f(x) = H_1 \left[1 - \exp\left\{ -H_2(x - U)^2 \right\} \right],$$

where U is the antimode and $H_1 > 0$ and $H_2 > 0$ are parameters.

We use a constant improper density throughout. The value of the density is denoted by D. The value of this constant cannot be estimated and has to be set by trial and error, matching the anticipated proportion of exceptional observations. The improper component has label 0, and is denoted by $0(K)$ in a model with K proper components. For Austria in 2010, with a single normally distributed (proper) density, we start by setting $D = 10^{-d}$, $d = 0.5, 0.6, \ldots, 1.0$. For $d = 0.5$ ($D \doteq 0.32$), the estimate of the fraction belonging to the improper component is $100\hat{p}_0 = 9.4\%$, more than the percentage of exceptional values that one could reasonably expect. For greater d (smaller D), the fraction is smaller, but it remains positive even for very small D:

$100\hat{p}_0 = 0.06\%$ for $d = 6$ and $100\hat{p}_0 = 0.05\%$ for $d = 7$. Reasonable choices for d are around 1.0 (that is, $D \doteq 0.1$), for which $100\hat{p}_0 \doteq 1.1\%$.

The estimated mean of the proper component increases with d up to $d = 0.8$ and then it decreases; it is 9.916 for $D = 10^{-0.5}$, 9.919 for $D = 10^{-0.8}$ and 9.914 for $D = 10^{-7}$. The estimated standard deviation increases with d, from 0.412 for $D = 10^{-0.5}$ to 0.499 for $D = 10^{-7}$. With increasing negative exponent d (decreasing D), first the households with very high income and then the households with very low income dominate the recruitment into the improper component. As fewer extreme observations are won by the improper component the fitted variance of the proper component increases.

This can be confirmed by inspecting the fitted probabilities. In R, we collate the probabilities $\hat{r}_{i,0(d)}$, $d = 0.5, \ldots, 0.9, 1, \ldots, 7$, into a $n \times p$ matrix called AT10r, and then evaluate

```
rng <- apply(sign(apply(AT10r, 1, diff)), 2, range)
```

which summarises the within-household differences of consecutive values of $\hat{r}_{i,0(d)}$ by the minimum and maximum of their signs, formatted as a $2 \times n$ matrix. It suffices to check that $+1$ is not the maximum for any household. More detail is obtained by tabulating the first row of rng against the second, obtaining for Austria in 2010 the table

```
        -1     0
   -1   27  6159
    0    0     2
```

For the two households in cell $(0, 0)$, which have the smallest values of eHI (67 and 216 Euro), $\hat{r}_{i,0} = 1.0$ for all the exponents d. For the next three households in the ascending order of eHI (453.60, 562.00 and 868.00 Euro), $\hat{r}_{i,0(d)}$ equals to 1.0 up to a certain value of d and then it decreases. For 27 households counted in cell $(-1, -1)$, $\hat{r}_{i,0}$ decreases with each increment of d. These households have values of eHI up to 3281 Euro (14 households) and above 122750 Euro (13 households). For the remaining 6159 households, which have values of eHI in the range $(3821, 122\,750)$, $\hat{r}_{i,0}$ is positive for $d = 0.5$, decreases with d and reaches zero for $d = 6$ or earlier.

The association of eHI and the probabilities $\hat{r}_{i,0(d)}$ is described in Figure 4.8. The curves are plotted by connecting the points with coordinates $(x^{(i)}, \hat{r}_{i,0(d)})$, where $x^{(i)}$ is the ith smallest value of eHI. The curves have edges because of the relatively isolated values at which they are evaluated. It is essential to plot the curves on both linear and multiplicative scales (right-hand panel), to obtain high resolution at different regions of the support of eHI. The values of eHI are marked at the bottom of each panel by dots for values outside the range $(2000, 80\,000)$ Euro; the range is marked by a gray segment. The diagram confirms that the improper component is most competitive for extremely small and large values of eHI and that its fitted probability decreases with d (increases with D) for every value of eHI.

The fits of the model with two proper and one improper component are summarised in Table 4.3. The improper component has fitted probability

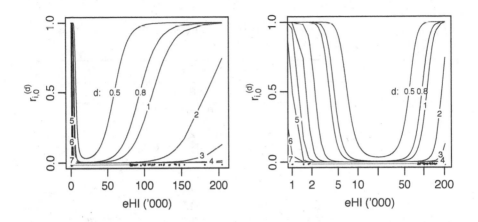

FIGURE 4.8
Fitted probabilities $\hat{r}_{i,0(d)}$ as functions of eHI.
The negative exponent of the improper density, $d = -\log_{10}(D)$, is marked on the curves.

around 0.01 for the negative exponent d around 0.75. The probability of the minority proper component 2(2) increases with D. The standard deviation of the majority component decreases with D very slowly. The standard deviation of the minority component also decreases, but at a much faster rate, because with increasing D most of the new recruits to the improper component are from the minority component—apart from the improper component, exceptional values are more attractive for the minority component. The changes in the fitted means reflect the balance of the new recruits from among the households with the smallest and largest values of eHI. The number of iterations is much greater than for the model with a single proper component (5–30). It increases as the two proper components become more similar and it is more difficult to distinguish them.

The solution with three proper components (see Table 4.1) is very similar to the solution for $d = 0.75$. The improper component plays a role similar to the proper component 3(3) that has a large variance. Convergence is achieved only after more than 5000 iterations. The fitted probability of the improper component is 0.006 for $d = 0.15$ and 0.019 for $d = 0.2$. The fitted distributions for $d = 0.15$ are $\mathcal{N}(9.90, 0.43^2)$, $\mathcal{N}(10.88, 0.03^2)$ and $\mathcal{N}(8.52, 0.43^2)$, with respective probabilities 0.95, 0.03 and 0.01. This is substantially different from any of the solutions in Table 4.1. For $d = 0.20$, we obtain the fitted distributions $\mathcal{N}(9.92, 0.33^2)$, $\mathcal{N}(9.96, 0.56^2)$ and $\mathcal{N}(9.24, 0.05^2)$, with respective probabilities 0.51, 0.44 and 0.03. This fit closely resembles the solution in Table 4.1 for $K = 4$ proper components, with the highly dispersed com-

TABLE 4.3
Fit of the model with two normally distributed (proper) components and one improper component with constant value $D = 10^{-d}$.

	d						
	2.0	1.0	0.9	0.8	0.75	0.7	0.6
$100\hat{p}_0$	0.01	0.21	0.38	0.74	1.12	1.84	4.37
$100\hat{p}_1$	93.01	87.28	84.08	79.82	75.71	66.24	49.90
$\hat{\mu}_1$	9.92	9.92	9.92	9.91	9.91	9.90	9.90
$\hat{\sigma}_1$	0.43	0.42	0.41	0.40	0.40	0.39	0.39
$100\hat{p}_2$	6.98	12.50	15.53	19.44	23.17	31.93	45.74
$\hat{\mu}_2$	9.79	9.90	9.92	9.94	9.95	9.95	9.97
$\hat{\sigma}_2$	1.07	0.85	0.78	0.71	0.66	0.59	0.49
Iterations	557	1543	1629	2226	3254	6364	5643

ponent 4(4) replaced by the improper component. The other three standard deviations are slightly smaller in the fit with the improper component.

There is no universal criterion for selecting one of the fitted models as the best. The models are approximations to the density of the observations, and their features gain credibility only when they appear in the fits of several alternative models. The best model, whichever way it is selected, is subject to uncertainty. Its worth is undermined when the second-best or another runner-up model yields a substantially different fit, and yet might be the 'best' in an hypothetical replication. In the analysis of several countries or years, it is advantageous to compromise on the selection of models by leaning toward the same model specification (bases, numbers of components and the presence of an improper component) for all the countries.

Banfield and Raftery (1993) proposed mixture models with a highly dispersed component for exceptional (outlying) observations. For more details about improper components, see Longford and D'Urso (2011).

4.5 Components as Clusters

Having concluded an analysis with a set of fitted components, we may want to decide whether they, or some of them, are distinct clusters. The first step in this process is a definition of the term 'cluster'. We commonly understand the term as delineating a subsample or a subpopulation which is well-separated or stands out from the rest of the sample or population. Being a cluster is a mutual relationship of a set of groups; it is a property of a group only in relation to a set of other groups.

We can detach the discussion from the context of mixture models and define a component generally as a pair (p, D) of a probability, $p \in (0,1)$, and a distribution D. We will regard components A and B as clusters if a randomly selected observation that belongs to one of them but we do not know to which, would be assigned to that component with high probability; that is, if the two components would be confused infrequently. For the assignment of an observation, we use the version of Equation (4.2) for two components. We assign observation x to component A with probability

$$r_{A|AB}(x) = \frac{p_A\, f_A(x)}{p_A\, f_A(x) + p_B\, f_B(x)}, \tag{4.5}$$

where $p_B = 1 - p_A$. This definition can be extended to the case in which $p_A + p_B \neq 1$, because $r_{A|AB}(x)$ depends on p_A and p_B only through the ratio p_B/p_A. The probability in Equation (4.5) is conditional on x. The subscripts of r indicate that the observation originates from A or B. The marginal probability of assignment to A of an observation that originates from B, for which we know only that it originates either from A or B, is obtained by integrating the probability in Equation (4.5) over the distribution of B:

$$r_{A|B} = \int \frac{p_A\, f_A(x) f_B(x)}{p_A\, f_A(x) + p_B\, f_B(x)}\, \mathrm{d}x. \tag{4.6}$$

We refer to $r_{A|B}$ as the index or rate of confusion of A by B. Components A and B are said to be confused if both $r_{A|B}$ and $r_{B|A}$ are greater than an a priori set threshold T. Thus, we would get the assignment defined by Equation (4.5) wrong quite often. Components A and B are said to be well-separated, and are clusters, if $r_{A|B} < T$ and $r_{B|A} < T$. The expression for $r_{A|B}$ is symmetric in A and B, except for the factor p_A in the numerator in Equation (4.6). We have the identity

$$p_B\, r_{A|B} = p_A\, r_{B|A}. \tag{4.7}$$

The rates $r_{A|B}$ and $r_{B|A}$ coincide only when $p_A = p_B$.

For classifying more than two components, we have three variants. A set of components are *pairwise* clusters if every pair of them are clusters. A component is a cluster with respect to the other components (*jointly*) if the component and its complement (union of the other components) are clusters. A set of components is *mutually* confused if the probability that a randomly selected unit from their mixture is assigned to the correct component is smaller than T. For a set of K components, this probability is

$$\int \frac{p_1^2 f_1^2(x) + p_2^2 f_2^2(x) + \cdots + p_K^2 f_K^2(x)}{p_1 f_1(x) + p_2 f_2(x) + \cdots + p_K f_K(x)}\, \mathrm{d}x. \tag{4.8}$$

It is easy to show that this integral is smaller than or equal to unity, and is equal to unity only when $f_{k_1}(x)f_{k_2}(x) = 0$ for all x and all pairs $k_1 \neq k_2$, that is, when the component densities have nonoverlapping supports. A subset of the components may satisfy a more stringent condition than all the components. A component with a small probability may have only a small impact on the probability in Equation (4.8), so good separation of some pairs of components may 'hide' poorer separation of some others.

Since in general $r_{A|B} \neq r_{B|A}$, the two rates can be separated by T. Then A and B are neither confused nor well-separated. Component A is said to be a satellite of another component B if $r_{A|B} < T$ and $r_{B|A} > T$. If A is a satellite of B, then necessarily $p_A < p_B$, because $p_A \geq p_B$ implies that $r_{A|B} \geq r_{B|A}$.

An essential property of a confusion index is that, for fixed probabilities p_A and p_B, it attains its maximum when A and B have identical densities. This is surprisingly easy to prove for continuous distributions by realising that the integrand in Equation (4.6) is related to the weighted harmonic mean,

$$W(x) = \left[\frac{1}{p_A + p_B} \left\{ \frac{p_B}{f_A(x)} + \frac{p_A}{f_B(x)} \right\} \right]^{-1}.$$

This is smaller than or equal to the corresponding arithmetic mean,

$$\frac{1}{p_A + p_B} \left\{ p_B \, f_A(x) + p_A \, f_B(x) \right\},$$

which integrates to unity. Therefore $\int W(x)\,\mathrm{d}x \leq 1$ and equality occurs only when $f_A(x) = f_B(x)$ for all x. Since the integrand in Equation (4.6) is equal to $p_A W(x)$, we have the inequality $r_{A|B} \leq p_A$, in which equality holds only when $f_A \equiv f_B$.

The integral in Equation (4.6) can be evaluated only by simulations for most of the common distributions. We draw a random sample $x_1, x_2, \ldots, x_m$ of large size ($m = 100\,000$) from the distribution of component B and approximate $r_{A|B}$ by the average of the values of the integrand $r_{A|AB}(x_i)$.

By way of example, we evaluate the confusion index for component $\{0.9, \mathcal{N}(0,1)\}$ by $\{0.1, \mathcal{N}(0.5, 0.5)\}$. The index is equal to $r_{A|B} = 0.8701$. By reversing the roles of the two components, we obtain the value of the index $r_{B|A} = 0.0967$. The two values satisfy Equation (4.7), subject to the uncertainty entailed in m replications. There is a small advantage in evaluating $r_{A|B}$ instead of $r_{B|A}$ when the distribution of B has smaller dispersion. The evaluation is quite fast, taking about 0.05 sec. for one index, so we do not have to be sparing with the number of replications m.

The confusion index is invariant with respect to the same strictly monotone transformation of both components. Therefore the index for a pair of components with lognormal distributions can be evaluated with reference to their normal generating distributions. Further, one of the components can be standardised to $\mathcal{N}(0,1)$. Figure 4.9 displays the values of the index $r_{A|B}$ for $\{0.9, \mathcal{N}(0,1)\}$ by $\{0.1, \mathcal{N}(\mu, 1)\}$ as function of μ, and for $\{0.9, \mathcal{N}(0,1)\}$ by $\{0.1, \mathcal{N}(0, \sigma^2)\}$ as function of σ. No smoothing was used in drawing the curves

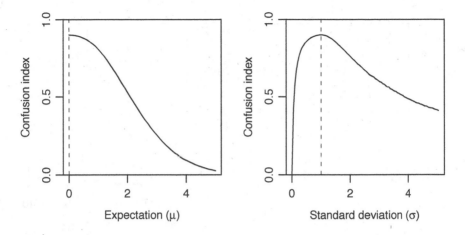

FIGURE 4.9

Confusion index $r_{A|B}$ of the standard normal distribution ($p = 0.9$) by a normal distribution with different expectation and variance.

in the diagram. Assuming that their exact versions are smooth, the roughness of the plotted curves would indicate poor approximation by the simulation.

As a function of μ, that is, of the standardised difference in expectation (with both variances equal to 1.0), the confusion index attains its maximum of $p_A = 0.9$ at $\mu = 0$, when the two distributions coincide, and it decreases, slowly at first and then much more rapidly, until it smoothly approaches zero; see the left-hand panel. For $\mu = 5$ the index attains the value 0.0264. As a function of σ (with identical expectations), the confusion index attains its maximum of $p_A = 0.9$ for $\sigma = 1$ and remains substantial for σ much greater than 1.0 (right-hand panel); for example, its value for $\sigma = 5$ is 0.413. For $\sigma = 0$, the index is equal to zero—the degenerate distribution (with all its mass at zero) is perfectly well-separated from any continuous distribution. For small $\sigma > 0$, the index increases rapidly.

Owing to the Equation (4.7), we do not have to simulate the values of the confusion index for different probabilities p_A and p_B. As a function of μ the index drawn in the left-hand panel is symmetric around $\mu = 0$. For the function $g(\sigma)$ in the right-hand panel, we have the identity $g(\sigma) = g(1/\sigma)$, so g is symmetric on the log scale, as a function of $\log(\sigma)$.

To complete the definition of a cluster, we have to set the threshold T. We could avoid this by quoting the two confusion indices, such as $r_{A|B} = 0.8701$ and $r_{B|A} = 0.0967$ in an earlier example. The two components are confused if $T < 0.0967$ and B is a satellite of A if $0.097 < T < 0.870$. We can set about fixing the value of T by the following exercise. We draw the scaled densities for two (normal) components and evaluate their confusion indices. We assess the components by their scaled densities whether we would like them to be

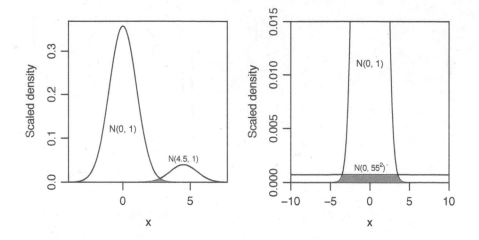

FIGURE 4.10
Pairs of scaled densities on the borderline of being clusters for $T = 0.05$.

regarded as clusters or not. By adjusting a parameter of one distribution (the other could be the standard normal), we find the borderline between being and not being clusters. The corresponding value of $\min(r_{A|B}, r_{B|A})$ is suitable for T. Figure 4.10 gives two examples of (normal) scaled densities which are on the borderline of being clusters with $T = 0.05$. In both, the majority component is standard normal with $p_A = 0.9$. The minority component in the left-hand panel is $\{0.1, \mathcal{N}(4.5, 1)\}$; its distribution differs only by the mean. The minority component in the right-hand panel is $\{0.1, \mathcal{N}(0, 55^2)\}$; its distribution differs only by the variance.

The confusion index can be interpreted as the area under both densities, highlighted in each panel by shading. When the distributions differ only in their expectations, the area is a small wedge at around $x = 2.8$ (left-hand panel). When the distributions differ only in their variances, the variance has to be so large that the density is effectively constant and the relevant area is a narrow horizontal strip. The vertical axis in the right-hand panel has a reduced scale; otherwise the density of the widely dispersed distribution could not be discerned from zero.

4.5.1   Confusion Matrix

For a set of $K \geq 2$ components, the confusion matrix is defined as the $K \times K$ matrix of confusion indices for the row by the column components. For example, the confusion matrix for the mixture model fit with $K = 4$ components for Austria in 2010, given in Table 4.1, is

$$
\begin{pmatrix}
0.500 & 0.675 & 0.860 & 0.675 \\
0.285 & 0.500 & 0.723 & 0.748 \\
0.035 & 0.071 & 0.500 & 0.290 \\
0.009 & 0.025 & 0.099 & 0.500
\end{pmatrix}. \tag{4.9}
$$

Thus, component $4(4)$, $\mathcal{N}(8.60, 1.34^2)$ with $p_4 = 0.01$, is a satellite of all the other three components, unless we choose a threshold T smaller than 0.099, in which case $3(4)$ and $4(4)$ are confused. The small-dispersion component $3(4)$, $\{0.028, \mathcal{N}(9.24, 0.05^2)\}$, is a satellite of components $1(4)$ and $2(4)$ for $T > 0.071$. For $0.035 < T < 0.071$, it is confused with $2(4)$. Components $1(4)$ and $2(4)$ are confused for $T < 0.28$.

We do not have to conclude with a categorical statement about clusters among the components. It suffices to state that components $3(4)$ and $4(4)$ come closest to being a pair of clusters, if we choose $T > 0.29$; both components $3(4)$ and $4(4)$ are satellites of both components $1(4)$ and $2(4)$ for $T > 0.071$, and for $T > 0.10$ component $4(4)$ is also a satellite of component $3(4)$. As an alternative, we may agree on a range of plausible values of T, such as $0.06 - 0.12$. Then all pairs of components are classified unequivocally, except for pairs $(2, 4)$ and $(4, 3)$ in which the first is a satellite of the second for T up to 0.071 and 0.099, respectively, and the pairs are confused otherwise.

The confusion index $r_{A|B}$ cannot exceed $p_A/(p_A + p_B)$. Therefore, components with small probabilities p_A are bound to be satellites of components with large p_B, unless they are well-separated, when even $r_{B|A}$ is small. For more details and examples related to confusion, see Longford and Bartošová (2014).

4.A   Appendix. Programming Notes

4.A.1   EM Algorithm for Mixtures of Normal Distributions

The function EUmix2 for fitting a mixture of normal distributions by the EM algorithm has the following arguments:

```
EUmix2 <- function(x, wei=1, sta=list(), K=0, maxit=20, tol=8)
{
###     EM algorithm for univariate mixtures (with sampling weights)

##  x      The outcome variable
##  wei    The vector of weights  (or scalar for equal weights)
##  sta    The initial solutions
##  K      The number of components (used only when length(sta)==0)
##  maxit  The maximum number of iterations
##  tol    The convergence criterion
```

The initial solution can be given as a list in which each element corresponds to a component and is a vector of length 3, comprising probability ($\hat{p}_k$), mean ($\hat{\mu}_k$) and standard deviation ($\hat{\sigma}_k$). The solution can also be given as a matrix with three columns for the corresponding sets of estimates. If an initial solution is given, then the number of components is set to the length of the list or to the number of rows of the matrix in the argument sta. The default is an empty list; if it applies, then the number of components has to be given by the argument K. If sta is not empty, then the value of K is ignored. The objects maxit and tol are positive scalars related to the iterations of the EM algorithm. They have roles similar to their namesakes in function EUnewton; see Appendix 3.A.2.

The number of observations is defined by the length of the first argument x. The vector of weights has to have the same length; the only exception permitted is for equal weights, which can be specified by a positive scalar argument or as an empty vector, c(). Negative values of wei are changed to zero, which amounts to discarding the corresponding observations.

```
##  The number of observations
n <- length(x)

##  Check on the sampling weights
if (length(wei) < 2) wei <- rep(1, n)

if (length(wei) != n)
    stop(paste("Incompatible lengths of  x  and  wei.",
               n, "vs.", length(wei)))

##  Negative weights are changed to zero
wei[wei < 0] <- 0

##  The total of the weights
swe <- sum(wei)

##  The effective sample size
nb <- sum(wei)^2/sum(wei^2)
```

If not given, the initial solution is set next; otherwise it is arranged in a format convenient for further use. The initial solution is set by a method with no sophistication, but it works for all countries, years and numbers of components up to six.

```
##  Check on the number of components
if (length(sta) + K == 0)
    stop("Cannot ascertain the number of components.")

##  The sample mean and standard deviation
MV <- MeVa(x)

##  If initial solution not given
```

```
if (length(sta) == 0)
estN <- cbind(pp=2*seq(K)/K/(K+1),
         mu=seq(MV[1] - MV[2], MV[1] + MV[2], length=K),
         sig=MV[2] * seq(0.5, 2.5, length=K) )

## If initial solution given
else
{
## Organise the initial values into a matrix
if (is.list(sta))
    estN <- matrix(sapply(sta, ExtrE, seq(3)), ncol=3, byrow=T)

if (is.matrix(sta)) estN <- sta[, 1:3]

## The number of components
K <- nrow(estN)
} ## End of the condition over the initial solution
```

The function MeVa returns the mean and standard deviation of a vector. The function ExtrE extracts an element (or a set of elements) of a vector; see Appendix 1.A.4.

The matrix estN contains the current (provisional) estimates of the marginal probabilities p_k, the means μ_k and the standard deviations σ_k as its columns. The deviance is defined as the -2-multiple of the log-likelihood. Before the first iteration, it is evaluated with the weights assumed to be equal.

```
## The initial deviance
devN <- n*(1+log(2*pi)) + (n-1)*var(x)
```

Before the iterations, a scalar that indicates whether further iterations are necessary (dist) and the iteration counter (iter) are initialised. Also, the matrix Conver is initialised for storing the results of the iterations.

```
## For the iterations
dist <- 0
iter <- 0

## The convergence monitor
Conver <- matrix(c(estN, devN, dist), 1)
```

The iterations of the EM algorithm are performed until their number reaches its specified maximum, or a convergence criterion, specified below, is satisfied.

```
while (iter < maxit & dist < tol)
{
## The iteration counter
iter <- iter + 1
```

The value of dist is established at the end of the iteration. The current values of the parameter estimates and of the deviance are copied, so that when they are updated the magnitude of the changes can be established.

```
##   Store the current fit
estP <- estN
devP <- devN
```

The scaled densities are evaluated in a loop over the components and stacked as the columns of matrix dns.

```
##   Initialisation for the scaled densities
dns <- c()
```

```
##   The current scaled densities
for (i in seq(K))
dns <- cbind(dns, EUdns(x, estN[i, ]))
```

The function EUdns evaluates a scaled density on a vector of points

```
EUdns <- function(Pts, Prs)
###     Evaluation of a scaled normal density
```

```
##   Pts    The evaluation points
##   Prs    The parameters (p, mu, sig)
```

```
Prs[1] * dnorm(Pts, Prs[2], Prs[3])
```

Braces, { and }, are not necessary to delimit this function because it has only one evaluation. It does not have to be assigned to an object because, being the only, it is also the last evaluation.

The matrix of probabilities $\hat{r}_{i,k}$ is obtained from the matrix of values of the densities dns by applying the function Prob on its rows. The function standardises a vector to have unit total.

```
##   The conditional probabilities   --   the E step
pro <- apply(dns, 1, Prob)
```

In the following expressions, the current deviance is evaluated and the M step is executed. The estimated means and standard deviations are evaluated with weights equal to the product of the current probabilities $\hat{r}_{i,k}$ and the sampling weights w_i. The within-component totals related to $\hat{p}_k$ are calculated first and are adjusted (standardised) later.

```
##   The deviance
devN <- -2*sum(wei * log(apply(dns, 1, sum)))/swe * nb
```

```
##   The M step
ppN <- pro %*% matrix(wei)
muN <- pro %*% matrix(wei * x) / ppN
```

```
estN <- cbind(p=ppN/swe, mu=muN,
              sig=sqrt((pro %*% matrix(wei * x^2))/ppN - muN^2) )
```

Each iteration is concluded by evaluating a summary of the changes made by the iteration and by storing the new solution. The scalar dist evaluates the negative $\log_{10}$ of the average of the mean squared differences of the estimated means, standard deviations and marginal probabililites and the squared change of the deviance. Large values of `dist` correspond to small changes. The iterations are stopped when `dist` exceeds `tol`; `tol` can be loosely interpreted as the number of decimal places of precision. Also, the iterations are stopped if an estimated probability or a standard deviation are very small.

```
##  The convergence criterion
dist <- -log(mean((estN - estP)^2) + (devN-devP)^2, base=10)/2


##  Check on the singularity
sng <- (min(estN[, -2]) < 0.0001)


##  Crash out from the iterations
if (sng) dist <- tol+1


##  Updating the convergence monitor
Conver <- rbind(Conver, c(estN, devN, dist))
}  ##  End of the iterations (while)
```

See Appendix 3.A.2 for similar use of objects `dist`, `tol`, `iter`, `maxit` and `Conver`. The result of applying `EUmix2` is formatted as a list comprising `Conver`, suitably labelled, and the estimated probabilities $\hat{r}_{i,k}$ in the matrix `pro`. A warning is issued if the maximum number of iterations is exhausted without reaching convergence.

```
if (iter == maxit & dist < tol)
  warning(paste("Convergence not reached after", iter, "iterations."))


if (sng)
  warning(paste("Singularity reached after", iter, "iterations."))


##  Labelling
dimnames(Conver) <- list(seq(0, iter), c(
    paste(rep(c("p", "mu", "sig"), rep(K, 3)), rep(seq(K), 3), sep=""),
        "Deviance", "Conv.crit"))


##  The results
list(Convergence=Conver, Probs=pro)
}  ##  End of function  EUmix2
```

The following code is for an example of fitting four components with lognormal distributions.

```
##  The indicator of a household record (restriction to uniques)
whi <- !duplicated(AT04[,"hid"])


##  Application
ATmix4 <- EUmix2(log(1+AT04[whi, "hx090"]), wei=AT04[whi, "db090"],
          K=4,
##        sta=list(c(0.35, 9.73, 0.35), c(0.25, 9.51, 1.1),
```

```
##                 c(0.30, 9.80, 0.35), c(0.10, 9.00, 0.2)),
###        sta=rbind(c(0.35, 9.73, 0.35), c(0.25, 9.51, 1.1),
###                 c(0.30, 9.80, 0.35), c(0.10, 9.00, 0.2)),
           maxit=1500)
```

(The object AT04 is the individual-level dataset for Austria in 2004.) By commenting out the line with K=4 and uncommenting the two lines that follow, or as an alternative the next two lines (with three hashes #), we can fit the model with the provided initial solution.

A compact version of the results is extracted by the function

```
EUmixF <- function(res, dgt=3)
{
### Extracting a compact version of the mixture model fit

## res    The results (list)
## dgt    The number of digits in rounding

## The result as a vector
rsv <- res[[1]][nrow(res[[1]]),]

list(Estimate=matrix(Dround(rsv[seq(length(rsv)-2)], dgt), nrow=3,
     byrow=T,
     dimnames=list(c("p", "mu", "sigma"), seq((length(rsv)-2)/3))),
     Fit=round(rsv[length(rsv)-c(1,0)], dgt), N.iter=nrow(res[[1]])-1)
} ## End of function  EUmixF
```

The first element of the resulting list is a character matrix, suitable for direct insertion into a LaTex document. The other two elements are related to the fit: the deviance, the value of the convergence criterion, and the number of iterations.

The following application extracts the results for a sequence of model fits; EUmix2r*, with * = 2, 3, 4, and 5, are the (long-form) results of fitting two to five components to household-level eHI in Austria in 2010.

```
ATmixFr <- list(M2=ATmixF(ATmix2r2), M3=ATmixF(ATmix2r3),
                M4=ATmixF(ATmix2r4), M5=ATmixF(EUmix2r5))
```

For example, ATmixFr$M4, the condensed result for four components, is

```
$M4
$M4$Estimate
          1     2     3     4
p     0.667 0.009 0.286 0.028
mu    9.927 8.600 9.984 9.237
sigma 0.361 1.338 0.656 0.048

$M4$Fit
 Deviance Conv.crit
 7033.402     8.003
```

```
$M4$N.iter
[1] 841
```

The function EUmixH2 plots the marginal and scaled component densities for a set of fitted models. It is advisable to use it only for two models; otherwise the diagram is too cluttered.

```
EUmixH2 <- function(x=c(), res, pts, npt=201, ltp=c(2,1), ylm=c())
{
###   Graph of the results of several mixture model fits

## x      The data (optional)
## res    List of result lists
## pts    The evaluation points
## npt    The number of evaluation points
## ltp    The line types for the densities
## ylm    The limits of the vertical axis
```

The vector of the data, x, is optional. If provided, its histogram is plotted in the background. Each model fit, obtained by the function EUmix2, is a list, and the argument of **res** is meant to be a list of two such lists. The densities are evaluated on a grid of points. Either the grid is provided as the value of **pts**, or only its limits; in the latter case, the value of **npt** is used as the number of evenly spaced evaluation points on the grid. The densities for the components are drawn by distinct line types specified by the argument **ltp**. For detailed exploration, the densities may be plotted in a particular range specified by **ylm**, set by trial and error. The extent of the horizontal axis is controlled by **pts**.

The relevant parts of the results are extracted first and suitably formatted. The numbers of components of the models are established and the grid of values for evaluating the density is set.

```
##  Extract the compact version of the results
res <- lapply(lapply(lapply(res, EUmixF, dgt=6), ExtrL, 1), as.numeric)

##  The numbers of components
nc <- sapply(res, length)/3

##  Formatting
for (i in seq(length(res)))
res[[i]] <- matrix(res[[i]][seq(3*nc[i])], nrow=3)

##  The evaluation points
if (length(pts) < 10) pts <- seq(min(pts), max(pts), length=npt)
```

The scaled densities (**dnt**) and their totals (the marginal densities) are evaluated in a loop over the models:

```
##  The component densities
```

```
dns <- list()

for (k in seq(length(res)))
{
dnt <- c()

##  The scaled densities
for (i in seq(nc[k]))
dnt <- rbind(dnt, EUdns(pts, res[[k]][,i]))

##  The marginal density
dns[[k]] <- rbind(dnt, apply(dnt, 2, sum) )
}  ##  End of the loop over the model fits (k)
```

The extent (range) of the vertical axis is set by default, unless it is given as the value of `ylm`. If the data is given in argument `x`, then its histogram is drawn; otherwise the empty shell of the plot is drawn

```
if (length(ylm) != 2) ylm <- range(unlist(dns))

if (length(x) > 0)
##  Histogram of the data (if data given; no weights)
hist(x, nclass=101, freq=F, xlim=range(pts), ylim=ylm, main="",
     lwd=0.5, border="gray50", cex.lab=0.9, cex.axis=0.9)

else

##  Empty shell of the plot
plot(range(pts), ylm, type="n", xlab="x", ylab="f(x)",
     ylim=ylm, cex.lab=0.9, cex.axis=0.9)
```

Finally, the densities are drawn with model-specific line types. The marginal densities are drawn with line thickness 1.4 (argument `lwd`) and the scaled densities of the `nc[k]` components of model k with thickness 0.4.

```
for (k in seq(length(res)))
{
for (i in seq(nc[k]+1))
lines(pts, dns[[k]][i,], lwd=0.4+(i>nc[k]), lty=ltp[k])
}
res
}  ##  End of function  EUmixH2
```

The function returns the compact version of the model fits.

The legend is added in left-hand panel of Figure 4.4 by the expression

```
legend(7.15, 0.95, lty=c(-1,2,1), c("Mixt. densities",
    paste(c(2,3), "components")), cex=0.6)
```

It is placed outside the function because its setting is specific to the panel.

Some fine-tuning is required in setting the arguments of **par** and EUmixH2 as well as those of `legend`. It is practical to comment out first the application of function `Figure` and the specification of its arguments, and inspect the diagram in an R graphics window. However, when the function `Figure` is applied, some rescaling takes place and further adjustment may be necessary.

The fitted conditional probabilities $\hat{r}_{i,k}$ are tabulated by the function `EUtab`. It operates on the results generated by `EUmix2` for models with any number of components, but one or two components have to be selected; otherwise no result is produced.

```
EUtab <- function(res, i=1, sbs=c())
{
###    Tabulation of the conditional probabilities

## res    The results list
## i      The number of bins or decimal places
## sbs    The components (rows of res[[2]]) selected for tabulation

## Sort out decimals vs. number of bins
if (i < 5) i <- 10^i
```

By the second argument, `i`, one can specify the number of categories (bins), or the number of decimal places in rounding that will be used for tabulation. For example, `i=1` is for ten bins, 0, 1, ..., 9, and `i=2` for 100. The minimum number of bins is five. Only integer values of `i` should be used. The value of `sbs` should be left at the default for two-component models. For models with more than two components, one or a pair of components (their row numbers in `res[[2]]`) should be given; `sbs` can also be set to a logical vector of the same length as `nrow(res[[2]])`.

```
## Extract the probabilities, expand and round them
res <- floor(i * res[[2]])

## Just one row for two components
if (nrow(res) == 2) res <- matrix(res[1, ], 1)

## Reduce to the selected components
if (length(sbs) > 0)  res <- res[sbs, ]

## The number of selected components
nc <- nrow(res)
```

The first expression above generates the assignment to the bins. If the model has only two components, the fitted probabilities for the second component are discarded. Depending on the number of components left after selection, either a one- or a two-way table is constructed.

```
## Separate evaluations for one and two vectors of probabilities
if (nc == 1)        table(res)
else if (nc == 2)   table(res[1, ], res[2, ])
else                stop("Too many components.")
} ## End of function  EUtab
```

Examples of one- and two-way tabulation with several degrees of coarseness are

```
##  Initialisation
ATtab22 <- ATtab23 <- list()

##  Two components
for (i in c(10, 20, 100))
ATtab22[[as.character(i)]] <- EUtab(ATmix2r2, i)

##  Three components
for (i in c(10, 20))
ATtab23[[as.character(i)]] <- EUtab(ATmix2r3, i, sbs=c(1,2))
```

These results (counts) can be converted to fractions and percentages by applying the respective functions `Prob` and `Perc`.

The ternary plot is implemented in the function `EUmixT`. Its arguments are the $(3 \times n)$ matrix of probabilities and the standard deviation of the noise added to the plotted points.

```
EUmixT <- function(prb, std=0.005)
{
### Ternary plot of the probabilities of mixture components

## prb  The (3 x n) matrix of probabilities
## std  The standard deviation of the jitter

## The number of observations
n <- ncol(prb)
```

The coordinates are set without drawing the axes. Then the triangle is drawn and the vertices are labelled.

```
## The empty shell of the plot;  no axes
plot(c(-1,1), c(0, sqrt(3))+0.03*c(-1,2), type="n", axes=F,
   xlab="", ylab="")

## The coordinates of the vertices
pts <- list(c(-1,1,0), c(0,0,sqrt(3)))

## The triangle
polygon(pts[[1]], pts[[2]], lwd=0.5, border="gray50")

## A small gap
shr <- 0.05

## Labelling the vertices
text(pts[[1]]+pts[[1]]*shr, pts[[2]]+c(-1,-1,1.5)*shr,seq(3),cex=0.85)
```

The points are drawn by a transformation of the rows of `prb`, with random noise added to both coordinates.

```
points(1 - prb[3,] - 2*prb[1,] + rnorm(n)*std,
       sqrt(3)*prb[3,] + rnorm(n)*std, pch=16, cex=0.2)
} ## End of function EUmixT
```

4.A.2 Improper Component

The EM algorithm for fitting mixtures with an improper component, function EUmixB, is implemented by a few changes of the function EUmix2 for the original EM algorithm. There is an additional argument, f0, the value of the (constant) improper density. The initial setting for the probabilities has to reflect that a probability is associated also with the improper component. It is set inflexibly to 0.01, but in the iterations it is estimated. In the iterations, the value of the scaled improper density is attached to the scaled proper densities as the first column:

```
## The scaled densities
dns <- pN[1] * f0

for (i in seq(K))
dns <- cbind(dns, EUdns(x, c(pN[i+1], meN[i], sdN[i])))

## The marginal density (denominator)
dnsT <- apply(dns, 1, sum)
```

Two additional optional arguments control the output: dgt specifies the number of digits in rounding, and long is for display of the entire convergence monitor Conver (if set to TRUE) or for a compact listing (the same layout as returned by function EUmixF) otherwise. The last evaluation of the function is

```
## The detail of the output
if (long)
lapply(list(Convergence=Conver, Probs=rik), round, dgt)
else
{
Res <- t(cbind(pN[-1], meN, sdN))
dimnames(Res) <- list(c("p", "Mean", "StD"), paste("C",seq(K), sep=""))

lapply(list(Improper.probability=pN[1], Proper.Comps=Res,
       Iterations=iter, Probs=rik), round, dgt)
} ## End of the output options
} ## End of function EUmixB
```

Applying the function EUmixB entails some fine-tuning to find appropriate values of the improper density. A grid of values (or their decimal logarithms) may be defined first, and then narrowed down to the range of interest, where the fitted probability is in the target range. The following expressions are for fitting a set of models with two proper and one improper component, for several values of the latter.

```
##  The values for the improper log-density
f0s <- c(seq(5, 9)/10, seq(7))

##  For two proper components
AT10Bt <- list()

for (fs in f0s)
AT10Bt[[paste("E-",fs,sep="")]] <- EUmixB(dt19, wei9, sta=list(), K=2,
        f0=10^(-fs), maxit=5000, tol=8, dgt=6, long=F)
```

4.A.3   Confusion Index

The confusion index for two normally distributed components is evaluated
by simulations. Each component is specified as a vector of length 3, with the
probability, expectation and variance as its elements. A large random sample is
drawn from the second component, the probabilities of assigning its elements
to the first component are evaluated, and the average of these probabilities is
established.

```
EUconf <- function(cmp1, cmp2, nre=100000)
{
###   The confusion index for two univariate normal components

##  cmp1    Component  A  (p, mean, variance)
##  cmp2    Component  B  (p, mean, variance)
##  nre     The number of replications

##  A sample from distribution  B
smp <- rnorm(nre, cmp2[2], sqrt(cmp2[3]))

##  The scaled densities
scd <- c()

for (cmp in list(cmp1, cmp2))
scd <- cbind(scd, cmp[1] * dnorm(smp, cmp[2], sqrt(cmp[3])))

mean(scd[, 1] / (scd[, 1] + scd[, 2]) )
}  ##  End of function  EUconf
```

The following example

```
EUconf(c(0.9, 0, 1), c(0.1, 0.5, 0.5), nre=100000)
```

yields the value 0.870 in 0.05 seconds. You do not have to be sparing with the
number of replications **nre**.

The confusion matrix is evaluated by the function **EUcnfM**. It evaluates its
off-diagonal elements in a double loop. Its argument **perm** is for permuting the
components, e.g., in the descending order of the marginal probabilities.

```
EUcnfM <- function(res, perm, nre=100000)
```

```
{
### Evaluation of the confusion matrix

## res    The result of the mixture model fit
## perm   The permutation of the components
## nre    The number of replications

## Extract the relevant details
res <- apply(EUmixF(res, dgt=8)[[1]], 2, as.numeric)[, perm]

## The number of components
K <- ncol(res)
```

The evaluation is speeded up by using Equation (4.7).

```
## The diagonal of the confusion matrix
mat <- diag(rep(0.5, K))

## The pairwise evaluations
for (k1 in seq(K-1))
{
for (k2 in seq(k1+1, K))
{
## The confusion index one way
mat[k1, k2] <- EUconf(res[, k1], res[, k2], nre)

## The confusion index in reverse
mat[k2, k1] <- mat[k1, k2] * res[1, k2] / res[1, k1]
}} ## End of the loop over the elements of a matrix  (k1, k2)

## Labelling
dimnames(mat) <- list(colnames(res), colnames(res))

list(Components=res, Confusion=mat)
} ## End of function  EUcnfM
```

Code could be added to check that **perm** is indeed a permutation.

```
if (mean(sort(perm) == seq(length(perm)) < 1) )
    stop("The argument perm is not a permutation.")
```

5

Regions

5.1 Introduction

An index defined for a country can be applied also to its regions. This is particularly relevant for the largest European countries, which have many regions and where substantial differences among the regions are likely to be present. When a region forms a substantial part of the country and is represented in the survey by a substantial part of the national sample, we simply apply the methods and procedures devised for a country. One important exception in this is that the national standard (e.g., a given percentage of the national median equivalised household income (eHI) used as the poverty threshold) is applied instead of a standard specific to the region (based on the median eHI for the region). When a region is represented in the survey by a relatively small sample, for which the estimates would be associated with large standard errors, we seek some improvements by means of small-area methods elaborated in the next section. In their established form, they are relevant in the European Union Statistics on Income and Living Conditions (EU-SILC) up to 2010 only for Spain and France, for which, respectively, 19 and 22 regions are defined in the data. For other populous countries, no regions are defined for the UK and Germany and only five regions for Italy and six for Poland. Czech Republic has eight regions; the other countries have at most four regions identified in the data. Several of the smallest countries form a region on their own (Cyprus, Estonia, Iceland, Luxembourg, Malta and Slovenia).

The sample sizes (numbers of individuals and households) of the countries and regions in EU-SILC in 2010 are presented graphically in Figure 5.1. Each country is represented by two horizontal segments with vertical ticks delimiting the sample sizes of its regions. The top segment (black) is for individuals and the bottom (gray) for households. In relation to individuals, the segment for households is extended 2.5 times, so that the two segments for a country have similar lengths and their average household sizes can be related to the reference value of 2.5. For countries and regions with average household size smaller than 2.5, the individual-level (black) segment is shorter (e.g., Austria, Germany and the UK). The average household size is much greater than 2.5 in Poland, Slovenia, Slovakia and Spain. The sample sizes for the regions are listed for countries with between two and eight regions. The national sample sizes are printed at the right-hand margin (with the heading 'Sample'), fol-

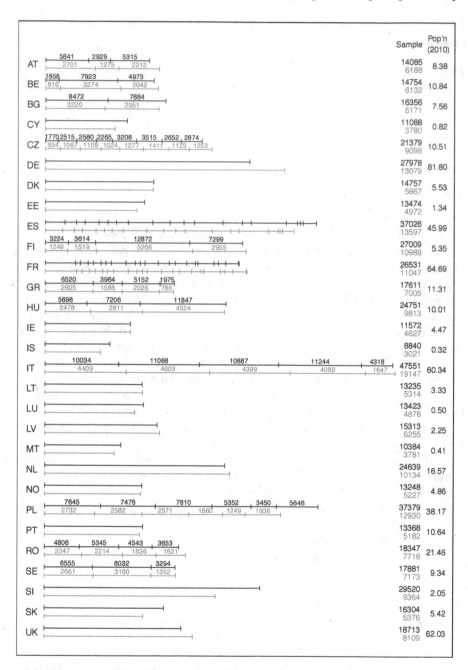

FIGURE 5.1

Sample sizes for the countries and their regions in EU-SILC in 2010.
The rightmost column gives the countries' population sizes, in millions. Source:
Eurostat. The countries are listed in Appendix 1.A.7.

lowed by the population size. The two countries with many regions, Spain and France, are studied in detail in respective Sections 5.5 and 5.6.

The national sample sizes are not strongly related to the population sizes of the countries. For example, the sample sizes for France and the UK, both with population in excess of 60 million, are smaller than for Slovenia, with population of just over two million.

5.2 Analysis of Regions

Austria comprises three regions labelled AT1, AT2 and AT3 in its datasets. The regions are Eastern Austria (AT1, comprising the states of Burgenland, Lower Austria and the City of Vienna), Southern Austria (AT2: Carinthia and Styria) and Western Austria (AT3: Upper Austria, Salzburg (state), Tyrol and Vorarlberg). We refer to the regions by the numerals, 1, 2 and 3. The estimates of the poverty rates for the regions in 2004 and 2010 are drawn in the left-hand panels of Figure 5.2. The sets of three curves for the regions refer to the common annual threshold of $\tau\%$ of the national median cHI, estimated separately for individuals and households; $30 \le \tau \le 80$. The right-hand panels display their scaled versions, the ratios of the region-level and national rates. The rates are evaluated for both individuals (black lines) and households (gray). Note that different scaling is applied to the individual- and household-level rates, by the national rates at the corresponding level. Household-level rates tend to be higher because poverty is more common in single-member households. The scaled versions of the two sets of rates are close to one another. The estimated poverty rates in 2010 are uniformly lower in region 3 than in the other two regions. In 2004, its poverty rates were higher than for regions 1 and 2 for the lowest thresholds.

The curves in the left-hand panels inform us about the magnitudes of the rates, but because these are in a wide range, they do not convey the details of the differences among the regions. The scaled curves in the right-hand panels are much better suited for this, but they do not inform about the (absolute) magnitudes of the rates.

We evaluate the scaled quantiles for each region and the seven years, 2004–2010, applying the same scaling by the annual national median. The scaled quantiles estimated at the individual level are displayed in Figure 5.3. The connected sets of seven horizontal segments represent the whole country (AT, at the top) and the three regions (1, 2 and 3), and each segment extends from the second to the 98th scaled percentile for the indicated (sub-)sample. The segments are for the respective years 2004–2010 (from top to bottom). They are connected at the corresponding scaled percentiles 2, 5, 10, 25, 50, 75, 90, 95 and 98. The quartiles (25th and 75th percentiles) are connected, both horizontally and vertically (year a to year $a + 1$), by thicker lines. Region 1

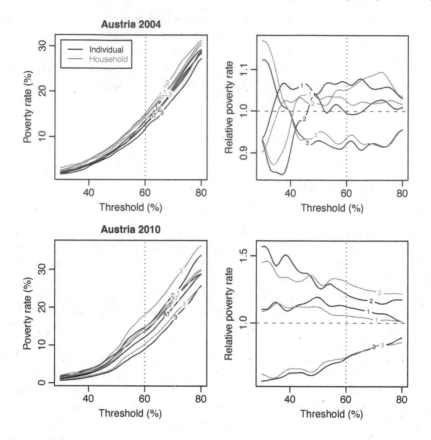

FIGURE 5.2

Estimates of the poverty rate functions for Austria and its regions in 2004 and 2010.

The right-hand panels display the functions scaled by the national rate function.

has higher and region 2 lower estimated median eHI than the national median throughout the seven years, but only by a small percentage every year. The scaled quartiles change only slightly over the years, without any discernible trend. Both quartiles for region 2 are uniformly smaller than their counterparts for the other regions. The extreme percentiles change a lot from one year to the next. We show below that the cause of this is their large sampling variation.

The top left-hand panel of Figure 5.4 displays the estimated Gini coefficients for Austria and its regions in years 2004–2010. The coefficient for region 1 is greater and for 2 and 3 smaller than for the country (labelled AT). Whether such a separation applies also to the underlying (population) values can be established by bootstrap; see Section 2.3. In its application, we have to bear in mind that the estimators for the regions are correlated with the

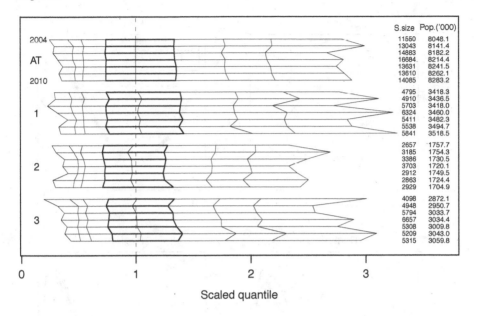

FIGURE 5.3
Scaled quantiles for Austria and its regions in 2004–2010.
Individual-level estimates. The sample and population sizes (individuals) are
listed at the right-hand margin.

estimator for the country, because each region's sample forms a substantial
part of the national sample.

The other three panels compare the Lorenz curves for the regions (black
lines) with the national Lorenz curves (gray) for the seven years. The transformation $p^2 - L(p)$ for the curves is essential; without it, the curves for a
setting would be indistinguishable from both one another and between the
country and each region. The annual Lorenz curves for region 1 are below
their national counterparts (above them after the transformation), whereas
for region 2 they are above and for region 3 there is some overlap. However,
fine details are difficult to discern even with this arrangement. We can gain
more insight by plotting the differences 'region vs. country', because they are
in a narrow range. Details are omitted.

Table 5.1 in Appendix 5.A.2 displays the estimated poverty rates, linear,
squared and log-poverty gaps (individual level) for Austria and its regions
for the seven years. The values of the poverty gap with linear and quadratic
kernels are not scaled. The table may serve as a template for a 'report card'
for a country and its regions. The Laeken indicators (Eurostat, 2005, and
Marlier *et al.*, 2007) include numerous indices evaluated for regions and other
subpopulations.

The next example, with the eight regions of Czech Republic, explores the

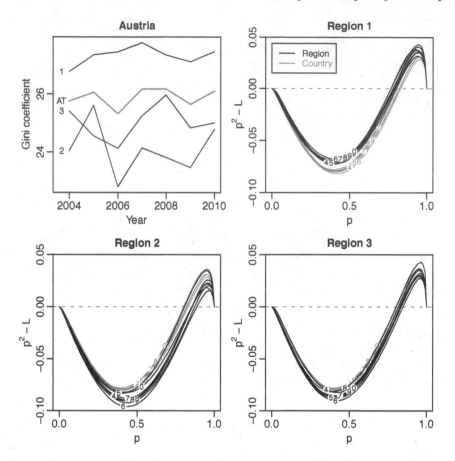

FIGURE 5.4
Gini coefficients and transformed Lorenz curves for Austria and its regions,
2004–2010; individual level.

general problem of making inferences about a country's regions further. Data
is available for years 2005–2010. Except for 2005, the annual sample sizes
for Czech Republic are much greater than for Austria, but are split to more
subsamples. The regions of Czech Republic have smaller subsample sizes than
regions 1 and 3 for all the six years in common.

The plot of scaled quantiles of eHI in Figure 5.5 shows that region 1 (Praha,
Prague) has much higher median income than the other regions. For the national samples we can observe a trend of (near) constant percentiles 5, 10 and
25 and decreasing high percentiles (reduction of inequality), except for 98%
in 2010. But this trend is not present for any of the regional estimates. We
can identify two possible reasons for this: genuine differences in the progression (trends) among the regions and sampling variation. Only the latter can

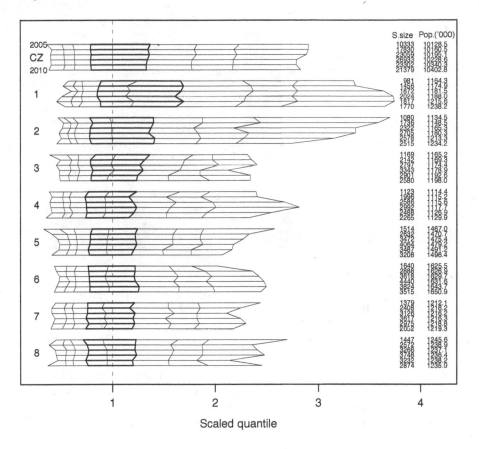

FIGURE 5.5
Scaled quantiles of eIII for Czech Republic and its regions in 2005–2010.
The regions are: 1: Praha, 2: Střední Čechy, 3: Jihozápad, 4: Severozápad, 5:
Severovýchod, 6: Jihovýchod, 7: Střední Morava and 8: Moravskoslezsko.

be explored, by bootstrap. The wild variation of the high percentile suggests
that sampling variation is the more likely explanation.

We estimate the sampling variances of the percentile estimators by using
1000 bootstrap replications and present the results for the regions of Czech
Republic in 2010 in Figure 5.6. The values of the estimates are accompanied by
the estimated standard errors in parentheses. The magnitude of the standard
error is represented by the length of the black segment underneath. (Longer
segments are a bit thinner.) Notable are the large standard errors for the
highest percentiles. For them, the standard errors are strongly associated with
the magnitude of the estimates. We cannot therefore assign much importance
to the sharp drop or increase of the estimate of the 98th percentile in respective
regions 3 and 4 in 2008, nor to the zig-zag pattern for regions 7 and 8.

	2%	5%	10%	25%	50%	75%	90%	95%	98%
CZ	35.7(0.7)	49.1(0.7)	61.8(0.4)	79.3(0.3)	100.0(0.0)	130.8(0.6)	172.4(1.0)	209.5(2.0)	272.1(4.2)
1	51.8(2.0)	63.4(1.6)	77.6(1.3)	98.7(0.8)	126.7(1.8)	171.4(2.7)	237.7(4.4)	285.6(7.8)	476.2(50.0)
2	31.0(3.5)	49.2(2.4)	63.9(1.3)	82.6(0.9)	108.6(1.7)	146.4(2.0)	187.1(3.7)	227.0(7.2)	278.3(17.1)
3	40.9(1.9)	55.1(2.1)	65.2(1.2)	81.2(0.8)	100.3(1.0)	125.5(1.2)	162.1(3.9)	197.3(4.8)	246.6(12.7)
4	28.7(1.3)	39.1(2.3)	54.6(1.4)	72.5(1.1)	91.7(0.7)	118.3(1.6)	149.5(2.6)	182.0(4.8)	244.0(12.3)
5	37.8(1.5)	52.1(1.4)	64.0(0.9)	78.5(0.7)	98.7(0.8)	124.6(1.3)	158.3(2.3)	176.7(3.2)	234.7(10.7)
6	39.9(2.7)	50.1(1.4)	60.7(1.3)	77.8(0.7)	97.6(0.8)	128.1(1.3)	162.8(3.5)	197.6(3.7)	249.2(11.4)
7	35.8(1.0)	44.9(2.0)	59.5(1.3)	75.8(1.1)	95.1(0.8)	121.0(1.7)	153.7(2.2)	184.8(4.3)	227.0(8.7)
8	29.0(1.0)	38.6(1.8)	56.4(1.9)	75.5(1.1)	92.0(0.8)	117.4(1.3)	151.2(3.3)	180.5(3.5)	224.2(13.4)

FIGURE 5.6
Bootstrap estimates of the standard errors of the scaled percentiles for the regions of Czech Republic in 2010.
All values are multiplied by 100. Based on 1000 replications.

We conclude that the scaled percentiles can be reliably compared for percentiles in the range 2 – 90, but the standard errors for percentiles 95 and 98 are excessive. However, the differences of the estimates are also quite large. In particular, the estimates for Prague are much greater than for the country; most of the country's high income is in the capital. Also, the household sizes tend to be smaller there. Their average in the sample is $1770/834 = 2.12$ for Prague and 2.37 for the rest of the country; see Figure 5.1. The bootstrap evaluation for one year takes about 140 sec.

5.3   Small-Area Estimation

In congenial circumstances, when the subdomain of interest, such as a region, is represented in the dataset by a sufficiently large sample, a summary of a variable in the domain would be estimated with negligible bias and small sampling variance, using only the values of the variable from the domain. Such estimators are called *direct*. All the estimators used in the previous sections are direct. They do not use any auxiliary information, such as the values of other variables from the same domain or the values of the same and related variables from other domains. Examples of other domains that could be exploited as auxiliary are not only other regions, but even the same region in the past, a

different subpopulation of the same region and a prima facie-related variable
from the same (or another) domain.

We address first estimation of a region-level index θ_d of a variable Y,
using no other variables than Y. The starting point are (unbiased) direct
estimators $\hat{\theta}_d$, $d = 1, \ldots, D$. We assume that the national value of the index,
θ, is estimated by $\hat{\theta}$ without bias. Suppose the sampling variances $v = \mathrm{var}(\hat{\theta})$
and $v_d = \mathrm{var}(\hat{\theta}_d)$ are known. For some regions d, v_d is too large, but v is
small. We seek estimators that draw on information from outside the target
region. Such an estimator is usually biased, and its bias is not known. But if
its variance is much smaller than for the direct estimator, compensating for
the bias even in the most pessimistic scenario, then direct estimation is not
efficient.

The region-level estimates $\hat{\theta}_d$ of an index θ_d differ for two reasons: sampling variation, as a result of which $\hat{\theta}_d \neq \theta_d$, and genuine differences among
the regions, as a result of which the values $\theta_1, \theta_2, \ldots, \theta_D$ are not identical.
When θ_d for a particular region d is estimated with low precision we might
contemplate the national value of the index θ, or its estimate $\hat{\theta}$, as an alternative to the direct estimator $\hat{\theta}_d$. This strategy is pursued by Gonzalez and
Hoza (1978).

The mean squared error (MSE) of $\hat{\theta}$ for θ_d is $v + (\theta_d - \theta)^2$. In our case,
$v \ll v_d$. If we were confident that θ_d does not differ substantially from θ, θ_d
would be estimated more efficiently by $\hat{\theta}$ than by $\hat{\theta}_d$. If we have no information
about the deviation $\theta_d - \theta$ specific to a region d, then we might define the
following rule: for regions d with v_d up to a certain critical value v^*, use $\hat{\theta}_d$;
for regions with $v_d > v^*$, use $\hat{\theta}$. You may not like the discontinuity involved.

Further improvement can be attained by combining the estimators $\hat{\theta}_d$ and
$\hat{\theta}$, defining

$$\tilde{\theta}_d = (1 - b_d)\,\hat{\theta}_d + b_d\,\hat{\theta}, \tag{5.1}$$

with b_d to be set, separately for each region. Such an estimator is called
composite—it is a composition of some *basis* estimators, $\hat{\theta}_d$ and $\hat{\theta}$ in our case.
The basis may comprise more than two estimators. When considering $\tilde{\theta}_d$ as a
function of b_d, we write $\tilde{\theta}_d = \tilde{\theta}_d(b_d)$. The coefficient b_d is set with the aim to
minimise the MSE of $\tilde{\theta}_d$. As a function of b_d, this MSE is

$$
\begin{aligned}
m_d(b_d) &= (1 - b_d)^2\, v_d + b_d^2 v^2 + 2b_d(1 - b_d)\,\mathrm{cov}\!\left(\hat{\theta}_d, \hat{\theta}\right) + b_d^2(\theta_d - \theta)^2 . \\
&= b_d^2\left\{ v_d + v - 2c_d + (\theta_d - \theta)^2 \right\} - 2b_d(v_d - c_d) + v_d ,
\end{aligned}
$$

where $c_d = \mathrm{cov}(\hat{\theta}_d, \hat{\theta})$. Denote also $B_d = \theta_d - \theta$. The minimum of $m_d(b_d)$, a
quadratic function of b_d, is attained for

$$b_d^* = \frac{v_d - c_d}{v_d + v - 2c_d + B_d^2} . \tag{5.2}$$

When $\hat{\theta}_d$ are mutually independent and $\hat{\theta}$ is a linear combination of the $\hat{\theta}_d$,

$\hat\theta = (u_1\hat\theta_1 + \cdots + u_D\hat\theta_D)/u_+$ for suitable scalars (constants) $u_1, \ldots, u_D$ and $u_+ = u_1 + \cdots + u_D$, then

$$c_d = \frac{u_d}{u_+}\, v_d\,.$$

Otherwise this identity holds only approximately and with some changes. The presence of B_d in Equation (5.2) is an inconvenience. If its value were available, θ_d would be estimated efficiently by $B_d + \hat\theta$, and the composition in Equation (5.1) would be of no use. We remove the dependence of b_d^* on B_d by substituting for each B_d^2 their average over the regions,

$$\sigma_{\mathrm B}^2 \;=\; \frac{1}{D}\sum_{j=1}^{D} B_j^2\,, \qquad\qquad (5.3)$$

or its estimate. We refer to $\sigma_{\mathrm B}^2$ as the region-level variance, and to the replacement of B_d^2 by $\sigma_{\mathrm B}^2$ or $\hat\sigma_{\mathrm B}^2$ as *averaging*. Estimation of $\sigma_{\mathrm B}^2$ is described in Appendix 5.A.1. Note that $\sigma_{\mathrm B}^2$ is the variance of the D quantities θ_j when θ is their mean.

When $u_d \ll u_+$, then $v_d \gg v$ and also $c_d \ll v_d$. We can then drop the terms c_d and v from Equation (5.2), and by substituting $\sigma_{\mathrm B}^2$ for B_d^2 we obtain a simpler expression,

$$b_d \;=\; \frac{v_d}{v_d + \sigma_{\mathrm B}^2}\,, \qquad\qquad (5.4)$$

for which the composite estimator in Equation (5.1) has an appealing interpretation. If we ignore the sampling variation of $\hat\theta$, then $\sigma_{\mathrm B}^2$ is the averaged precision of $\hat\theta$ for estimating θ_d for a region d. Therefore, the coefficients in the composite estimator $\tilde\theta_d$ are proportional to the precisions of the basis estimators $\hat\theta_d$ and $\hat\theta$ for θ_d. The coefficient in Equation (5.4) is not suitable for a region d that forms a substantial subsample of the data, such as 25% or more, because for such a region $\mathrm{cor}(\hat\theta_d, \hat\theta)$ is large and c_d cannot be ignored. Further, estimation of the region-level variance $\sigma_{\mathrm B}^2$ is problematic when there are only a few regions in the country (and in the data), or when a few regions account for most of the sample (e.g., 75% or more).

The role of θ or $\hat\theta$ can be described as being a pivot for the region-level estimators $\hat\theta_d$. It may be advantageous to replace θ with the average $\bar\theta = (\theta_1 + \theta_2 + \cdots + \theta_D)/D$. With B_d redefined as $\theta_d - \bar\theta$, the average of the squares B_d^2 is the finite-population variance of the quantities $\theta_1, \theta_2, \ldots, \theta_D$. An obvious estimator of $\bar\theta$ is the average of the direct estimators $\hat\theta_d$, but its precision may be unduly influenced by the precision for the region with the smallest representation in the sample. Some fine judgement is therefore called for, especially when $\hat\theta$ is a complex statistic.

Setting these issues aside, suppose we obtain the exact values of the coefficients b_d^*. Then the MSE for region d is reduced by

$$v_d - m_d\left(b_d^*\right) \;=\; \frac{(v_d - c_d)^2}{v_d + v - 2c_d + B_d^2}\,,$$

a positive quantity for every region, because $c_d < v_d$ owing to the inequalities $v < v_d$ and $|c_d| \leq \sqrt{v\,v_d}$. Also, $v_d = m_d(0)$, so $m_d(b_d^*)$ is bound to be smaller. When we apply averaging and substitute σ_B^2 for each B_d^2 in the denominator in Equation (5.2), we fare somewhat less well. The difference of the MSEs of $\hat\theta_d$ and $\tilde\theta_d(b_d)$, with b_d based on σ_B^2, is

$$
\begin{aligned}
m_d(0) - m_d(b_d) &= 2b_d\,(v_d - c_d) - b_d^2\,(v + v_d - 2c_d + B_d^2) \\
&= b_d^2\left\{2\,(v + v_d - 2c_d + \sigma_\text{B}^2) - (v + v_d - 2c_d + B_d^2)\right\},
\end{aligned}
$$

and therefore the composite estimator $\tilde\theta_d(b_d)$ is more efficient than $\hat\theta_d$ only when

$$
B_d^2 < 2\sigma_\text{B}^2 + v_d + v - 2c_d.
$$

Thus, composition will fail for a small fraction of regions that are extreme— that have either very small or very large values of θ_d. Ironically, the problem is more acute for more precisely estimated regions, for which v_d is small. However, for such regions b_d is small and the losses in efficiency are small. Also, large regions are less likely to be extreme than small regions, with the exception of (the region that contains) the capital. We may decide upfront to use only direct estimators for such regions. Since σ_B^2 is estimated, some further inefficiency is incurred. We can ameliorate this problem by increasing the value of the estimate $\hat\sigma_\text{B}^2$. As a result, the reduction of the MSE is smaller for some regions, but the composition is less efficient for fewer regions than it would be without this inflation.

The composite estimator $\tilde\theta_d(\hat b_d)$, with an estimate substituted for b_d in Equation (5.1), depends on the variance estimate $\hat\sigma_\text{B}^2$ only through the coefficient $\hat b_d$, in which it appears in the denominator of a fraction. Estimating σ_B^2 efficiently does not result in efficient estimation of b_d^*, because efficiency is not retained by a nonlinear transformation. (Neither is unbiasedness.) There is no single adjustment of $\hat\sigma_\text{B}^2$ that would make all $\hat b_d^*$ efficient for the corresponding b_d^*. Merely, we can explore the consequences of the error in estimating σ_B^2 or of substituting an adjustment of $\hat\sigma_\text{B}^2$ for σ_B^2.

Denote the denominator of b_d^* in Equation (5.2) by Q_d; $Q_d = v_d + v - c_d + B_d^2$. If we use a coefficient b_d different from the optimal b_d^*, the loss of efficiency is

$$
\begin{aligned}
m_d(b_d) - m_d(b_d^*) &= \left(b_d^2 - b_d^{*\,2}\right)Q_d - 2\,(b_d - b_d^*)\,(v_d - c) \\
&= (b_d - b_d^*)^2\,Q_d,
\end{aligned}
$$

so the loss is an increasing function of $|b_d - b_d^*|$. Suppose we substitute $B_d^{2\prime} = B_d^2 + \Delta$ instead of B_d^2. Then the difference of the corresponding coefficients b_d is

$$
b_d - b_d^* = -\frac{b_d^*\Delta}{Q_d + \Delta}.
$$

To compare the change in b_d resulting from increasing B_d and reducing it by Δ, we evaluate the total of these changes,

$$b_d^* \left(\frac{-\Delta}{Q_d + \Delta} + \frac{\Delta}{Q_d - \Delta} \right) = \frac{2b_d^* \Delta^2}{(Q_d + \Delta)(Q_d - \Delta)}.$$

This is positive, so the change by reducing B_d^2 by Δ is greater than by increasing it by Δ. So, understatement of B_d^2 leads to greater inflation of the MSE than overstatement by the same quantity Δ. We should therefore prefer to err on the side of overestimating B_d^2, or its substitute σ_B^2, the result of which is underestimation of b_d. This can be interpreted as leaning toward the direct estimator and relying less on the similarity of the regions. Indeed, if we choose a sufficiently small value of b_d we are almost certain to have an improvement for every region, but the improvements will be very small. This idea is distantly connected with ridge regression (Hoerl and Kennard, 1970). In contrast, if we overstate b_d we may obtain an estimator less efficient than the direct estimator for several regions. By a similar analysis we conclude that the consequences of substituting $B_d^2 \Delta$ are less serious than substituting B_d^2/Δ for any scalar $\Delta > 1$.

The estimator $\tilde{\theta}_d$ can be described as a *shrinkage* estimator, pulling the (original) direct estimator closer toward its pivot (or focus) $\hat{\theta}$. Shrinkage is stronger (b_d or $\hat{b}_d$ is greater) when $\hat{\sigma}_B^2$ is smaller because $\hat{\theta}$ is then a more credible alternative to $\hat{\theta}_d$, or contains more information about θ_d. Shrinkage is also stronger when v_d is greater, because $\hat{\theta}$ is then relatively more useful. The ideas of shrinkage were developed by Robbins (1955) and James and Stein (1961), and further elaborated by Efron and Morris (1972, 1973 and 1975).

5.4 Using Auxiliary Information

In the previous section, we combined a direct estimator for a region with its national version. The only information we used about the two estimators were their expectations and sampling variances, and that the direct estimator is unbiased. We sought to exploit the similarity of the regions, and in particular that B_d^2 is small in relation to v_d. This suggests that we could combine the direct estimator with any other estimator, or indeed with several estimators or some other quantities. With statistics that are in some way closer to θ_d we are likely to be more successful. An obvious choice in this regard is the value of $\hat{\theta}_d$ from the previous or an earlier year. After all, the index for a region could not have changed substantially from one year to the next.

In a general formulation, we consider a vector of auxiliary statistics $\hat{\boldsymbol{\xi}}_d$ which are unbiased (direct) estimators of some region-level quantities $\boldsymbol{\xi}_d$. Let $\hat{\boldsymbol{\theta}}_d = (\hat{\theta}_d, \hat{\boldsymbol{\xi}}_d^\top)^\top$ and $\hat{\boldsymbol{\theta}} = (\hat{\theta}, \hat{\boldsymbol{\xi}}^\top)^\top$, and let $\mathbf{e} = (1, 0, \ldots, 0)^\top$ be the indicator vector for the first element in $\boldsymbol{\theta}_d = (\theta_d, \boldsymbol{\xi}_d^\top)^\top$, so that our targets are $\theta_d = \mathbf{e}^\top \boldsymbol{\theta}_d$ and their direct estimators are $\mathbf{e}^\top \hat{\boldsymbol{\theta}}_d$. The multivariate composition is defined as

$$\tilde{\theta}_d = (\mathbf{e} - \mathbf{b}_d)^\top \hat{\boldsymbol{\theta}}_d + \mathbf{b}_d^\top \hat{\boldsymbol{\theta}},$$

with the vector of coefficients $\mathbf{b}_d$ to be set. The composition introduced in Equation (5.1) is obtained when $\boldsymbol{\xi}_d$ is empty and $\mathbf{e} = 1$. We refer to that composition as univariate.

The optimal vector $\mathbf{b}_d$ is found by minimising the MSE of $\tilde{\theta}_d$, using the matrix analogues of the operations in the previous section. First, the MSE of $\tilde{\theta}_d$ is

$$\begin{aligned}
m_d(\mathbf{b}_d) &= (\mathbf{e} - \mathbf{b}_d)^\top \mathbf{V}_d (\mathbf{e} - \mathbf{b}_d) + \mathbf{b}_d^\top \mathbf{V} \mathbf{b}_d + (\mathbf{e} - \mathbf{b}_d)^\top \mathbf{C}_d \mathbf{b}_d \\
&\quad + \mathbf{b}_d^\top \mathbf{C}_d^\top (\mathbf{e} - \mathbf{b}_d) + \left(\mathbf{b}_d^\top \mathbf{B}_d \right)^2,
\end{aligned}$$

where $\mathbf{V}_d = \mathrm{var}(\hat{\boldsymbol{\theta}}_d)$, $\mathbf{V} = \mathrm{var}(\hat{\boldsymbol{\theta}})$, $\mathbf{C}_d = \mathrm{cov}(\hat{\boldsymbol{\theta}}_d, \hat{\boldsymbol{\theta}})$ and $\mathbf{B}_d = \boldsymbol{\theta}_d - \boldsymbol{\theta}$ are the respective multivariate versions of v_d, v, c_d and B_d. Some diagonal elements of $\mathbf{V}_d$ and $\mathbf{V}$ may vanish. For example, an auxiliary variable may be extracted from a census or register. Note that we do not require the values for the elementary units (individuals or households), only their region-level summaries. Also, an auxiliary variable may be defined directly for the regions.

The minimum of $m_d(\mathbf{b}_d)$, a multivariate quadratic function of $\mathbf{b}_d$, is found by completing the square:

$$\begin{aligned}
m_d(\mathbf{b}_d) &= \mathbf{b}_d^\top \mathbf{Q}_d \mathbf{b}_d - \mathbf{e}^\top \mathbf{P}_d \mathbf{b}_d - \mathbf{b}_d^\top \mathbf{P}_d \mathbf{e} + \mathbf{e}^\top \mathbf{V}_d \mathbf{e} \\
&= \left(\mathbf{b}_d - \mathbf{Q}_d^{-1} \mathbf{P}_d \mathbf{e} \right)^\top \mathbf{Q}_d \left(\mathbf{b}_d - \mathbf{Q}_d^{-1} \mathbf{P}_d \mathbf{e} \right) + \mathbf{e}^\top \left(\mathbf{V} - \mathbf{P}_d^\top \mathbf{Q}_d^{-1} \mathbf{P}_d \right) \mathbf{e},
\end{aligned}$$

where

$$\begin{aligned}
\mathbf{Q}_d &= \mathbf{V}_d + \mathbf{V} + \mathbf{C}_d + \mathbf{C}_d^\top + \mathbf{B}_d \mathbf{B}_d^\top \\
\mathbf{P}_d &= \mathbf{V}_d - \mathbf{C}_d.
\end{aligned} \tag{5.5}$$

Hence the minimum of m_d is attained for $\mathbf{b}_d^* = \mathbf{Q}_d^{-1} \mathbf{P}_d \mathbf{e}$ and the minimum attained is the diagonal element of $\mathbf{V} - \mathbf{P}_d^\top \mathbf{Q}_d^{-1} \mathbf{P}_d$ that corresponds to $\hat{\theta}_d$. We substitute for each $\mathbf{B}_d \mathbf{B}_d^\top$ its region-level expectation, the matrix

$$\boldsymbol{\Sigma}_\mathrm{B} = \frac{1}{D} \sum_{j=1}^{D} (\boldsymbol{\theta}_d - \boldsymbol{\theta})(\boldsymbol{\theta}_d - \boldsymbol{\theta})^\top.$$

Further, the matrices $\mathbf{V}_d$, $\mathbf{V}$, $\mathbf{C}_d$ and $\boldsymbol{\Sigma}_\mathrm{B}$ have to be estimated, although $\mathbf{C}_d$ is usually derived directly from $\mathbf{V}_d$, and $\mathbf{V}$ from $\mathbf{V}_1, \mathbf{V}_2, \dots, \mathbf{V}_D$. Estimation of the elements of $\boldsymbol{\Sigma}_\mathrm{B}$ by moment matching is described in Appendix 5.A.1.

The matrix $\boldsymbol{\Sigma}_\mathrm{B}$ is estimated elementwise and therefore the estimate $\hat{\boldsymbol{\Sigma}}_\mathrm{B}$ may have one or several negative eigenvalues, so that it is not a variance matrix. This problem is resolved by adding a positive constant to the diagonal of the original matrix $\hat{\Sigma}_B$. Even an estimated variance $\hat{\sigma}_B^2$ may be negative, although such instances are rare. We would resolve it by truncating it at zero or at a positive (default) value, the smallest variance that we would entertain. In the multivariate setting, we add to the diagonal of $\hat{\Sigma}_B$ the positive scalar for which all the eigenvalues are positive and the ratio of the largest and smallest eigenvalue is 1000. This ratio is called the condition number. It is easy to show that such a scalar is unique when some (but not all) eigenvalues are negative or the condition number is greater than the threshold (1000). Alternatives to this approach include multiplying each covariance in $\hat{\Sigma}$ by a positive constant smaller than unity.

For more details of the method and examples from other surveys, see Longford (2005, Part II). The original paper on composition for small-area estimation is Longford (1999). Lohmann (2011) compares various aspects of the data in EU-SILC surveys and in administrative registers. Longford (2004) illustrates how register data can be exploited for small-area estimation even when the definitions of the target variable (used in the survey) and the variable used in the register differ. Rao (2003) is another established reference to small-area estimation; it expands on the review by Ghosh and Rao (1994). Fay and Herriott (1979) is generally acknowledged as the beginning of modern small-area estimation.

5.5   Regions of Spain

Spain is represented in the EU-SILC samples by all its 19 regions (17 autonomous communities and two autonomous cities) in years 2006–2010. In 2004 and 2005, the survey was not conducted in one of the regions, 64 (Melilla), a city-enclave on the north African coast. The regions vary in their population sizes from around 80 000 each in Ceuta and Melilla, the two autonomous cities, to several million in Andalusia (8.4), Catalonia (7.5), Madrid (6.5) and Valencia (5.1); the total population of Spain is around 47 million (www.ine.es; figures for January 2010).

The sample sizes of the regions are presented in Figure 5.7. Each region is associated with seven pairs of horizontal segments, one for each year of the survey, descending from 2004 at the top to 2010 at the bottom of the septet. The black segments connect the effective and the factual sample sizes for individuals and the gray segments their household-level counterparts. The horizontal coordinates of the latter are multiplied by 2.5, to have lengths similar to the corresponding segments for individuals. The average household size is greater than 2.5 in the subsamples for all regions and years, by the widest margin in Ceuta, Melilla and Murcia (region 62). The horizontal axis is

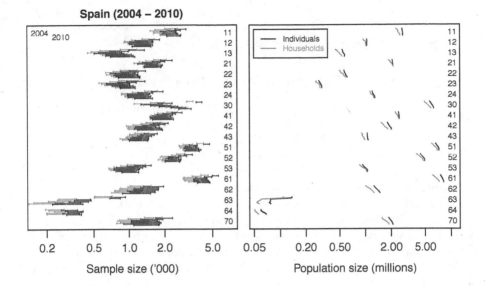

FIGURE 5.7

Sample sizes and estimated population sizes of the Spanish autonomous communities and cities; 2004–2010.

The regions are: 11: Galicia, 12: Asturias, 13: Cantabria, 21: País Vasco (Basque Country), 22: Navarra, 23: La Rioja, 24: Aragón, 30: Madrid (city and environs), 41: Castilla y León, 42: Castilla-La Mancha, 43: Extremadura, 51: Cataluña (Catalonia), 52: Valencia, 53: Illes Balears (islands), 61: Andalusia, 62: Murcia, 63: Ceuta (city), 64: Melilla (city) and 70: Canarias (islands).

on the multiplicative scale which is more appropriate for making comparisons by means of ratios (or percentages) rather than by differences.

The right-hand panel displays the totals of the weights, which are the estimated population sizes. Assuming that by poststratification and calibration of the weights the total of the weights was matched with the available (or estimated) population size, the displayed estimates are exact, or nearly so. For each region and level (individual or household), the seven annual estimates are connected. The vertical coordinates of these estimates are the same as for the corresponding summaries in the left-hand panel. Among the most populous regions, Andalusia (61) has a greater average household size than Catalonia (51) or Valencia (52), where in 2010 it was close to the reference of 2.5 members per household. The population has been increasing unevenly; the increase over the six years was only slight in Galicia (11), Asturias (12), País Vasco (21) and Castilla y León (41), and it was fastest in the most populous regions and in Balears (53). In general, the numbers of households have been increasing faster than the numbers of individuals. There is an anomaly in region 63 (Ceuta), where the estimated population size dropped substantially from around 135 000 in 2004 and 2005 to 76 000 in 2006. We conjecture that the figures for 2004 and 2005 are the totals for Ceuta and Melilla.

5.5.1   Composite Estimation of the Poverty Rates

The direct estimates of the poverty rates in the regions are between 9% and 37% in years 2004–2006 and are somewhat more spread in the later years. Except for 2004, the Basque Country (21) and Navarra (22) have the two lowest estimated poverty rates. Extremadura (43) and Ceuta (63) have the two highest estimated rates in every year, except for 2009, when the highest estimate is for Melilla. The standard errors for the rates, estimated by bootstrap with 1000 replications, are around 1% for Catalonia (51), and somewhat higher for the other two most populous regions, Madrid (30) and Andalusia (61). The standard errors have been increasing over the years for several of the least populous regions, including the two autonomous cities on the north African coast, Ceuta and Melilla, in concert with the sample sizes declining over the years.

To illustrate small-area estimation, we analyse first each year's data separately. The estimated national poverty rate has been around 20% (19.8– 20.2%) in 2004–2009, and it increased to 21.0% in 2010. These rates are estimated with high precision, with standard errors around 0.4%. This does not support the claim that the poverty rate has been constant over the years 2004–2009, although changes of large magnitude in the population rates can be ruled out.

The region-level variance σ_B^2 is estimated by 72.1, 55.5, 73.0, 74.9, 76.3, 77.8 and 61.7%^{2} for the respective years 2004–2010, so the standard deviation of the regional poverty rates in the studied years is about 8%. Figure 5.8 displays the direct and (univariate) composite estimates of the poverty rates for each year 2004–2010, representing the results of seven annual analyses. The standard errors of the composite estimators are themselves estimated, but they are in general biased. We see that composite estimates differ from the direct estimates only slightly for most regions, and substantial differences occur only for three regions, Extremadura (43) and the two north African cities (63 and 64).

The composite estimator is rather ineffective because the regions' poverty rates differ so much. With a large region-level standard deviation $\hat{\sigma}_B$, combining the direct and national estimators is not useful because $\hat{\theta}$ is likely to have a bias for θ_d that is large in relation to the standard error of the direct estimator $\hat{\theta}_d$. For most regions, $v_d \ll \hat{\sigma}_B^2$.

The key condition for effectiveness of a composite estimator is *similarity* of the target quantities, the poverty rates. We conjectured earlier that poverty rates have the property of inertia—they are not altered substantially in the course of a few years, and when some changes take place, they tend to be similar in extent and are in the same direction for most of the regions. Even

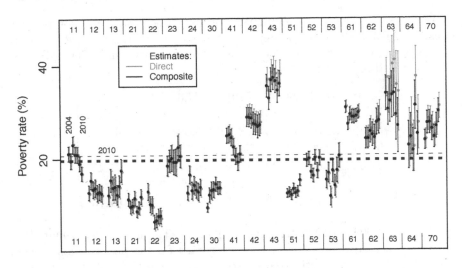

FIGURE 5.8

Direct and univariate composite estimates of the poverty rates in the regions of Spain, 2004–2010.

The national poverty rates are represented by horizontal dashes. Each vertical segment extends between $\hat{\theta}_d - \hat{s}_d$ and $\hat{\theta}_d + \hat{s}_d$, where $\hat{\theta}_d$ is the estimate (marked by a knot) and $\hat{s}_d$ the estimated standard error.

the increase in the national poverty rate by about 1% from 2009 to 2010 is small when related to the changes in the estimates from one year to the next for most regions. Figure 5.8 indirectly corroborates our hypothesis of inertia. The annual estimates for the regions with smaller standard errors (large samples) differ only slightly (regions 12, 21, 24, 30, 51 and 61), much less than the estimates with large standard errors (regions 43, 63, 64 and 70). This suggests that sampling variation is the principal reason for the year-to-year differences of the estimated rates within the regions.

We exploit the inter-year similarity of the rates by including the rates in the previous year(s) in the basis of the composition. Such an estimator can be motivated by the following scenario. Suppose the annual poverty rates in the regions are constant and so are the effective sample sizes, and the annual samples are mutually independent. Then by pooling the data across the years we would increase the effective sample size for every region about sevenfold, and reduce the standard error accordingly. In reality, the precision is increased much less because the rates differ somewhat, but for most regions the bias due to these differences is well worth incurring in exchange for variance reduction (using much more data).

We seek the optimal combination of the direct estimators of poverty rates for 2010 with their counterparts for earlier years. Thus, we set $\mathbf{e} = (0, 0, 0, 0, 0, 0, 1)^\top$ when using the estimates for each year since 2004. The estimated variance matrix is

$$\hat{\boldsymbol{\Sigma}}_{\mathrm{B}} = \begin{pmatrix} 72.08 & 51.61 & 60.82 & 57.07 & 57.53 & 54.89 & 54.73 \\ 51.61 & 55.50 & 55.63 & 52.90 & 52.88 & 51.12 & 52.12 \\ 60.82 & 55.63 & 73.03 & 78.20 & 79.32 & 74.38 & 69.52 \\ 57.07 & 52.90 & 78.20 & 74.93 & 80.58 & 77.61 & 72.27 \\ 57.53 & 52.88 & 79.32 & 80.58 & 76.30 & 75.59 & 71.45 \\ 54.89 & 51.12 & 74.38 & 77.61 & 75.59 & 77.76 & 72.84 \\ 54.73 & 52.12 & 69.52 & 72.27 & 71.45 & 72.84 & 61.75 \end{pmatrix} \tag{5.6}$$

for the years 2004–2010 in the ascending order. The matrix is estimated elementwise, see Appendix 5.A.1, with no regard for the positive definiteness of the target $\boldsymbol{\Sigma}_{\mathrm{B}}$. The estimated matrix in Equation (5.6) is not a variance matrix; some of its 'correlations' exceed unity and it has three negative eigenvalues, equal to -4.0, -5.4 and -5.9. We inflate the diagonal of $\hat{\boldsymbol{\Sigma}}_{\mathrm{B}}$ so that all the eigenvalues would become positive and the condition number would be equal to 1000. The matrix $\hat{\boldsymbol{\Sigma}}_{\mathrm{B}} + 6.38\mathbf{I}$, where $\mathbf{I}$ denotes the (7×7) identity matrix, has this property. Note that we require positive definiteness only for each $\hat{\mathbf{Q}}_d$ given by Equation (5.5), not for $\hat{\boldsymbol{\Sigma}}_{\mathrm{B}}$. However, efficient estimation of the elements of $\boldsymbol{\Sigma}_{\mathrm{B}}$, or of the matrix as a unit, does not convert to efficient estimation of $\mathbf{b}_d$ or θ_d, because they involve nonlinear transformation of $\hat{\boldsymbol{\Sigma}}_{\mathrm{B}}$, namely the inverse of $\hat{\mathbf{Q}}_d$. That is why we prefer to err on the side of stability and inflate the variances in $\hat{\boldsymbol{\Sigma}}_{\mathrm{B}}$ a bit more than necessary for positive definiteness of $\hat{\mathbf{Q}}_d$.

Figure 5.9 presents the direct and several composite estimates and the associated estimated standard errors in a compact form. The composite estimates use the data from 2009 (bivariate composite), from 2008 and 2009 (trivariate composite), and so on, from 2004–2009 (seven-variate composite) as auxiliary to the direct estimates for 2010. The estimates for a region are connected, starting with the direct on the left and continuing with the composite estimates of increasing complexity (from univariate to seven-variate). The diagram shows that the direct and univariate composite estimates differ substantially for a few regions (43, 63, 64 and 70), as discussed earlier, but there are nontrivial differences among the composite estimates for a few other regions, such as 11, 13, 53 and 62.

The standard errors of the direct and univariate composite estimators, displayed in the bottom panel, differ only slightly for all regions but, as we anticipated, they are much smaller for bivariate than univariate composition for most regions. For trivariate to seven-variate composition, the estimated standard errors are increased slightly for most regions. If the variance matrix $\boldsymbol{\Sigma}_{\mathrm{B}}$ were known such increases (reversals) could be ruled out. They occur with $\hat{\boldsymbol{\Sigma}}_{\mathrm{B}}$ because this matrix would not be positive definite without an adjustment

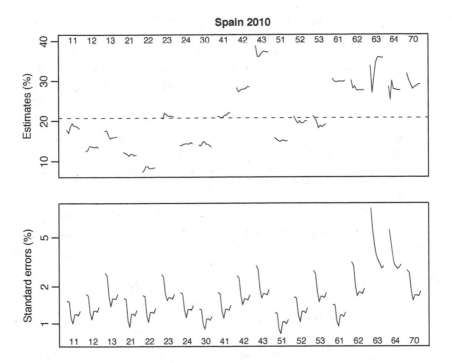

FIGURE 5.9

Direct and univariate to seven-variate composite estimates of the poverty rates in the regions of Spain, 2004–2010 and estimated standard errors.
The eight quantities (estimates or standard errors) for a region are connected by solid lines. Their order (left to right) is: direct, univariate composite, bivariate composite (using the direct estimates from 2009 as auxiliary), ..., seven-variate composite (using the estimates from all six previous years, 2004–2009, as auxiliary information).

(of the diagonal) that not only deals with negative eigenvalues, but ensures that $\hat{\Sigma}_B$ is not too close to singularity and the inverses $\hat{Q}_d^{-1}$ do not have any large elements. The estimated standard errors are uniformly largest for the direct estimator, followed by the univariate composite estimator. It is smallest for the trivariate estimator for all regions except the two smallest, Ceuta (63) and Melilla (64), for which the more complex estimators are preferred— seven-variate for Ceuta and five-variate for Melilla (recall that Melilla is not represented in the survey in 2004 or 2005).

Stability of the estimates can be assessed not only by the small differences among the multivariate composite estimators, but also by inspecting the coefficients $\hat{b}_d$. We expect the direct estimates from more recent years to be more useful. Therefore their coefficients should be greater (in absolute value), and should all have the same sign.

In the bivariate composition, the current year 2010 is associated for every region with a positive coefficient and the previous year with a negative coefficient with slightly smaller absolute value, except for the small regions 63 and 64, for which the differences are greater, 0.33 and 0.24, respectively. In trivariate composition, the coefficients for the current year are somewhat greater, and the coefficients for 2008 and 2009 are negative for all regions, the former slightly greater (closer to zero) than the latter, except for regions 63 and 64. The within-region totals of the coefficients, $\hat{\mathbf{b}}_d^\top \mathbf{1}_3$, where $\mathbf{1} = (1,1,1)^\top$, are all positive, in the range $(0.04, 0.10)$, except for regions 63 and 64 $(0.24$ and 0.18, respectively). The order of these totals reflects the standard error of the direct estimator (smallest for Catalonia, followed by Madrid and Andalusia; largest for the Balears, Extremadura and Murcia). The coefficients are altered only slightly when the condition number is changed from 1000 to 10 000 or to 200. The coefficients become unstable when we do not apply the constraints on the positive definiteness of $\hat{\boldsymbol{\Sigma}}_B$. Then the absolute values of the coefficients are in several instances greater than unity and their signs change from one year to the next.

The coefficients for more complex estimators have a similar pattern, with positive values for the direct estimates and negative values for all the auxiliary estimates. The coefficients have small positive totals within regions $(\hat{\mathbf{b}}_d^\top \mathbf{1})$, but their magnitudes for the auxiliary estimates do not have the anticipated pattern. At this point, stability breaks down; these estimators are too complex.

Finally, we discuss the nature of the standard errors. First, they are estimated. The target of their estimation is the minimum MSE that would be attained in the congenial circumstances, with $\mathbf{V}_d$, $\mathbf{V}$ and $\mathbf{C}_d$ known *and* with $\mathbf{B}_d\mathbf{B}_d^\top$ replaced by $\boldsymbol{\Sigma}_B$, itself assumed to be known. Ignoring the uncertainty about v_d, v and c_d in univariate composition, the expression for the standard error is close to unbiasedness when $|\theta_d - \theta| \doteq \sigma_B$; Asturias (12) and Murcia (62) are likely examples. The poverty rates in La Rioja (23), Castilla y Leon (41) and Valencia (52) are probably close to the national rate. If that is the case, then the standard errors for them are overestimated. Their estimation could be improved by giving more weight to the national rate $\hat{\theta}$ in the composition. After all, if $\theta_d = \theta$, then $\hat{\theta}$ is an estimator of θ_d that could not be improved. At the other extreme, the poverty rate of Extremadura (43) is likely to be much greater than $\theta + \sigma_B$. The MSE of its estimator is likely to be underestimated. If we knew that θ_d is exceptional, we could improve $\tilde{\theta}_d$ by reducing the coefficient b_d to incur smaller bias. Multivariate composition involves a matrix inversion. The inverse of an efficient estimator of a matrix is not efficient, because the inverse is a distinctly nonlinear transformation. Further, the standard errors are affected by averaging (replacing $\mathbf{B}_d\mathbf{B}_d^\top$ by $\boldsymbol{\Sigma}_B$), so they are not unbiased, and in some cases the bias is not trivial.

Based on the data for a single year (2010), we cannot identify these regions with any level of certainty. If we have the direct estimates from the previous years at our disposal, then these improvements can be made, but multivari-

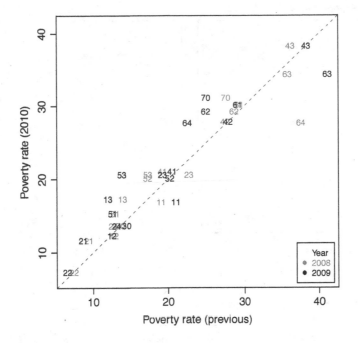

FIGURE 5.10
Direct estimates of the region-level poverty rates in Spain for 2008 and 2009 against the estimates in 2010.
The regions' codes for estimated rates in 2009 are printed in black and the codes for 2008 in gray.

ate composition is more effective. With multivariate composition, we have a similar problem with the nature of the standard errors, but our criteria for exceptionality have to be altered. Now they refer to the association of the targets θ_d with their counterparts from the previous years. Figure 5.10 displays the plot of the direct estimates of the poverty rates in 2008 and 2009 against the direct estimates in 2010. It suggests that regions 63 (Ceuta) and 64 (Melilla) are outliers, in discord with the association among the rates for the other regions. However, the standard errors for Melilla are so large that the evidence of exceptionality implied by the diagram is very weak. *If* Melilla were exceptional, by an unusually large change in the rate θ_d from 2009 to 2010, then the bivariate composition would be suboptimal, and may be even less efficient than univariate composite or even direct estimation. In summary, the composition is suboptimal for a few regions, but we do not know for which.

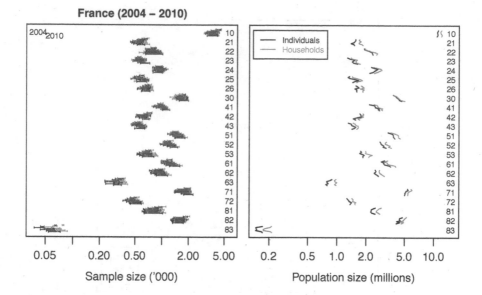

FIGURE 5.11

Sample sizes and estimated population sizes of the regions of France.
The regions are: 10: Île de France (including Paris), 21: Champagne-Ardenne, 22: Picardie, 23: Haute-Normandie, 24: Centre, 25: Basse Normandie, 26: Bourgogne, 30: Nord-Pas de Calais, 41: Lorraine, 42: Alsace, 43: Franche-Comté, 51: Pays de la Loire, 52: Bretagne, 53: Poitou-Charentes, 61: Aquitaine, 62: Midi-Pyrénées, 63: Limousin, 71: Rhône-Alpes, 72: Auvergne, 81: Languedoc-Roussillon, 82: Provence-Alpes-Côte de Azur, 83: Corse (Corsica).

5.6   Regions of France

To avoid repetitive discourse, we analyse the linear poverty gaps, not the poverty rates, in the regions of France. Figure 5.11 displays the sample and estimated population sizes of the regions of the country. One region, Île de France (10), has much greater sample size than the others, reflecting its much greater population size (nearly 11 million in 2009, 17.5% of the population of France) than the other regions (up to 5.1 million). Further, there is a region with exceptionally small sample size, Corse (Corsica, 83), in proportion with its exceptionally small population size (less than 200 000); the next least populous region is Limousin (63), with population of nearly 800 000. The five overseas regions of France, Guyane (French Guyana), Guadeloupe, Martinique, Mayotte and La Réunion, are not covered by the surveys.

    The twiddles in the right-hand panel suggest that the totals of weights are not equal to the corresponding population sizes, because such irregular changes in the population counts are implausible. For example, the estimated population of Corsica is 172 000 (individuals) in 2004, then slightly below 150 000 in each year 2005 – 2008, and around 170 000 in 2009 and 2010. The estimated average household size is smaller than 2.5 in every region, although only by a narrow margin in Nord-Pas de Calais (30) in 2004 – 2008.

The estimated national poverty gap was around 300 Euro in years 2004 – 2008, except for 2005 (266 Euro), and then it increased to 336 Euro in 2009 and 395 Euro in 2010. The direct region-level estimates are the highest for Corsica in every year, between 440 and 1600 Euro in the seven years. Five different regions have the smallest estimated poverty gap in a year, but the estimates for these regions in some other years are quite distant from the smallest. For example, Alsace (42) has the smallest estimated poverty gap in 2007, but only the 16th smallest in 2010, above the estimated national level. Such a change in the estimates has to be attributed mainly to the sampling variation (chance), because a similar change in the underlying (population) poverty gap is not plausible. This is confirmed by the standard errors estimated by bootstrap. They are 12 – 16 Euro for the national estimator, 26 – 42 Euro for Île de France, 50 – 100 Euro for a typical region, and 200 – 500 Euro for Corsica. For any given region, or the whole country, higher estimate of the poverty gap is associated with higher estimate of the standard error.

The direct and univariate composite estimates are plotted in Figure 5.12 using the same layout as Figure 5.8 for Spain. With the exception of Corsica (83), the estimates are tightly bunched around the annual national poverty gaps (266 – 395 Euro), although several regions have increased values in 2010 (e.g., Bourgogne, 26, Limousin, 63). The estimates of the region-level standard deviation σ_B are smaller than 100 Euro in 2004 and 2007 – 2009, and equal to 116 Euro in 2010. They are 210 and 263 Euro in the respective years 2005 and 2006, when the estimates for Corsica are much greater than in the other years. With the likely exception of Corsica, the regions have very similar values of the poverty gap. As a consequence, the coefficients b_d are substantial for most regions and the composite estimates are strongly pulled toward the national estimate. The shrinkage coefficient $\hat{b}_d$ for Corsica is 0.94—estimation relies principally on the national poverty gap; in contrast, its value for Île de France is 0.10. For most other regions $\hat{b}_d$ is in the range $(0.20, 0.45)$.

The multivariate composite estimates and their standard errors are presented in Figure 5.13. They indicate that the change of the estimate from the direct to the univariate composite is partly reversed for Corsica (83) by the multivariate composition. The panel for the standard errors at the bottom indicates that the (average) standard errors decrease with complexity of the composition—information from the past years is useful for estimating the poverty gap in 2010. For several regions there is a reversal with the inclusion of the direct estimates from 2008 and 2007, but that is more than offset in more complex composite estimators by using information from the earlier years. The estimated gains in precision are substantial, even for the most pop-

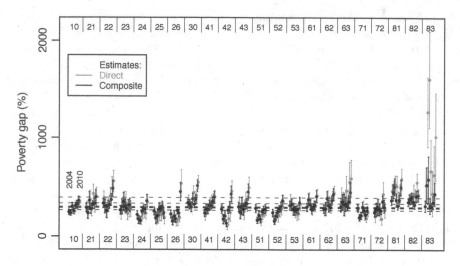

FIGURE 5.12
Direct and univariate composite estimates of the poverty gap for the regions
of France.

ulous regions, Île de France (10) and Rhône-Alpes (71), because of the strong
(estimated) similarity of the poverty gaps across the years.

The region of Corsica is a likely outlier, more clearly identified in the
analysis of poverty rates (not reported here). One might be tempted to exclude
it from the analysis and make special arrangements for its estimation. If it is
indeed an outlier, then the vector of coefficients in $\hat{b}_{83}$ is not appropriate
(or could be improved) and the assessment of the precision by the estimated
standard error is biased. In this case, exclusion of Corsica from the analysis
leads to only minor changes, indicating some small improvement, because
the estimated region-level variances are very small even with this apparently
exceptional region included.

5.7 Simulations

Based on Figure 5.9, we choose the bivariate composition for the poverty rates
in the regions of Spain, exploiting as auxiliary information only the data from
one previous year, 2009, and six-variate composition for linear poverty gaps
in the regions of France. The choice could be informed further by a simulation

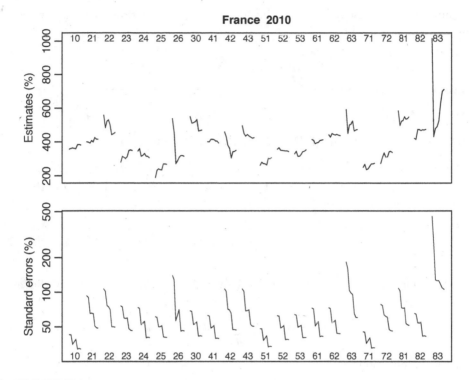

FIGURE 5.13
Multivariate composite estimates of the poverty gap for the regions of France.

study in which we generate artificial populations of the country and its regions for the seven years and replicate on it the processes of sample selection and evaluation of the eight alternative estimators. The principal difficulty in this process is not computing or data storage, but the construction of these populations and realistic implementation of the sampling process with some (or all) of its imperfections. We have to make several shortcuts to make a simulation study feasible. The annual populations can be replaced by the population rates θ_d and the sampling process by drawing simple random samples with the region-level effective sample sizes $n_{d,b}$. The population sizes are large enough for the distinction between sampling with and without replacement to be ignored, and the sample sizes are large enough for the normal approximation to the distributions of the sample rates $\hat{\theta}_d$ to be satisfactory.

We can anticipate a picture much more convoluted than that implied by Figures 5.9 and 5.13. Different compositions may be optimal for different regions, and so the choice of a single estimator for all the regions may entail some compromise. Further, the conclusions may differ as we introduce some changes in the frame (the population quantities and sample sizes) on which the simulations are based. Implementing the scheme and working out the details is left as a project for the reader.

5.A   Appendix

5.A.1   Estimation of Region-Level (Co-)Variances

We describe first estimation of the variance $\sigma_{\mathrm{B}}^2 = \sum_{d=1}^{D}(\theta_d - \bar{\theta})^2/D$ from a set of mutually independent direct estimators $\hat{\theta}_d$ of the targets θ_d; $\bar{\theta}$ is their mean. We use the statistic (sample summary)

$$S_{\mathrm{B}} = \frac{1}{D}\sum_{d=1}^{D}\left(\hat{\theta}_d - \hat{\bar{\theta}}\right)^2,$$

and obtain an unbiased estimator of σ_{B}^2 by adjusting S_{B} for its bias. We have

$$\mathrm{E}\left(S_{\mathrm{B}}\right) = \sigma_{\mathrm{B}}^2 + \frac{1}{D}\sum_{d=1}^{D}(v_d + v - 2c_d),$$

using the notation introduced in Section 5.3. Therefore

$$\hat{\sigma}_{\mathrm{B}}^2 = S_{\mathrm{B}} - v - \frac{1}{D}\sum_{d=1}^{D}(v_d - 2c_d)$$

is an unbiased estimator of σ_{B}^2. We can drop from this expression v and each c_d that is much smaller than the corresponding v_d; $c_d \ll v_d$ when the subsample for region d does not form a substantial part of the national sample. With v and all c_d dropped,

$$\hat{\sigma}_{\mathrm{B}}^2 = S_{\mathrm{B}} - \frac{1}{D}\sum_{d=1}^{D}v_d$$

and $\hat{\sigma}_{\mathrm{B}}^2$ can be interpreted as the naive estimator S_{B} adjusted by the average variance of the estimators $\hat{\theta}_d$. This is an example of a moment-matching estimator, obtained by matching the value of a statistic with its moment, in this case its expectation. The covariance c_d accounts for the dependence of $\hat{\theta}_d$ and $\hat{\bar{\theta}}$. It can be ignored when there are many regions and none of them is a large part of the country, nor has a dominant subsample in the data.

To estimate a covariance in the region-level variance matrix $\boldsymbol{\Sigma}_{\mathrm{B}}$, we form the cross-product statistic

$$S_{\mathrm{B}} = \frac{1}{D}\sum_{d=1}^{D}\left(\hat{\theta}_d - \hat{\bar{\theta}}\right)\left(\hat{\xi}_d - \hat{\bar{\xi}}\right),$$

where $\hat{\theta}_d$ and $\hat{\xi}_d$, $d = 1, \ldots, D$, are pairs of estimates in the basis of the composition. If these pairs are based on different (independent) data sources, then they are independent, and

$$\mathrm{E}\,(S_\mathrm{B}) = \frac{1}{D} \sum_{d=1}^{D} \left(\theta_d - \bar{\theta}\right) \left(\xi_d - \bar{\xi}\right),$$

equal to the corresponding region-level covariance. Thus, this covariance is estimated naively without bias.

5.A.2   Report Card for Austria and Its Regions

This section presents Table 5.1 in which the estimates of the various poverty indices for Austria and its regions, and the associated (estimated) standard errors, are collected.

5.B   Programming Notes

The functions for the evaluations described in Section 5.2, related to direct estimation, are compiled by editing the functions used in Chapters 1 and 3. The key changes are the isolation of the data for the separate regions and formatting (organisation) of the region-level results. The functions have an additional argument reg; its value is the name of the variable that indicates the region or, more generally, the subdomain. In all applications, its value is "db040". It is useful to retain this argument so that the function could be used for analysis of partitions other than regions, such as age groups, household types and similar.

The function EUread, introduced in Section 1.A.1, has to be adapted for input of the region-variable db040, because its values are character strings, such as AT1, AT2 and AT3 for Austria. For other countries that have regions defined in the data, the codes comprise the two letters of the country's acronym and one or two digits. Exceptions are Italy, with codes ATC, ..., ATG for its five regions, and Finland, which has a region FI1A.

The function EUreadR is developed from EUread by adding the argument Reg for the name of the region-variable.

```
EUreadR <- function(Dire, cou, Year, Ext, VRS, Reg, IH=1)
```

Expressions are added in EUreadR to check that the variable declared by Reg indeed has the country's acronym in its first two characters, to remove these two characters, and then convert the dataset to numeric.

```
...
Dat <- read.csv(paste(Dire, cou, EUYear(Year), ".", Ext, sep=""),
```

TABLE 5.1
Report card for Austria and its regions, 2004–2010 (individual level).

	Poverty rate		Poverty gap		Squared pov. gap		Log-poverty gap	
2004								
AT	12.79	(0.56)	0.346	(0.020)	1.683	(0.131)	0.944	(0.042)
AT1	13.65	(0.86)	0.354	(0.028)	1.592	(0.174)	1.007	(0.063)
AT2	12.71	(1.27)	0.335	(0.042)	1.514	(0.266)	0.942	(0.095)
AT3	11.83	(0.97)	0.344	(0.033)	1.895	(0.241)	0.871	(0.071)
2005								
AT	12.29	(0.54)	0.313	(0.019)	1.512	(0.126)	0.900	(0.039)
AT1	13.01	(0.81)	0.386	(0.032)	2.013	(0.219)	0.973	(0.060)
AT2	13.07	(1.30)	0.297	(0.036)	1.290	(0.237)	0.951	(0.094)
AT3	10.99	(0.90)	0.238	(0.027)	1.062	(0.182)	0.784	(0.064)
2006								
AT	12.55	(0.50)	0.323	(0.019)	1.535	(0.129)	0.926	(0.038)
AT1	14.53	(0.82)	0.440	(0.035)	2.351	(0.254)	1.099	(0.062)
AT2	12.62	(1.09)	0.243	(0.029)	0.888	(0.155)	0.889	(0.078)
AT3	10.28	(0.79)	0.236	(0.024)	0.983	(0.151)	0.753	(0.058)
2007								
AT	12.01	(0.50)	0.322	(0.018)	1.539	(0.119)	0.888	(0.037)
AT1	13.94	(0.81)	0.375	(0.030)	1.848	(0.216)	1.031	(0.060)
AT2	12.55	(1.18)	0.330	(0.041)	1.539	(0.252)	0.924	(0.087)
AT3	9.50	(0.73)	0.258	(0.024)	1.188	(0.142)	0.705	(0.054)
2008								
AT	12.36	(0.53)	0.295	(0.019)	1.234	(0.114)	0.907	(0.040)
AT1	13.94	(0.88)	0.333	(0.028)	1.342	(0.165)	1.035	(0.065)
AT2	11.26	(1.10)	0.293	(0.046)	1.395	(0.316)	0.832	(0.084)
AT3	11.16	(0.90)	0.253	(0.026)	1.015	(0.144)	0.802	(0.066)
2009								
AT	12.02	(0.57)	0.337	(0.022)	1.652	(0.152)	0.896	(0.043)
AT1	13.50	(0.98)	0.375	(0.035)	1.884	(0.264)	0.990	(0.072)
AT2	13.35	(1.20)	0.396	(0.047)	1.931	(0.313)	1.022	(0.093)
AT3	9.55	(0.78)	0.261	(0.027)	1.229	(0.172)	0.716	(0.059)
2010								
AT	12.12	(0.50)	0.330	(0.019)	1.480	(0.121)	0.911	(0.038)
AT1	13.66	(0.83)	0.373	(0.029)	1.614	(0.173)	1.032	(0.064)
AT2	14.62	(1.12)	0.443	(0.047)	2.168	(0.336)	1.124	(0.087)
AT3	8.96	(0.77)	0.218	(0.024)	0.943	(0.151)	0.654	(0.056)

Notes: The units of the indices: Poverty rate: %; Poverty gap: thousands of Euro; Squared poverty gap: millions of Euro^2; Log-poverty gap: log-Euro. The values of the poverty gap are not scaled. All estimates are direct. Standard errors (in parentheses) are based on 1000 bootstrap replications.

```
                header=T, sep=",")[, VRS]
...
```

```
##  Check the  region indicator
if (sum(substring(Dat[, Reg], 1, 2) != cou) > 0)
    stop(paste("Problem with the region indicator; variable ",
        Reg, ".", sep=""))
```

```
##  Discard the country acronym from the region indicator
Dat[, Reg] <- substring(Dat[, Reg], 3)
```

```
##  Arrangement for Italy (regions C, D, E, F, G)
if(sum(is.na(match(sort(unique(Dat[, Reg])), LETTERS)))==0)
        Dat[, Reg] <- match(Dat[, Reg], LETTERS)
```

```
##  Conversion to numeric
apply(Dat, 2, as.numeric)
}  ## End of function  EUreadR
```

For Italy, the character retained after removing the first two (IT) is converted to a digit by matching it with the alphabet (C – 3, ..., G – 7). Finland has regions FI13, FI18, FI19 and FI1A. The last region is recoded to FI11, so that the indicator of the region, after stripping the first two characters, can be a numerical variable.

```
Dat[Dat == "1A"] <- 11
```

Alternatives to this way of dealing with categorical variables include formatting the data in a (structured) list of variables, called **data.frame**, which is a list of vectors, in which each element (vector) has the same **names**, or creating a list of two elements, one for numeric and one for categorical data.

Within a function for estimating regional quantities, we form first the vector of codes of the regions:

```
Rgs <- sort(unique(dat[, reg]))
```

Then in the loop in which we evaluate the region-level summaries, we use a logical variable to indicate inclusion of subjects in a particular analysis, indicated below by the function EUfun.

```
for (rg in Rgs)
{
whi <- dat[, reg] == rg

EUfun(dat[whi, "hx090"], ...)
}
```

For some simple evaluations, the function **tapply** can be used. For example,

```
tapply(dat[, "db090"], dat[, reg], sum)
```

generates the region-level totals of the sampling weights, suitably labelled, not requiring the vector Rgs. The flexibility of this expression can be greatly enhanced by replacing the argument sum with a user-defined function.

The results are formatted in a list. Its first element is the set of results for the country (based on the entire sample), and the other elements are the results for the regions. The elements can themselves be lists. It is important for the elements to have identical attributes (e.g., matrices of identical sizes, with identical dimnames). A specific part of the results from every element is extracted by applying functions the Extr* functions ExtrR, ExtrC, ExtrE and ExtrL; see Appendix 1.A.4.

Small-area estimation starts with direct estimation of the indices for the regions.

```
EUregA <- function(Cou, Yea, dire="~/Splus/EUsilc12/DATA/", ext="dta",
              STD=60, div=1000, reg="db040")
{
### Poverty statistics for the regions of a country

## Cou   The country (two-leter acronym)
## Yea   The year (digit or 10)
## dire  The data directory
## ext   The extension of the dataset
## STD   The threshold (single value)
## div   The divisor (to avoid very large numbers)
## reg   The variable that indicates the region

## Check on the threshold -- has to be a (single) fraction
STD <- STD[1]
while (max(STD) > 1.5) STD <- STD/100
```

The data is input for the country and year at the household level and eHI is truncated at zero. The weighted median of eHI is then evaluated (at the individual level) and the shortfall calculated, together with its transformations for the various kernels (identity, linear, quadratic and log). The values of the shortfall are divided by the argument div, set to 1000, to avoid an output full of large numbers.

```
## Data input (incl. region) -- household level
dat <- EUreadR(dire, Cou, Yea, ext, VRS=c("hx040", "hx090", "db090"),
            reg, IH=2)

## Truncation at zero
dat[dat[, "hx090"] < 0, "hx090"] <- 0

## The weights and their total
weiP <- dat[, "db090"] * dat[, "hx040"]
Swe <- sum(weiP)

## The median, multiplied by the threshold
```

```
MD <- QuantW(dat[, "hx090"], weiP) * STD

## The shortfall
SHF <- (MD - dat[, "hx090"])/div
SHF[SHF < 0] <- 0

## The transformations of the shortfall
OUT <- cbind(SHF>0, SHF, SHF^2, log(1+div*SHF))
```

The national and regional weighted means are evaluated by matrix multiplication as $w_+^{-1} \mathbf{Y}^\top \mathbf{w}$ and $w_{d+}^{-1} \mathbf{Y}_d^\top \mathbf{w}_d$, where $\mathbf{Y}$ is the matrix of the values transformed from the shortfall (OUT), $\mathbf{w}$ the vector of weights (wei) and w_+ their total (Swei); the subscript d denotes the restriction of each object to region d.

```
## The national estimates
EST <- t(OUT) %*% matrix(weiP) / Swe

## The regions
Rgs <- sort(unique(dat[, reg]))

## Evaluation for each region
for (rg in Rgs)
{
## The households in the region
whi <- dat[, reg] == rg

## The region-level estimates
EST <- cbind(EST, t(OUT[whi, ]) %*% matrix(weiP[whi]) / sum(wei[whi]) )
}

## Labelling (kernels x regions)
dimnames(EST) <- list(c(seq(0, 2), "log"), c("AT", Rgs))

t(EST)
} ## End of function EUregA
```

The function is applied to Austria for years 2004–2010 by an expression with initialisation of a list and a loop over the years.

```
AT410Ar <- list()

## Annual analyses
for (yr in seq(4, 10))
AT410Ar[[as.character(2000+yr)]] <- EUregA("AT", yr)
```

The object AT410Ar is a list of the annual results, each in the form of a matrix of regions by kernels, with the whole country included in the first row, labelled AT. For example, round(AT49Ar[["2010"]], 3) is

```
            0     1     2    log
AT   0.121 0.330 1.480 0.911
1    0.137 0.373 1.614 1.032
2    0.146 0.443 2.168 1.124
3    0.090 0.218 0.943 0.654
```

The standard errors associated with these estimates are estimated by boot-strap in function EUregB. It has the same set of arguments as EUregA, except for one addition, nre=1000, for the number of bootstrap replicates. The code of EUregB is adapted from EUregA by including the evaluation of the expressions between the median (MD) and the results (EST) as the expression-argument of the function replicate, with the draw of a bootstrap sample added:

```
## The weights
wei <- dat[, "db090"]
weiP <- dat[, "db090"] * dat[, "hx040"]

replicate(nre,
{
## Bootstrap sample
Sbj <- EUbts(wei)

## The bootstrap median, multiplied by the threshold
MD <- QuantW(dat[Sbj, "hx090"], wei[Sbj]) * STD

...   ## code omitted

## The national estimates
EST <- t(OUT)%*%matrix(dat[Sbj,"hx040"]) / sum(dat[Sbj,"hx040"])

## The regions
Rgs <- sort(unique(dat[Sbj, reg]))

## The regional estimates
for (rg in Rgs)
{
whi <- dat[Sbj, reg] == rg

EST <- cbind(EST, t(OUT[whi,])%*%matrix(dat[Sbj,"hx040"][whi]) /
             sum(dat[Sbj, "hx040"][whi]) )
}
} ## End of the expression in the replication
) ## End of replicate
} ## End of function EUregB
```

The output is a $4(D+1) \times M$ matrix, where D is the number of regions and M the number of bootstrap replications. The following application generates a list of annual results.

```
AT410Br <- list()
```

```
for (yr in seq(4, 10))
AT410Br[[as.character(2000+yr)]] <- EUregB("AT", yr, nre=1000)
```

This evaluation takes about 80 seconds of CPU time for Austria. The boot-strap replicate estimates are summarised by the function MeVa, which evalu-ates the means and standard deviations, see Appendix 1.A.4.

```
ES410Bs <- lapply(ES410Br, apply, 1, MeVa)
```

This list can be formatted further, e.g., by extracting the means and standard deviations and reshaping them to suitable matrices.

```
ES410Bt <- list(Expectations=lapply(lapply(ES410Bs, ExtrR, 1), matrix,
      ncol=4, byrow=T),
               St.errors=lapply(lapply(ES410Bs, ExtrR, 2), matrix,
      ncol=4, byrow=T) )
```

```
##  Labelling
for (yr in names(ES410Bt[[1]]))
dimnames(ES410Bt[[1]][[yr]]) <- dimnames(ES410Bt[[2]][[yr]]) <-
    dimnames(ES410Ar[[yr]])
```

The function EUssz collects the sample sizes and estimated population sizes (weight totals) for the regions of a country. After the input of the data, a single expression generates all the counts: the numbers of individuals and households in the sample, the effective sample sizes and the estimated popu-lation sizes. The effective sample sizes are evaluated by the function EUeff; see Appendix 1.A.3.

```
EUssz <- function(Co, Yr)
{
###   The sample sizes (regions)  of  country  Co  in  year  Yr

dat <- EUreadR("~/Splus/EUsilc12/DATA/", Co, Yr, Ext="dta",
              VRS=c("db090", "hx040", "hid"),"db040", IH=2)


##  The household sizes
hhsz <- dat[, "hx040"]

RES <- rbind(c(sum(hhsz), EUeff(rep(dat[, "db090"], hhsz)),
    nrow(dat), EUeff(dat[, "db090"]),
    sum(dat[, "db090"]*hhsz), sum(dat[, "db090"])),
    cbind(tapply(hhsz, dat[, "db040"], sum),
    tapply(rep(dat[, "db090"], hhsz), rep(dat[, "db040"], hhsz),
    EUeff),
    table(dat[, "db040"]), tapply(dat[, "db090"], dat[, "db040"],
    EUeff),
    tapply(rep(dat[, "db090"], hhsz), rep(dat[, "db040"], hhsz),sum),
    tapply(dat[, "db090"], dat[, "db040"], sum)))

dimnames(RES) <- list(c(Co, sort(unique(dat[, "db040"]))),
```

```
          c("n-Ind", "nb-Ind", "n-Hhld", "nb-Hhld", "P-Ind", "P-Hhld"))
RES
} ## End of function EUssz
```

The function is applied to generate the list of counts for Spain in the seven
years of the survey by the expressions

```
ES410aD <- list()

for (yr in seq(4,10))
ES410aD[[as.character(2000+yr)]] <- EUssz("ES", yr)

## Add a row for the empty region
for (i in seq(2))
{
ES410aD[[i]] <- rbind(ES410aD[[i]], ES64=0)
ES410aD[[i]] <- ES410aD[[i]][sort.list(rownames(ES410aD[[i]])), ]
}
```

which also insert zeros for region 64 in 2004 and 2005 when no data was
collected in the region.

5.B.1   Composite Estimation

The function EUregC for univariate composition uses the outputs of EUregA
and EUregB:

```
EUregC <- function(Dest, Ster, Essz, mnSB=0)
{
###   Small-area estimation  --  univariate composition

##   Dest   The direct estimates (country and the regions)
##   Ster   The estimated standard errors (country and the regions)
##   Essz   The effective sample sizes
##   mnSB   The smallest permissible variance (truncation)
```

The argument mnSB is used for truncating the moment-matching estimate of
the region-level variance σ_B^2. In the evaluations in the function, it is assumed
that the annual direct estimates are mutually independent, and the linear
approximation is used for the covariances c_d (Cd).

```
## The regions in common in Dest and Essz
rgs <- c(names(Dest), names(Essz))
rgs <- rgs[duplicated(rgs)]

## Estimation reduced to the regions in common
Dest <- Dest[rgs]
Essz <- Essz[rgs]

## From standard errors to sampling variances
Ster <- Ster^2
```

```
##  The relative standard errors
Rssz <- Essz[-1] / Essz[1]

##  The covariances
Cd <- Ster[-1] * (1 - Rssz)
```

Next, the region-level variance is estimated, with truncation if applicable.

```
SB <- mean((Dest[-1] - Dest[1])^2) - Ster[1] -
      mean(Ster[-1] * (1 - 2*Rssz))
SB <- max(c(SB, mnSB))
```

The coefficients b_d are then estimated and the composition of the direct region-level and the national estimators evaluated. The standard errors are estimated by the appropriate substitutions.

```
##  The coefficients  (the denominator  den  is used later)
den <- Ster[-1] * (1 - 2*Rssz) + Ster[1] + SB
bd <- Ster[-1]*(1-Rssz) / den

##  The composition
Est <- (1 - bd) * Dest[-1] + bd * Dest[1]

##  Estimated standard error
SteC <- sqrt(Ster[-1] - (Ster[-1]*(1 - Rssz))^2 / den)

list(Estimates=rbind(Composite=Est, St.Err.C=SteC, Direct=Dest[-1],
     St.Err.D=sqrt(Ster[-1]), Shrinkage=bd), Variance.B=SB)
}  ##  End of function  EUregC
```

The output is a list comprising two elements: a matrix of estimates, standard errors and coefficients $\hat{b}_d$ (D columns), and the estimated variance $\hat{\sigma}_B^2$.

The set of seven annual analyses for Spain are conducted by the code

```
##  The numbers of regions within years
Nrg <- c(rep(19,2), rep(20,5))
names(Nrg) <- names(ES410Ar)

##   Initialisation
ES410Cr <- list()

##  Using auxiliary information starting in year  yr
for (yr in as.character(2000+seq(4, 9)))
ES410Cr[[yr]] <- EUregC(ES410Ar[[yr]][, 1],
     ES410Bs[[yr]][2, seq(Nrg[yr])*4-3], ES410Cr[[yr]][, 6])
```

Note that the number of regions (+1 for the national values) is declared separately for each year, to take care of the absence of one region (Melilla) in the first two years.

5.B.2   Multivariate Composition

Multivariate composition is implemented by the function EUregD. It has three
mandatory arguments, lists lstE, lstV and lstN, for the respective direct
estimates, standard errors and population sizes of the regions, and several
optional arguments used for indicating the setting of the analysis and for its
fine-tuning. The data input is in the form of annual lists generated in the evaluation of direct estimates and their standard errors by the function EUregA,
and the (annual) population sizes of the regions. Data lists were generated for
several poverty indices (kernels); the argument vr indicates which kernel is to
be used (which estimates and standard errors are to be extracted), and the
argument cl identifies the column that contains the population sizes in the
elements of lstN (cl=5 for individuals and cl=6 for households). The argument dfl is for the default setting of the estimate and standard error for a
region not represented in lstE and lstN, because it was not represented in
the related survey. The argument DD specifies the maximum condition number
of the estimated variance matrix $\hat{\Sigma}_B$ and tol the negative exponent of the
smallest permitted value of an eigenvalue of $\hat{\Sigma}_B$. If the conditions associated
with DD and tol are not satisfied, the variances in $\hat{\Sigma}_B$ are inflated.

```
EUregD <- function(co, lstE, lstV, vr=1, lstN, cl=6, dfl=c(0.20,100),
          DD=1000, tol=8)
{
###     Small-area estimation   --   multivariate composition

##    co       The country acronym
##    lstE     The estimates (in the annual lists)
##    lstV     The standard errors (in the annual lists)
##    vr       The variable (column) in the annual lists
##    lstN     The population sizes (in the annual lists)
##    cl       The column (in the annual lists)
##    dfl      The default values for estimate and st.error
##             when they are missing
##    DD       The conditioning for the variance matrix
##    tol      For checking for a small eigenvalue

##  Check on the tolerance
if (tol > 0.1)  tol <- 10^(-tol)

##  All the areas
Rgs <- sort(unique(as.vector(unlist(sapply(lstE, rownames)))))
nrg <- length(Rgs)

##  The years
yrs <- names(lstE)
nyr <- length(yrs)
```

The relevant data from the three data-lists is extracted and formatted into
three matrices, MNS, VRS and PSZ, for the estimates, sampling variances and

population sizes, respectively. Their initialisation and the loop over the years are necessary to deal with regions not represented in a data-list.

```
##  Initialisation
MNS <- matrix(dfl[1], nrg, nyr)
VRS <- matrix(dfl[2]^2, nrg, nyr)
PSZ <- matrix(0, nrg, nyr)
dimnames(MNS) <- dimnames(VRS) <- dimnames(PSZ) <- list(Rgs, yrs)


##  The matrices of the relevant data
for (yr in yrs)
{
nms <- rownames(lstE[[yr]])

MNS[nms, yr] <- lstE[[yr]][, vr]
VRS[nms, yr] <- lstV[[yr]][, vr]^2
PSZ[, yr] <- lstN[[yr]][, cl]
}


##  The relative population sizes
PSZ <- PSZ %*% diag(1/PSZ[co,])
dimnames(PSZ) <- list(Rgs, yrs)
```

The elements of Σ_B are estimated separately for variances (diagonal of Σ_B) and covariances (off-diagonal). Note that the respective first rows of MNS and VRS contain the national means and sampling variances.

```
##  The region-level variances
SB <- c()

for (yr in yrs)
{
##  Discard the national statistics and 'no data' regions
whi <- c(PSZ[, yr] > 0 & rownames(PSZ) != co)

##  The estimate of the region-level variance
SB <- c(SB, mean((MNS[whi, yr] - MNS[1, yr])^2) - VRS[co, yr] -
           mean((VRS[whi, yr] * (1 - 2 * PSZ[whi, yr]))) )
}  ##  End of the loop over the years  (yr)


##  The variance matrix, initialisation for the covariances
SB <- diag(SB)

for (yr1 in seq(nyr-1))
{
for (yr2 in seq(yr1+1, nyr))
{
##  Direct estimates available for both years (and not the country)
whi <- (PSZ[, yr1] > 0) & (PSZ[, yr2] > 0) & (rownames(PSZ) != co)
```

```
##  The estimate of the covariance
SB[yr1, yr2] <- SB[yr2, yr1] <- cov(MNS[whi, yr1], MNS[whi, yr2])
}}  ##  End of the double loop (covariances)
```

Next, the variances are inflated if necessary. The scalar vAdd is the value added to each variance.

```
##  Dealing with negative eigenvalues
vAdd <- range(eigen(SB)[[1]])

##  The variance inflation
vAdd <- (vAdd[2] - vAdd[1]*DD)/(DD-1)

##  Truncation at zero
vAdd <- vAdd*(vAdd > 0)
diag(SB) <- diag(SB) + vAdd
```

Estimation for each region is carried out separately, in a loop. The objects V, Vd, Cd, Qd, bd and ee are the R versions of $\mathbf{V}$, $\mathbf{V}_d$, $\mathbf{C}_d$, $\mathbf{Q}_d$, $\mathbf{b}_d$ and $\mathbf{e}$ in Section 5.4. The first element of Rgs is All, indicating the national estimates. It is skipped in the loop.

```
##  The national variance and the indicator vector
V <- diag(VRS[1, ])
ee <- c(rep(0, nyr-1), 1)

##  Initialisation for the results
RES <- c()
bds <- c()

##  Separate estimation for each region (except 'All')
for (rg in Rgs[Rgs != co])
{
Vd <- diag(VRS[rg, ])
Cd <- diag(VRS[rg, ]) * diag(PSZ[rg, ])

Qd <- Vd + V - 2*Cd + SB
bd <- solve(Qd, Vd - Cd) %*% matrix(ee)

##  Collect the vectors of coefficients
bds <- cbind(bds, bd)

if (min(eigen(Qd)[[1]]) < tol)
    warning(paste("Small/negative eigenvalue for region ", rg, ".",
            sep=""))

##  The composite estimate and standard error for the region
RES <- cbind(RES, c(sum((ee - bd) * MNS[rg, ]) + sum(bd * MNS[co, ]),
            sqrt(sum(diag(Vd)*ee) - sum(ee * ((Vd - Cd) %*% bd)) ) ))
}  ##  End of the loop over the regions
```

```
##  Labelling
dimnames(RES) <- list(c("Estimate", "St.error"), Rgs[Rgs != co])
dimnames(bds) <- list(names(lstE), Rgs[Rgs != co])

list(RES, c(vAdd, eigen(SB)[[1]]), bds)
}  ##  End of function EUregD
```

The output is a list of three elements: a $2 \times D$ matrix of the estimates and estimated standard errors, the eigenvalues of $\hat{\Sigma}_B$ preceded by the scalar used in variance inflation and the $r \times D$ matrix of the coefficient vectors $\hat{\mathbf{b}}_d$. The weights assigned to the current year are given in its bottom row labelled "2010".

The collection of the bivariate to seven-variate composite estimates for the regions is obtained by the following expressions:

```
##  The number of years
nyr <- length(ES410Ar)

ES410Ds <- list()

for (yr in seq(nyr-1))
{
##  nyr-yr+1  is the dimension of the composition/shrinkage
ES410Ds[[as.character(nyr-yr)]] <-
EUregD("ES", ES410Ar[seq(yr, nyr)], ES410Bt[[2]][seq(yr, nyr)], vr=1,
             ES410aD[seq(yr, nyr)], cl=5, DD=1000)
}
```

The loop-variable yr indicates the first year of the auxiliary information, counted from 2004, so nyr-yr+1 is the dimension of the composition. The key change for estimating the poverty gaps is to set vr=2; the argument dfl does not have to be altered.

The results in ES410Dr can be reformatted by applying the Extr* functions. For example, the following expression forms a list of the matrices of estimates and standard errors, with the results of the univariate composition and direct estimation added as the last two columns.

```
##  The estimates and the standard errors separated
ES410Dt <- list(Estimates=cbind(
       sapply(lapply(ES410Ds, ExtrL, 1), ExtrR, 1),
       C=ES410Ds[[6]][[1]][1, ], D=ES410Bt[[1]][[6]][-1, 1]),
     St.errors=cbind(sapply(lapply(ES410Ds, ExtrL, 1), ExtrR, 2),
       C=ES410Ds[[6]][[1]][2, ], D=ES410Bt[[2]][[6]][-1, 1]))
```

5.B.3   Graphics

The collection of estimates for the regions and their estimated standard errors is plotted by the function EUregG.

```
EUregG <- function(lst, fct=0.75, mlt=100, abl=c(), ttl)
```

```
{
###    Graphics for direct and composite estimates

##    lst       The (condensed) result of EUregD
##    fct       The factor for plotting
##    mlt       The multiplier for the estimates
##    abl       The national mean (to mark it in the diagram)
##    ttl       Title for the plot
```

The labels of the regions are extracted. The numbers of estimates are used for setting the horizontal locations of the estimates in the plot (`Kest`). The results in the list `lst` are multiplied by `mlt`, for instance, from fractions to percentages, to make the axis labels more attractive.

```
##   The regions
Rgs <- rownames(lst[[1]])
nrg <- length(Rgs)

##   The estimates and the offsets
nEst <- ncol(lst[[1]])
Kest <- (seq(nEst) - (nEst+1)/2)/nEst * fct

for (i in seq(2))
lst[[i]] <- mlt*lst[[i]]
```

Now the empty shell of the plot is drawn. The extent of the x-axis is set exactly (`xaxs="i"`), and it is not labelled (`xaxt="n"`):

```
plot(c(0.25,nrg+0.75), range(lst[[1]])+c(0,1.5), type="n", xaxs="i",
    xaxt="n", xlab="", ylab="Estimates (%)", main=ttl,
    cex.lab=0.9, cex.axis=0.9, cex.main=0.9)
```

The contents of the top panel (estimates), that is, the connected sets of estimates for each region, the labels of the regions and the national rate, are drawn by the following expressions

```
##   The twiddles
for (rg in seq(nrg))
lines(rg+Kest, rev(lst[[1]][rg, ]), lwd=0.7)

##   The regions' labels
text(seq(nrg), max(lst[[1]])+1.5, Rgs, cex=0.7)

##   The national rate
abline(h=abl, lty=2, lwd=0.5)
```

A similar set of expressions takes care of the bottom panel of the diagram. Figure 5.9 is generated by the expression

```
Figure("ES3", EUregG(ES410Dt, 0.75, abl=100*ES410Ar[["2010"]][1,1],
    ttl="Spain 2010"),  4.5, c(2, 1), c(0.05, 3.5, 1.75,1), 0.6)
```

6

Transitions

Apart from the poverty status and its extent in any particular year, as quantified by the shortfall and summarised by a poverty index, the changes from one year to the next are also of interest. We refer to them as transitions, and the 'catchy' term associated with them is poverty *dynamics* (e.g., Duncan *et al.*, 1993, Stewart and Swaffield, 1999, and Jenkins, 2000). A household that is classified as poor in one year and as not poor in the next is regarded as a *success*, and a household with a change in the other direction as a *failure*. The balance of successes and failures, that is, of exits from and entries to poverty, is an overall measure of the change. The rate (percentage) of transitions in either direction is a measure of the mobility of households between the two poverty states.

This chapter defines indices for the changes households make from one poverty status to the other, and studies their properties. Such a change is referred to as a transition. The starting point is the percentages of households that make the transition from one year to the next in either direction, but we soon find these summaries wanting because their values are bounded by the related constituencies; only the poor in year 1 can exit poverty and only those who are not poor in year 1 could possibly enter poverty in the following year 2. This is resolved by defining the relative rates of transition. Next we highlight the issue that a large fraction of the transitions may be a result of relatively small changes in the equivalised household income (eHI), such as from the 58th percentile in year 1 to the 61st in year 2. We address this problem by defining substantial and partial transitions, scoring each transition by the magnitude of the relative change from one year to the next. An alternative based on multiple imputation is proposed and applied in a study of successive transitions across a sequence of years.

6.1 Panel Data

In general, we can learn about changes in poverty status in a population from one year to the next only by observing (recording) such changes for a sample of the population. If we want to learn about changes in not only one pair but several consecutive pairs of years, we have to maintain over the years contact

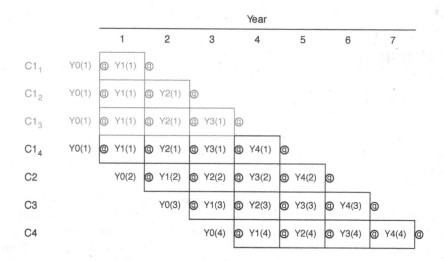

FIGURE 6.1

Rotating panel design.

The survey started in year 1 with four subcohorts $C1_k$, $k = 1, 2, 3$ and 4, each to be retired after completing the questionnaire in year k.

with a sufficiently large sample of subjects (individuals or households). The subjects cannot be retained in the sample for a very long time; some of them are retired from the survey every year, and the sample is replenished by new subjects. The rotating panel design has these features. An illustration of the design is given in Figure 6.1. Every year of participation is represented by a box with a reference to a cohort (horizontally) and the year (vertically). For example, box Y3(2) corresponds to the third year of cohort C2, which was introduced into the sample in year 2 and was retired from it after year 5. Questionnaires were administered to this cohort at the beginning of years 2, 3, 4 and 5, marked by encircled 'q', inquiring about the preceding year, in the first instance about year 1 at the beginning of year 2.

We assume that the survey started in year 1 by collecting information about the preceding year 0. The sample may start with a large cohort that will be retired en masse at the end of year 4. Then the panel would be suddenly depleted and would require many new subjects. We would have no capacity to learn about the changes between years 4 and 5, because no transitions would be recorded for these two years.

An alternative, represented by gray colour in the diagram, avoids this problem by splitting the year-1 cohort C1 to four subcohorts $C1_k$, $k = 1, 2, 3$ and 4, which will be retired at the end of year k. In this way, continuity of the panel is maintained from the very beginning. Cohorts C2 and C3 could also be split to subcohorts that would be retired in different years. From cohort

C4 on, the panel is settled to four subcohorts being present in every year, one of them introduced and one about to be retired, having been on the panel for the three previous years.

Appreciating that participation in a survey is a burden, we do not want to keep subjects on the panel (in the sample) for many years. That is why every subject is retired after at most three years, which involve four rounds (administrations) of the questionnaire. Also, the likelihood that households change their composition or move to another location increases, and keeping track of these changes becomes difficult. The sampling designs in the longitudinal part of the European Union Statistics on Income and Living Conditions (EU-SILC) are four-year rotating panels for every country.

Table 6.1 lists the household-level sample sizes of the panels for the 24 countries in the longitudinal part of EU-SILC in 2010. The three blocks list the numbers of all households, of households with data in every (planned) round of the survey, and of the households that have the same value of the equivalised household size (eHS, introduced in Section 1.1) in every year. The latter households are called intact. Within each block, the entries are for the cohorts (rotational groups) started in 2007, 2008 and 2009. The three cohorts have similar sizes only for a few countries: Austria, Spain, Finland, Italy, Romania and the UK. Refusals and other forms of nonresponse interfere with the plans to have a settled symmetric design over the years. The changes in the sizes of the groups for Bulgaria (increase), France and Norway (reductions) are extreme. The panel for Luxembourg was not refreshed in 2008, and a substantially smaller rotational group was introduced in 2009 than in 2007. Note that the cohort introduced in 2007 was extremely large, comprising 4355 households with more than 12 000 individuals. That is a sample of about 1:40. (The population of Luxembourg in 2007 was nearly half a million.) The national panels in 2010 also contain new cohorts, intended for 2010−2013; their summaries are not listed in the table.

For some analyses, only households with records in every year are used. The second block of Table 6.1, with the counts of households that have no records missing, indicates that in some countries many households do not cooperate fully or are not followed up. For example, in the rotational group that started in Austria in 2007, only 1119 households out of 2186 (51%) have records in every year. The rate of full cooperation is higher in the other two rotational groups; however, further losses in 2011 (and in 2012 for the group started in 2009) can be expected. Denmark appears to have no missing records. This is likely to be an artefact of how the data was submitted, omitting all households with incomplete responses, or extracting the data from an administrative register. Norway has no missing records in the rotational group 2009−2010 and, together with Romania, exceptionally high rate of compliance with the survey protocol. The rate of compliance apparent from the submitted records is lowest in the UK; only 1486 households have complete records out of 3284 participating households in rotational group 2007−2010 (45%); it is followed by Austria (51%) and Belgium (56%). Of course, this comparison is tainted

TABLE 6.1

Household-level sample sizes of the national panels in EU-SILC in the 2007–2010 database.

	All households			Full participation			Intact households		
	7–10	8–10	9–10	7–10	8–10	9–10	7–10	8–10	9–10
AT	2186	1901	2045	1119	1260	1688	765	993	1497
BE	2119	1618	1853	1183	1107	1460	790	846	1245
BG	1069	1696	2020	776	1550	1926	494	1283	1763
CZ	2713	2099	2590	2264	1717	2373	1705	1427	2179
DK	970	1094	1252	970	1094	1252	626	790	1037
EE	1611	1514	1235	1087	1133	1109	703	813	955
ES	4054	4010	4013	2715	3029	3396	1753	2279	2856
FI	1830	1829	1757	1394	1432	1528	876	1018	1333
FR	8249	1953	1874	5142	1558	1660	3513	1211	1465
GR	1714	2526	2305	1151	1855	2013	816	1474	1826
HU	3215	2578	3361	1871	1947	2806	1355	1555	2489
IS	639	719	766	501	589	645	226	332	476
IT	6324	6218	5922	3987	4439	4625	2931	3682	4245
LT	1589	1252	1323	1234	1119	1283	901	888	1152
LU	4355	0	1122	2429	0	813	1507	0	706
LV	1593	1928	1932	1137	1419	1616	743	1006	1354
NL	3731	3621	3079	2071	2340	2431	1532	1868	2176
NO	2857	541	614	2433	458	614	1406	321	496
PL	3930	3866	3492	2909	3076	3109	1855	2311	2695
PT	1368	1364	1649	989	1105	1443	711	871	1299
RO	2031	1933	1975	1894	1855	1943	1605	1691	1863
SE	2164	1868	1836	1556	1405	1524	970	1020	1253
SI	2952	3390	3242	1592	2167	2542	1006	1559	2170
UK	3284	3034	2938	1486	1651	1945	1054	1272	1709

Note: Records with missing values on variables on which eHI is based are not counted.

by how the records are compiled (preprocessed), and reflects neither on the general disposition of households toward participating in EU-SILC nor on the success in retaining households on the panel. In a principled way of compiling a database, all the households that were included in the survey would be listed, even if they refused to participate outright. However, as the database contains records for individuals, an ambiguity would arise for households for which the number of its members was not established. In some instances, even the existence of the selected household may be in doubt, so the principled approach could be implemented only with some compromise.

The level of compliance with the survey protocol can be described in greater detail by tabulating the patterns of presence. For each household and year in which it is a subject in the survey, we define the indicator of presence as unity if the household has a record in the data, and as zero otherwise. For a household in the rotational group 2007–2010, the pattern of its presence is defined by 'gluing' the four indicators together. For instance, 1111 is the code for full compliance and 0110 for having records only for 2008 and 2009. Of the $2^4 = 16$ possible quartets comprising zeros and unities, the pattern 0000 does not occur because households that take no part in the survey, even though they have been selected, are not represented in the database. Such losses include not only refusals, but also failure to contact the household, selecting a dwelling that is empty or that does not exist (with the details sometimes not established), and other contingencies that include errors in the sampling frame. Not all the other 15 patterns occur for every country. In particular, the pattern 1001, for having records only in 2007 and 2010, occurs only in three countries, Bulgaria and Romania once each and in Hungary for 101 households. We conjecture that after the failures in 2008 and 2009 no attempts would be made to contact a household in its concluding year 2010 in most countries.

Table 6.2 displays the tabulations of the patterns for the countries in the rotational groups 2007–2010. Pattern 1001, fully described above, is omitted from the table. The patterns can be classified into the following groups:

F.	Full compliance:	1111;
D.	Drop-out:	1000, 1100, 1110;
L.	Late entry:	0001, 0011, 0111;
1M.	One missing:	1011, 1101;
1P.	One participation:	0100, 0010;
2O.	Other:	0101, 0110, 1001, 1010.

In some longitudinal studies, drop-out is the only incomplete pattern, and analysis of such studies is much simpler than when all patterns are present (Diggle *et al.*, 2002; Little and Rubin, 2002). Following full compliance, the counts are largest for the drop-out patterns. Several countries have no households with patterns 0xxx (Finland, Iceland, the Netherlands, Sweden and Slovenia, in addition to Denmark). It appears that no attempt was made in these countries to contact any households in 2008–2010 that did not cooperate (or were not contacted) in 2007. In Norway, the late entries 0011 and 0111 are the only incomplete patterns. In Finland and the Netherlands, all incomplete records are drop-outs. The incomplete patterns that start and end with unity, 1xx1, that is, 1001, 1011 and 1101, can be interpreted as successful recoveries of households' participation in the survey. Belgium, France and Italy have many such records.

Figure 6.2 displays a graphical summary of the patterns of presence for Austria and Belgium. Each year is represented by a vertical bar comprising layers that correspond to the patterns. Presence in the survey is indicated by shading. The thickness of a layer is proportional to the number of households

TABLE 6.2
Patterns of presence of households in the EU-SILC rotational groups 2007–2010.

	L	1P	L	1P	2O	2O	L	D	2O	1M	D	1M	D	F
	1	10	11	100	101	110	111	1000	1010	1011	1100	1101	1110	1111
AT	10	5	21	5	0	1	20	682	0	0	232	0	91	1119
BE	30	13	18	2	2	4	24	351	21	72	190	47	162	1183
BG	4	1	12	3	0	0	10	210	0	0	25	0	27	776
CZ	23	0	21	3	0	1	15	205	2	13	105	0	61	2264
DK	0	0	0	0	0	0	0	0	0	0	0	0	0	970
EE	30	6	22	5	0	4	18	154	2	11	186	15	71	1087
ES	61	17	70	16	1	6	55	408	14	69	322	51	249	2715
FI	0	0	0	0	0	0	0	195	0	0	125	0	116	1394
FR	98	30	128	49	9	40	159	891	39	94	759	101	710	5142
GR	18	1	11	4	0	0	7	237	0	0	180	0	105	1151
HU	24	4	12	3	0	11	13	520	18	43	334	30	231	1871
IS	0	0	0	0	0	0	0	45	2	9	33	8	41	501
IT	47	20	59	13	4	7	59	698	47	93	516	104	670	3987
LT	3	0	3	2	0	0	5	184	2	31	84	9	32	1234
LU	25	5	26	165	0	64	268	581	0	0	493	0	299	2429
LV	9	0	2	0	0	3	2	216	5	13	105	21	80	1137
NL	0	0	0	0	0	0	0	838	0	0	444	0	378	2071
NO	0	0	259	0	0	0	165	0	0	0	0	0	0	2433
PL	21	6	34	6	2	5	26	391	3	15	311	30	171	2909
PT	11	3	16	2	2	1	9	106	9	30	68	21	101	989
RO	2	0	6	0	0	1	1	50	1	9	32	5	30	1894
SE	0	0	0	0	0	0	0	257	3	8	145	3	192	1556
SI	0	0	0	0	0	0	0	711	4	20	378	10	237	1592
UK	21	16	11	9	0	4	17	920	0	0	517	1	282	1486

Notes: To fit the table on the page, pattern 1001 is omitted. The only nonzero entries in the omitted column are for Bulgaria (1), Hungary (101) and Romania (1). The leading zeros of the patterns in the second line of the headings are dropped.

with the pattern. The most frequent patterns are marked at the right-hand margin and the counts of households with the pattern at the left-hand margin. We conclude from the diagram that the rate of full cooperation is higher in Belgium and there is a higher rate of households with patterns 1M in Belgium. In Austria, most of the households with incomplete records have pattern D; the other incomplete patterns account for only 2.8% of the households, compared to 11.0% in Belgium. A lighter shade of gray between the bars for two consecutive years indicates continuity, that transitions between the years can be observed.

As an alternative, the thickness of the layers could be set proportional to the total weight within the pattern. The households have year-specific weights, so we would have to settle on the year on which the weights are to be based.

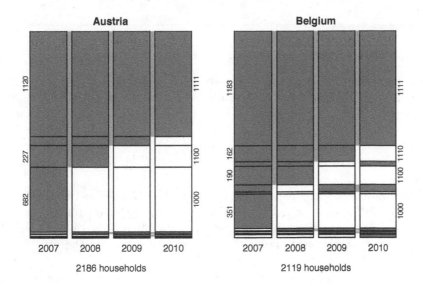

FIGURE 6.2
Patterns of presence on the panel for Austria and Belgium; rotational groups
2007 – 2010.

Note the discrepancy of the count of units with pattern 1111 in Austria
between the diagram and Tables 6.1 and 6.2 (1120 vs. 1119). The reason for
it is a household with missing value of eHS in 2010, caused by no cooperation from two adult members of a household that had more members in the
previous years. For the tables, listwise deletion was applied with respect to
all the variables used, including eHS, and then all the counts established. For
the diagram, listwise deletion was applied, but eHS is not among the variables used. A discrepancy arises also for the pattern 1100. Four households
with pattern of presence 1110 have missing values of eHS in 2009 because of
no cooperation of one member of each household. These four households are
counted in pattern 1110 in Figure 6.2, but as pattern 1100 in Table 6.2. These
glitches can be fixed by replacing the single listwise deletion with several, each
time deleting incomplete records with respect to the minimal list of variables.
However, eHS is rarely missing; when it is missing eHI, the key variable, is
not available either. In the data for Belgium, the variable eHS has no missing
values.

6.2   Absolute and Relative Rates of Transition

In any single year, a household has two possible poverty states, being poor
and not being poor. For a pair of consecutive years 1 and 2, there are four
patterns of poverty: exit from poverty (poor in year 1 and not poor in year 2),
entry to poverty (not poor in year 1 and poor in year 2), staying not poor (not
poor in either year) and staying poor (poor in both years). The former two
patterns indicate a change of status (moving) and the latter two no change
(staying).

The absolute rate of a pattern for years 1 and 2 is defined as the percentage
of individuals in year 1 who have this pattern. The latter reference to year 1
in this definition is important, because the population of a country changes
from one year to the next. For estimating this rate, we use the weights for
year 1.

The poverty status can be altered substantially by the departure of a
member from the household or the arrival of a new member (by birth, death,
marriage or the like). We regard a household as intact, or as having the same
composition, if it has the same value of eHS in the two years. Thus, a household
is not intact if one member is classified as a child in one year and as an adult
in the next, but is regarded as intact if one adult is replaced by another.

We reduce our attention to intact households. The dataset for Austria
has 3641 households that are intact between 2009 and 2010. Another 483
households have records for both years but have an altered value of eHS. Note
that the figures in the right-hand block of Table 6.1 are for intact households
within the cohorts. A household in the cohort 2007–2010 is not counted as
intact in the table if it had a change of eHS between 2008 and 2009, but
it may be intact in 2009–2010. The poverty status in each year is derived
by comparing the household's eHI with the estimated standard for the year.
The annual standard is estimated with the weights for the year involved. By
evaluating the standard for a range of thresholds, we obtain the four rates as
functions of the threshold.

Figure 6.3 displays these curves for the four patterns for Austria in years
2009 and 2010. The absolute rates of change are plotted in panel A. Black
colour is used for exit from poverty and gray for entry to poverty. Thin solid
lines connect the estimates for thresholds of 40, 41, ..., 80%, and thicker lines
connect their smoothed versions. With this layout, we highlight the smoothed
curves but give an indication of the extent of smoothing. The rates of switching
from either state increase with the threshold for exit from poverty, except for
an anomaly in the range $50 < \tau < 55\%$, where the rate of exit has a sudden
increase followed by a decrease, and the rate of entry has a small decrease
followed by an increase that catches up with the overall trend. The two curves
intersect several times. The rate of staying poor increases with the threshold
and the rate of staying in prosperity declines with τ. The curves formed from

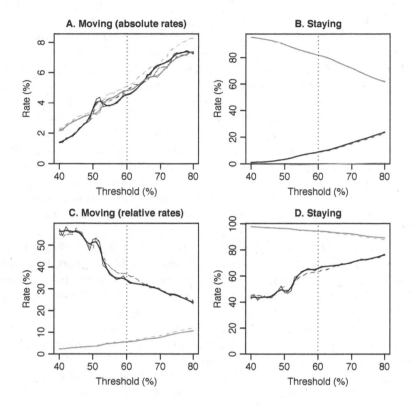

FIGURE 6.3

Estimated rates of transition between poverty states in Austria from 2009 to 2010.

Black colour indicates being poor in year 2009. Solid lines are for estimates based on intact households and dashes for all households with complete records for 2007–2010. Thin lines are for the estimates and thicker lines for their smoothed versions.

the estimates do not need any smoothing—the original curves are completely obscured by the smoothed curves in panel B.

Being poor in year 1 is a prerequisite for exiting from poverty in the next year. Therefore, the poverty rate in year 1, as an upper limit, is an important factor in the rate of exit from poverty. The *relative rate* of exit from poverty is defined as the percentage of those who exit poverty, using the number of poor in year 1 as the denominator (divisor). The relative rates of entry to poverty are defined similarly, using the households that are not poor in year 1 as the denominator. From now on, we refer to the rates displayed in panels A and B of Figure 6.3 as *absolute*. If all the poor in year 1 (say, 15% of the households) changed their status the next year, the relative rate of exit from poverty would be 100%, whereas the absolute rate would be 15%, the highest possible. The relative rates are better suited for comparisons, although any comparison of the rates of transition for two settings (countries or years) with very different (static) rates of poverty is problematic.

The relative rates are plotted in panels C and D. For exit from poverty, they differ radically from the corresponding absolute rates because the denominator in their evaluation, the poverty rate in 2009, changes substantially between the thresholds of 40% and 80%, from 2.4% to 31.3%. Thus, its increase is about 13-fold. The denominator for evaluating the relative rate of entry to poverty is the absolute rate of prosperity in 2009, which is in the range 68.7%–97.6%. It declines across the thresholds we consider only about 1.4 times. The curves of the absolute and relative rates of entry to poverty (gray colour) have similar shapes, but this cannot be seen in panels A and C because they are on very different scales. The relative rates of staying poor are affected by the anomaly at around $\tau = 50\%$.

The estimated absolute rates of entry to and exit from poverty differ by between -1.1%, at $\tau = 51.5\%$, and 1.0% at $\tau = 42.5\%$. Apart from a few reversals, both functions are increasing. Discord between the two rate functions does not imply any accumulation or reduction of the number of poor households because new households are formed every year and others become members of the studied population (by forming new households or becoming Austrian residents), and some of them start in the state of poverty. Other households disappear (by death and emigration).

The rates of transition estimated for all households, including those with changed composition, are drawn in Figure 6.3 by thin dashes. Intact households have lower (absolute) rates of entry to poverty for all thresholds, by up to 1.0%. The estimated rates of staying poor differ from the rates for all households by up to 1.2%. The differences are negative for a few thresholds, but only by a narrow margin in all instances. The differences are greater for higher thresholds. The rates of staying not poor differ by between -0.5% and 0.6%. In relation to the high absolute rates, 61%–95%, such differences are quite small.

The relative rates drawn in panel C of Figure 6.3 appear to be smoother because they are presented with much less detail, with the vertical axis extending over 60% compared to only 8% for the absolute rates in panel A. The appearance (shape and magnitude) of the function for exit from poverty (black) is changed substantially by pro-rating, because the rates of poverty (the denominators) are small fractions that change a lot on the multiplicative scale over the plotting range of $40 < \tau < 80\%$. In contrast, the largest denominator for entry to poverty is only about 1.4 times greater than the smallest.

With data for several years, we can study the evolution of the rates of transition. Figure 6.4 displays the absolute and relative rates of transition for Austria for the three pairs of consecutive years from 2007 to 2010. The absolute rates of exit from poverty for $\tau > 60\%$ are lower for 2009–2010 than for the two earlier pairs of years. The rates of entry to poverty have declined

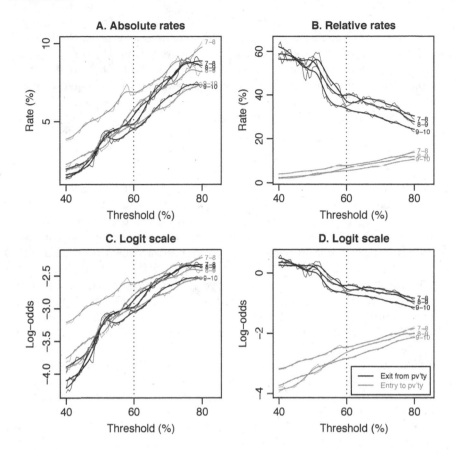

FIGURE 6.4
Estimated rates of transition between poverty states in Austria from 2007 to 2010; linear and logit scales.
Black colour indicates exit from poverty and gray colour entry to poverty.

over the three pairs of years, with the largest decline after 2008 for small thresholds τ. The relative rates have declined over the three pairs of years for entry to poverty for all thresholds and for exit from poverty for $\tau > 60\%$.

The decrease of the relative rate of entry to poverty from 14.5% to 10.8% ($\tau = 80\%$), by 3.7%, cannot be straightforwardly compared to the decrease of the rate of exit from poverty from 2007–2008 to 2009–2010, from 28.8% to 23.1%, by 5.7%, because changes took place at different percentages. In brief, the linear scale is not appropriate for comparing percentages. A low rate has a limit for how much it can be reduced, and similarly a high rate for how much it can be increased. The odds ratio resolves this iniquity. The odds for a probability p is defined as $p/(1-p)$. (See Section 1.2.1 for an application of the odds ratio.) We prefer to use the log-odds, $\mathrm{logit}(p) = \log(p) - \log(1-p)$

because it is antisymmetric; $\text{logit}(p) = -\text{logit}(1 - p)$. We refer to the unit on this scale as a *logit*. Ratios of two odds convert to differences on the logit scale. The bottom panels of Figure 6.4 reproduce the curves from the top panels on the log-odds (logit) scale. On this scale, the decrease from 14.5% to 10.8% is equal 0.33 logits, whereas the decrease from 28.8% to 23.1% is equal to 0.30, slightly less.

The inverse of the logit function is

$$\text{logit}^{-1}(x) = \frac{\exp(x)}{1 + \exp(x)} .$$

The logit function is very close to linearity for $0.25 < p < 0.75$, but for extreme (very high and very low) probabilities it is extremely nonlinear.

We can interpret the findings for Austria only by a reference to what is usual, and in the absence of such a norm or reference value, we compare the countries. Figure 6.5 presents the relative rates of transition to and from poverty for three pairs of consecutive years, from 2007 to 2010, for the 24 countries that have panel data for the four years. All the curves in the diagram are smoothed, yet they have a lot of irregularity. The relative rates of exit from poverty are more dispersed for low thresholds τ than for higher thresholds. The countries' curves intersect a lot, both owing to local irregularities and trends consistent across the thresholds. For example, Denmark has the eighth highest estimated rate of exit from poverty in 2007–2008 for $\tau = 40\%$ and the lowest estimated rate for $\tau = 80\%$. The three panels for exit from poverty have few consistent patterns. The rates for Austria and the UK are among the highest for all three years and the entire range of thresholds, and the rates for Norway and Romania are among the lowest. The rates for Iceland stand out for small thresholds in 2007–2008, but are unexceptional otherwise.

The relative rates of entry to poverty in the right-hand panels tend to diverge with increasing threshold. For low thresholds, the rates for Latvia stand out as the highest in 2008–2009, and for $\tau < 70\%$ in 2009–2010. Sweden and Norway have some of the lowest rates of entry to poverty in all three pairs of years. The rates for Bulgaria are the highest by a wide margin for thresholds $\tau > 60\%$ in 2007–2008, but are not exceptional in the other two pairs of years. The mound at $\tau = 50\%$ in 2009–2010 belongs to Romania. Without smoothing, it would be a spike reaching the rate of 45% at $\tau = 50.0\%$, but settling to unexceptional levels for lower and higher thresholds.

6.3   Substantial Transitions

We should be sceptical about any transition that is associated with a small change in eHI, because it amounts to a very small material change in the circumstances of the household and is to a great extent an accident of our

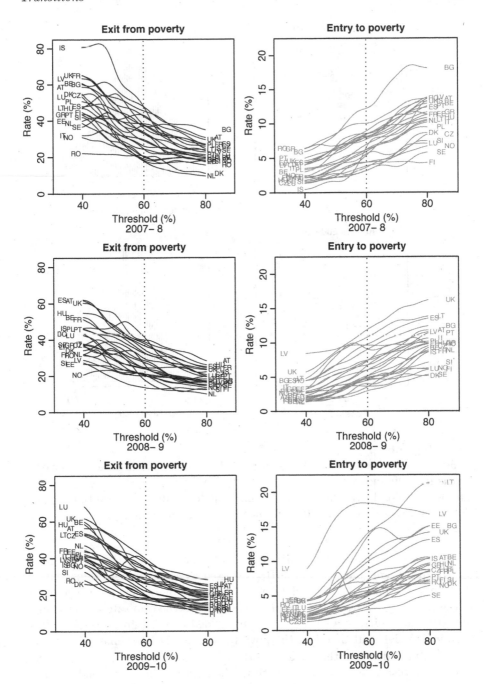

FIGURE 6.5

Relative rates of transition between poverty states in countries with panel data from 2007 to 2010; intact households.

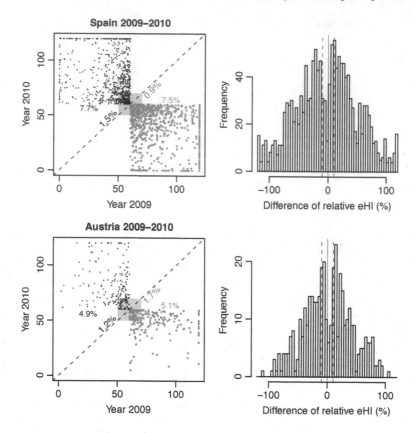

FIGURE 6.6

Relative equivalised household income of households with different poverty status in 2009 and 2010.

The households in the highlighted square made small (insubstantial) transitions. Values in excess of 120% and negative vales are truncated.

conventions. We define the relative eHI, and denote it by eHI%, as the ratio of eHI and the median value of eHI, expressed as a percentage. Thus, with a threshold τ, households with eHI% just below τ in one year and just above τ in the next year exit poverty, and households with the opposite pattern enter poverty. The arbitrariness is further compounded by sampling variation, since the median eHI, and therefore the poverty standard, are estimated.

Figure 6.6 displays the values of eHI% for households that have different poverty status in 2009 and 2010, with the households that have made a small (insubstantial) transition highlighted by the light-gray background, and the histogram of the differences of eHI% from 2010 to 2009. On the example of two countries, it shows that a large fraction of the transitions are 'accidental'. For both countries, the points representing households are densely packed in the vicinity of the threshold, within $\pm 10\%$ of the threshold $\tau = 60\%$ in particular (the lightly shaded squares in the left-hand panels and the vertical dashes in the histograms). In Spain, there are 1454 transitions among 9376 individuals (16.5%), and in Austria 394 transitions among 4124 individuals (9.6%). For Spain, nearly 20% of the exits from poverty are from eHI between the percentiles 50 and 60 in 2009 to percentiles $60-70$ in 2010 (1.5% out of 7.7%). In Austria such exits form nearly 25% of all the exits (1.2% out of 4.9%). In Spain 12% and in Austria 33% of the entries to poverty are for such insubstantial changes in the reverse direction.

It would be desirable to discount the transitions that occur as a result of small changes in eHI%. Instead of a single threshold, such as $\tau = 60\%$, we define a narrow range around it, such as $(55, 65)\%$, called the *transition zone*, and count only transitions in which the relative eHI of a household in one year is below the lower limit of the range, 55%, and in the other is above the upper limit, 65%. We refer to such transitions as *substantial*, and qualify them by the centre ($\tau = 60\%$ in this example) and half-width ($\nu = 5\%$) of the transition zone. The absolute and relative rates of substantial exit from and entry to poverty are defined in complete analogy with the original definitions of the transitions. These definitions correspond to $\nu = 0\%$; we refer to such transitions as *ordinary*.

A change from outside the zone to within, or in reverse, is not substantial, even if the centre of the zone is crossed in the process or the difference of the percentiles is large. Thus, a household could progress over two years from one poverty status to the other without a substantial transition, if its value of eHI% in the middle year is in the transition zone. Both changes could be large, say, from 40 eHI% to 62 eHI%, and then to 105 eHI%. That is an obvious drawback of the definition of substantial transition. Clearly, it is difficult to extend it to several pairs of years. We attend to this problem in the next two sections.

Instead of the transition rates for a single transition zone, we study the rates for a given centre and a range of half-widths. Figure 6.7 presents the rates of substantial transition from and to poverty in $2009-2010$ for the centre of $\tau = 60\%$ as functions of the half-width ν in the range $(0, 10)\%$. The left-hand limits of the curves are the ordinary transition rates at $\tau = 60\%$. These rates are absolute; their denominators are the totals of weights of all the individuals. The denominators for the relative rates are the totals of weights of all the individuals with eHI% in year 1 below $\tau - \nu$ for exit from poverty. For entry to poverty, the relative rates are based on those with eHI% above $\tau + \nu$.

The absolute rates for both exit from and entry to poverty are decreasing functions of the half-width ν because greater ν corresponds to a stricter condition for being counted as a substantial transition. The rates converge to zero with increasing ν, as the transition zone becomes wider and greater change in income is necessary to cross it. The countries' symbols are placed left to right after sorting their values for $\nu = 0$. This arrangement enables us to infer that Estonia has the third highest ordinary rate of exit from poverty, but much

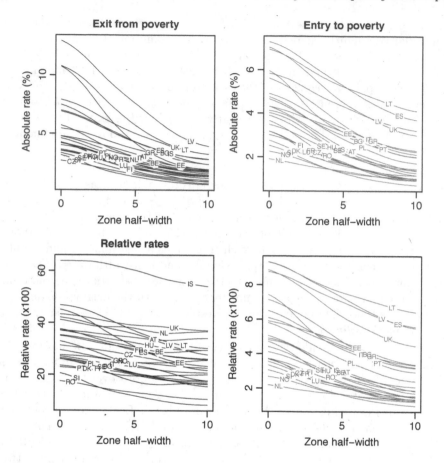

FIGURE 6.7
Absolute and relative rates of substantial transition from 2009 to 2010 for
centre $\tau = 60\%$ and half-widths in the range $(0, 10)\%$.

lower rates than most other countries for ν close to 10%. Further, Belgium,
Finland and Luxembourg have much lower rates at around $\nu = 5\%$ than what
their ordinary rates (or their ranks) would indicate. Latvia, Lithuania and the
UK have some of the highest estimated rates of transition in both directions.
Slovenia has one of the lowest absolute rates of substantial transition in both
directions for all half-widths. In general, the ranking of the countries for transition in one direction is similar to the ranking in the opposite direction, but
there are several exceptions, such as Norway.

A low absolute rate of exit from poverty has two complementary interpre-
tations. Either the poverty rate in the country is low, or poverty as a status is
difficult to exit, so that a lot of the existing poverty is persistent (long-term).
The bottom panels of Figure 6.7 display the relative rates of substantial tran-

sition. The absolute and relative rates of entry to poverty differ only slightly, mostly by the scale, because the relative rates are obtained by dividing the absolute rates by values close to 100%. In contrast, the relative rates of exit from poverty differ from the corresponding absolute rates a great deal, even more so than for the ordinary rates.

The curves for the relative rates of exit from poverty are close to constant for several countries. The rates are highest by far for Iceland, exceeding 50% for all half-widths. If the rates were this high for several consecutive pairs of years, then it would suggest that poverty is largely a transient (one-year) phenomenon for many households in Iceland. In fact, the high relative rates of exit from poverty arise because many households had small or no income as a consequence of the near collapse of the economy in Iceland in the preceding year, and in 2009 some recovery took place. In contrast, the relative rate of exit from poverty in Romania is below 20% throughout the range of half-widths; most poor households remain in poverty. The persistence of poverty cannot be studied through the transition rates from one year to the next, not even from the (substantial) transition rates for several pairs of years, because we have to link a household's transition between years 1 and 2 with its transition between years 2 and 3. For example, even in Iceland there could be a 'hard-core' stratum of households that remain in poverty for many years. Clearly, there are such strata in Romania and Slovenia and other countries with low relative rates of exit from poverty, because if the relative rates did not change from around 20% for several years, some households would be poor for at least five years in a row.

In an alternative definition of substantial transition, different transition zones are defined for the two types of transition. An exit from poverty is classified as substantial if eHI% in year 1 is below the centre τ and in year 2 exceeds $\tau + \nu$. An entry to poverty is substantial if eHI% in year 1 is above τ and in year 2 falls below $\tau - \nu$. This definition can be motivated by associating poverty with accumulated debts or postponed expenditure of ν% of the median eHI and prosperity with having some savings or being able to cope without expenditure of ν eHI%.

6.4 Partial Scoring of Transitions

By defining substantial transition we have raised the standard for what we regard as a transition, but we do not address the coarseness of the definition, that a pair of values of eHI (or eHI%) in two consecutive years is either counted or not counted as a transition. A substantial exit from poverty, say, from 30 eHI% to 95 eHI%, does not need distinguishing from another substantial exit, say, from 38 eHI% to 76 eHI%, because both of them are successes that need no qualification. However, a transition from 55 eHI% to 68 eHI%, although not substantial with centre set to $\tau = 60\%$ and half-width to $\nu = 10\%$, amounts to a nontrivial change very close to 'full' exit from poverty, very different from a transition from 58 eHI% to 61 eHI%.

We introduce a transition score by the following rules. A substantial transition, clearing the entire transition zone, is scored as unity. For transitions that are not substantial, we define a partial (fractional) score. For a household, denote by y_1 and y_2 the values of eHI% in the respective years 1 and 2. We set the household's score for exit from poverty to the product of two subscores. The subscore for year 1 is equal to unity if eHI% in that year is lower than $\tau - \nu$, equal to zero if eHI% is greater than τ, and is linearly interpolated in the range $(\tau - \nu, \tau)$:

$$s_{1+} = \frac{\tau - y_1}{\nu}.$$

The subscore for year 2 is defined by the mirror image of s_{1+}: it is equal to unity if eHI% in year 2 is greater than $\tau + \nu$, equal to zero if it is smaller than τ, and is linearly interpolated in the range $(\tau, \tau + \nu)$:

$$s_{2+} = \frac{y_2 - \tau}{\nu}.$$

For entry to poverty, the subscore 1 is equal to unity if eHI% in year 1 exceeds $\tau + \nu$, equal to zero if it falls short of τ, and when $\tau < y_1 < \tau + \nu$,

$$s_{1-} = \frac{y_1 - \tau}{\nu}.$$

The subscore for year 2 is equal to unity if eHI% in year 2 falls short of $\tau - \nu$, equal to zero if it exceeds τ, and when $\tau - \nu < y_2 < \tau$,

$$s_{2-} = \frac{\tau - y_2}{\nu}.$$

The (partial) score for exit from poverty is the product $s_+ = s_{1+} s_{2+}$ and for entry to poverty it is $s_- = s_{1-} s_{2-}$. See Figure 6.8 for an illustration. The panels in the first two rows depict the functions $s_{1\pm}$ and $s_{2\pm}$, and the panels at the bottom contain the contour plots of the scores s_+ and s_- as functions of eHI% in the two years, y_1 and y_2. The rates of transition (with partial scoring) are defined as the averages of these scores, converted to percentages.

Several variants of this approach can be devised. The linear subscore functions in Figure 6.8 are continuous, but not smooth, with sharp edges at τ and $\tau \pm \nu$. The edges can be removed by a smoothly increasing transformation f from $(0, 1)$ to $(0, 1)$ that has zero derivatives at both zero and unity. One class of such functions is $f_r(s) = s(2s)^{r-1}$ for $s \in (0, \frac{1}{2})$ and $f_r(s) = 1 - 2^{r-1}(1-s)^r$ for $s \in (\frac{1}{2}, 1)$, for $r > 1$. Its derivative at $s = \frac{1}{2}$ is equal to r. The scoring scheme in Figure 6.8 corresponds to $r = 1$. Although this is a very flexible class of transformations, it is of little practical use because for small r the averages based on them do not differ much from the averages based on linear

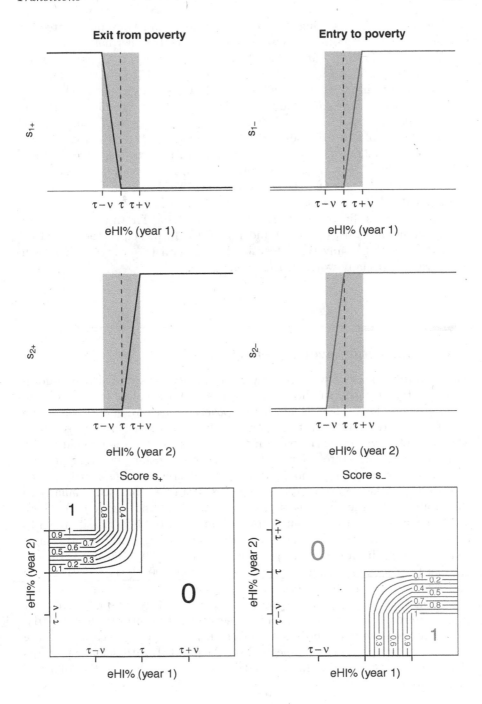

FIGURE 6.8
Subscores and scores for transitions.

scoring ($r = 1$). The scoring schemes for greater r can be approximated by schemes with $r = 1$ with a narrower transition zone (smaller half-width ν).

For Austria in 2009–2010, the estimated rates of exit from poverty with $\tau = 60\%$, $\nu = 5\%$ and $r \in (1,4)$ are in the range 4.132%–4.143% and for entry to poverty in the range 3.430%–3.464%. Thus, the exponent r has next to no influence on the transition rates. This result is not specific for Austria. For example, for Spain in 2009–2010 and the same values of τ and ν, the estimated rates of exit from poverty decline with the exponent r from 5.639% for $r = 1$ to 5.622% for $r = 4$, and for entry to poverty they decline from 6.392% to 6.355%. For large r, we obtain an approximation to the ordinary rate of exit from poverty for the transition zone $(57.5, 62.5)\%$. These rates are 5.603% and 6.320%, respectively.

The subscores illustrated in Figure 6.8 are partial for values of eHI% in two adjacent intervals, $(\tau - \nu, \tau)$ and $(\tau, \tau + \nu)$. In principle, they could be defined as partial in any two intervals. The interval $(\tau - \nu, \tau + \nu)$ for both subscores is especially worth considering.

6.5   Transitions over Several Years

The transition rates between poverty states in two consecutive years inform us about the turnover in the two categories of households. In this section, we study households that remain in one poverty status for several years; the two groups may be referred to as persistently poor and securely prosperous. More generally, we define the pattern of poverty states for a household by the sequence of annual states. Thus, 0000 is the pattern for persistent poverty over four years and 0101 and 1010 are the patterns of transition from one status to the other in every year. The patterns can be tabulated, similarly to the tabulation of the patterns of presence (Table 6.2 and Figure 6.2). There are 16 possible patterns for four years, so it may be useful to classify them to fewer groups. For example, one group may comprise the patterns of a single change (switch) from poverty to prosperity, that is, 1000, 1100 and 1110, and another for a single switch in the opposite direction, 0001, 0011 and 0111. The patterns 0101 and 1010 can form another group, and the remaining patterns can be in a single group or in two groups: in one status except for one year (four patterns), and two years each in either status (patterns 0110 and 1001). The group with four patterns can be split into two groups according to the status over the three of the four years. Note the parallels with defining patterns of presence in Section 6.1. In the analyses that follow, we use the following classification:

0000, persistent poverty;

1111, no experience of poverty;

→ P, switch to poverty   (patterns 1000, 1100 and 1110);

P →, switch from poverty (patterns 0001, 0011 and 0111);

 3P, poor in three years (patterns 0100 and 0010);

 1P, poor in one year   (patterns 1011 and 1101);

3SW, three switches   (patterns 0101 and 1010);

2&2, patterns 0110 and 1001.

For greater detail, we may define for a pair of years more than two states, such as poverty, gray area (close to the threshold), and prosperity. However, with three states there are 81 possible patterns for four years, far too many for a neat summary by means of a table. We have to dismiss the partial scores for transitions on similar grounds, even if we rounded or coarsened them, e.g., to 0, $\frac{1}{2}$ and 1.

The estimated rates of poverty patterns for the 24 countries with panel data for years 2007–2010 are listed in Table 6.3. Czech Republic, Hungary, Luxembourg, the Netherlands and Sweden have the highest estimated rates of the pattern 1111 (no experience of poverty), greater than 80%. At the other extreme, the frequencies of this pattern in Bulgaria, Estonia, Lithuania and Latvia are below 60%. Estonia, Greece and Latvia have the highest rates of persistent poverty (pattern 0000), in excess of 11%, and these rates are lowest in Iceland and the Netherlands. These two patterns have the highest frequencies for most countries. Notable exceptions are Bulgaria and the Baltic countries (Estonia, Latvia and Lithuania), with frequencies of a single switch of status (→ P to poverty or P → from poverty) exceeding 10%. The averages of the 24 percentages for the respective patterns 0000 and 1111 are 7.5% and 71.7%. The estimated frequencies for Belgium and Poland are close to these values.

The level of mobility can be summarised by the complement of the frequencies of the 'stable' patterns 0000 and 1111. In general, the level of mobility is lowest where the frequency of pattern 1111 is highest. For example, the level of mobility in Luxembourg is $100 - 3.8 - 85.1 = 11.1\%$; in Czech Republic it is also 11.1% and in the Netherlands it is 13.9%. In contrast, the level of mobility in Bulgaria and the Baltic countries exceeds 30%. Among the developed economies, only the UK has level of mobility greater than 30%. In the UK, single switches in either direction have frequencies close to 10% each (8.0% and 10.1%). The difference of the frequencies of switching in the two directions is also of interest, because it captures one aspect of the balance between poverty and prosperity. Notable are the large differences for Estonia (4.5% vs. 14.9%), Latvia (7.8% vs. 14.8%), Lithuania (7.1% vs. 16.5%) and the Netherlands (1.3% vs. 8.9%), in favour of exiting poverty in all four cases. The largest balance in favour of entering poverty is in Bulgaria (11.2% vs. 8.4%), Poland (8.6% vs. 5.6%) and Sweden (6.3% vs. 2.7%).

The other four patterns can be described as a single (exceptional) year in one of the two states (3P and 1P) and two years in each state (3SW and 2&2).

TABLE 6.3
Estimated rates of poverty patterns in 2007–2010.

	0000	1111	→P	P→	3P	1P	3SW	2&2	N
AT	5.3	76.7	3.6	6.9	1.6	3.5	1.1	1.3	765
BE	6.4	72.3	3.9	8.8	2.6	4.0	0.8	1.2	790
BG	9.8	59.1	11.2	8.4	1.1	6.1	1.1	3.2	494
CZ	3.6	85.2	3.5	3.7	0.8	2.4	0.3	0.5	1705
DK	8.5	77.2	4.2	7.1	0.6	1.0	0.5	0.9	626
EE	11.8	58.0	4.5	14.9	1.4	5.3	1.3	2.8	703
ES	8.9	64.9	7.0	8.0	3.0	4.5	1.9	1.8	1753
FI	8.9	74.3	3.0	5.8	1.0	4.2	0.8	2.1	876
FR	4.6	77.8	4.4	7.1	1.2	3.0	0.9	1.1	3513
GR	12.7	63.7	6.5	7.7	2.0	4.3	1.2	2.0	816
HU	3.4	82.3	2.3	6.4	1.0	2.8	0.6	1.1	1355
IS	2.2	76.6	2.9	8.0	0.6	3.9	2.8	3.0	226
IT	10.4	69.9	3.8	8.4	2.1	2.9	0.8	1.8	2931
LT	9.3	56.1	7.8	14.9	1.6	5.4	1.9	3.0	901
LU	3.8	85.1	1.7	5.7	0.4	1.4	0.6	1.3	1507
LV	12.6	49.0	7.1	16.5	1.3	6.5	1.9	5.1	743
NL	2.0	84.1	1.3	8.9	0.5	1.8	0.6	0.8	1532
NO	7.9	79.6	2.5	6.3	0.7	1.8	0.8	0.5	1406
PL	6.9	72.2	8.6	5.6	1.1	3.6	0.5	1.3	1855
PT	9.3	69.1	6.0	5.9	2.4	4.2	1.6	1.6	711
RO	10.3	67.3	6.5	8.9	2.6	2.3	1.0	1.0	1605
SE	4.4	81.4	6.3	2.7	0.6	2.5	0.8	1.3	970
SI	10.7	77.1	1.6	4.6	1.6	2.5	1.1	0.9	1006
UK	6.7	61.5	8.0	10.1	2.5	5.9	2.7	2.7	1054

These patterns account for small fractions of the households in most countries. Note that other instances of a single year in a status are covered by the single-switch patterns (0001, 0111, 1000 and 1110). The total of the frequencies of the four 'minority' categories of patterns, 3P, 1P, 3SW and 2&2, exceeds 12% only in Lithuania, Latvia and the UK. These totals are smaller than 4% in Denmark (3.0%), Luxembourg, the Netherlands and Norway (3.7% each) and Czech Republic (3.9%).

Many of the estimated rates we quote are associated with relatively large sampling variation, because the sample sizes, listed at the right-hand margin of the table, are rather modest, for Iceland (226) in particular. We want to point out that all the statements made in this section are about *estimates*. They do not carry over to the underlying population rates as a matter of course. For example, the standard error associated with the frequency of pattern 0000 in Belgium is approximately $100 \times \sqrt{0.064/1000} = 0.8\%$. The standard errors for more extreme frequencies are smaller. For instance, the standard error associated with the estimate of 2.0% for pattern 0000 in the Netherlands is around 0.36%, so there is no doubt that it is exceptionally small.

As we skim a large table, such as Table 6.3, we make several statements, each of them likely to be correct (when interpreted as a statement about the corresponding population quantities). But the likelihood that all the statements are correct is rather modest. Formal tests of hypotheses, or construction of confidence intervals, may generate the appearance of greater credibility of the statements. Such inferential statements, when made about the specific countries, are not valid because the countries have been selected by a purposeful search in the dataset(s) used for drawing the inferences. Integrity of the inferential statements could be restored by formulating the hypotheses to be tested in advance of data inspection. For example, we could formulate hypotheses based on the data currently available, and test them using EU-SILC samples that are (almost) independent of it, e.g., in cohort 2011 – 2014.

Even though we have inspected the data, we might conduct the exercise of formulating hypotheses that we would have (or should have) thought of prior to the exploratory analysis. By cheating in this exercise we would distort the impression of our insight and knowledge, without leaving any traceable grounds for accusations of misconduct. We should nevertheless resist the temptation and maintain our integrity.

6.6   Imputed Patterns

A drawback of the frequencies studied in the previous section is that they are *coarse*—they treat substantial transitions on par with insubstantial transitions. Partial scoring, which addresses this problem for a pair of consecutive years, is difficult to extend to a longer sequence of years because too many patterns would result, and they would be nigh impossible to study and concisely summarise. To work with patterns composed of zeros and unities, and yet make insubstantial transitions count less than substantial transitions, we borrow an idea from methods for missing values (Little and Rubin, 2002).

Missing values are a nuisance, both in senses of the commonplace and technical statistical terminology. When planning a study and all the operations that follow the data collection, we may compile a programme (e.g., an R function) to process the data further, or have the intention to do so. Suppose we anticipate a dataset in the form of a matrix, with all its entries recorded. If some of them are not recorded the execution of the programme will fail (reach an impasse) when it encounters an entry with its value missing. The result of any arithmetic operation that involves a missing value in a nontrivial way is 'missing', not a useful quantity for any analyst.

We can improvise and impute a value for each missing entry. Then the programme will perform without indicating any failure, but if its outputs are used for inferential statements, then their validity will be compromised. To see this, suppose the inferential statements would be valid if the dataset were complete—if all its entries were recorded as planned originally, using the prepared protocols. The dataset that we complete is called completed. As a random object, it differs from the (planned) complete dataset. After all, we could not have as much confidence in an inferential statement generated from a completed dataset than we would have from a complete one; the complete dataset would yield superior inferences because it is not tainted by our improvisation (imputation) which entails uncertainty additional to the uncertainty associated with sampling and planned data processing.

The method for analysing the complete data is an important asset of the analyst, and he or she would not like to abandon it as a failure and devise and implement an alternative method from scratch. Yet, no method of imputation (data completion) can yield a completed dataset with which a complete data method would work satisfactorily. The EM algorithm (Dempster, Laird and Rubin, 1977) offers an alternative viewpoint and prescription: Instead of estimating the values themselves, we should estimate the summaries of the missing values that are required in the complete-data method. When the EM algorithm is not feasible, multiple imputation is often a viable alternative. Instead of a single completion of the dataset, we generate several, in such a way that these alternative *plausible* datasets differ as much as our uncertainty about the missing items and their associations. The goals of efficient estimation of missing values and efficient estimation of population quantities are not in accord, and we are concerned solely with the latter goal.

In Chapter 4 we applied these ideas and implemented the EM algorithm for mixtures to a problem in which the items declared as missing were never intended to be observed. We depart once more from the original motivation of methods for missing values by grafting the approach to the problem at hand. We shall declare the poverty status of a household in a year as uncertain if its eHI% is in the transition zone $(\tau-\nu, \tau+\nu)$ and adopt the following model. Let y be the value of eHI% of a household and $\tau-\nu < y < \tau+\nu$. Then the probability that the household is poor is $(\tau+\nu-y)/(2\nu)$. The probability is proportional to the distance from the upper limit of the transition zone. The events of being poor are mutually independent for the households with eHI% in the zone, as well as for the distinct years. We impute a poverty status for each household and year when its eHI% is in the transition zone and analyse the resulting plausible poverty patterns. The outcome of this completed-data analysis is a table. We replicate this process of imputation several times and average the results of the analyses (the tables). The differences among these replicate tables, the results of the completed-data analyses, provide an indication of how much information is lost owing to the uncertainty about the poverty status of some households. This loss should not be regarded as a weakness of the analysis. On the contrary, the transparent and flexible representation of the uncertainty is its strength as it leads to more credible inferential statements. The flexibility is achieved by the choice of the transition zone.

Suppose there are M replicate completions (imputations) and completion $m = 1, \ldots, M$ yields the vector or table of estimated frequencies $\hat{\mathbf{p}}_m$ of patterns (or groups of patterns) with the vector of estimated sampling variances $\hat{\mathbf{v}}_m$. The variances are defined as $v_h = p_h(1 - p_h)/n_e$, where n_e is the effective sample size. Note that these variances refer to the complete dataset, with a dichotomous poverty status (0 or 1) assigned to every household in every year. Further, let $\tilde{\mathbf{b}}$ be the elementwise variances of $\hat{\mathbf{p}}_m$ and $\tilde{\mathbf{v}}$ the elementwise means of $\hat{\mathbf{v}}_m$. Then the set of sampling variances that refer to the incomplete data (with uncertainty about the poverty status in the transition zone), is estimated with at most small bias by

$$\tilde{\sigma}^2_{\mathrm{MI}} = \tilde{\mathbf{v}} + \frac{M+1}{M}\,\hat{\mathbf{b}}. \tag{6.1}$$

We refer to σ^2_{MI} as the multiple-imputation (MI) sampling variance, and to its square root as the MI standard error. The values underlying the between-imputation variances $\hat{\mathbf{b}}$ can be interpreted as the information lost because the poverty status is uncertain in the transition zone. The within-imputation variances $\mathbf{v}$ are the variances that would be present even if this source of uncertainty were removed. The factor $(M + 1)/M$ suggests that it is better to generate more sets of plausible values. But the returns diminish with increasing M, especially when $\mathbf{b}$ is smaller than $\mathbf{v}$. We generated $M = 100$ replications, which turned out to be an overkill because the entries of $\mathbf{b}$ are much smaller than their counterparts in $\mathbf{v}$. However, this does not involve excessive computing time. Also, accurate estimation of the sampling variance requires many more replications M than estimation of the frequencies. The latter is not handicapped by excessively large M.

For Austria, we obtained estimates (averages of $M = 100$ replicate completed-data analyses) of 4.57% for pattern 0000 and 76.75% for pattern 1111. The former differs somewhat from its counterpart in the analysis without considering a transition zone, 5.26%, see Table 6.3, but the latter does not (76.74%). For pattern 0000, the average within-imputation variance is 0.65 and the between-imputation variance is 0.08. Hence the MI standard error is $\sqrt{0.65 + 1.01 \times 0.08} = 0.85$. The number of imputations M has next to no impact on the result; it would be different only slightly even if we generated several times as many completions. For pattern 1111, the MI standard error is estimated by 1.67%; the between-imputation variance, 0.12, is about 22 times smaller than the within-imputation variance, 2.67(%²). For some smaller frequencies, the between-imputation variance is relatively larger. For example, the estimate of the rate of pattern 0001 is 1.34%, with MI standard error 0.49%. The respective within- and between-imputation variances are 0.20 and 0.04. If the latter were ignored, the standard error would be estimated by 0.44%. Details of the results are displayed in Table 6.4.

If we set the transition zone wider, say, by setting $\nu = 10\%$, the uncertainty about the poverty status and its pattern becomes greater and the estimates of the frequencies of the patterns are altered somewhat more. For Austria, the es-

TABLE 6.4

Multiple imputation analysis of poverty patterns for Austria, 2007–2010; $\tau = 60\%$ and $\nu = 5\%$.

	Pattern							
	0000	0001	0010	0011	0100	0101	0110	0111
Mean	4.57	1.34	0.80	1.26	0.86	0.65	0.71	4.15
Within-variance	0.65	0.20	0.12	0.19	0.13	0.10	0.11	0.60
Between-variance	0.08	0.04	0.03	0.04	0.04	0.03	0.05	0.07
MI st. error	0.85	0.49	0.39	0.47	0.41	0.36	0.40	0.82
	1000	1001	1010	1011	1100	1101	1110	1111
Mean	1.24	0.88	0.55	2.20	0.91	1.25	1.89	76.75
Within-variance	0.18	0.13	0.08	0.32	0.14	0.18	0.28	2.67
Between-variance	0.04	0.04	0.02	0.07	0.02	0.05	0.05	0.12
MI st. error	0.48	0.41	0.32	0.63	0.40	0.49	0.58	1.67

Notes: The units for the means and MI standard errors are percent (%); the units for the within- and between-imputation variances are $\%^2$. Based on 100 replicate completions.

timated frequency of pattern 1111 is reduced to 75.6% (from 76.7%) and the standard error is increased to 1.76% (from 1.67%). The within-imputation variance is changed only slightly, owing to the change of the frequency of the pattern and its estimation referring to a different complete dataset. The between-imputation variance is nearly trebled, from 0.12 to 0.32, reflecting the greater uncertainty about the states and patterns—many more assessments of the poverty status are in the transition zone, as its width is doubled from 10% to 20%. However, the impact of the between-imputation variance on MI sampling variance remains insubstantial. For example, the frequency of pattern 0000 is estimated as 3.60%, with standard error 0.81%, even smaller than with $\nu = 5\%$. Since the (estimated) frequency of the pattern is smaller, so is the within-imputation variance, 0.52. The increase of the between-imputation variance, from 0.078 to 0.132, although substantial in relative terms, leads to MI variance that is still smaller than with a narrower transition zone.

We note that the estimators of the rates of patterns for a country are negatively correlated because they are constrained by being a composition—their total is 100%. However, the correlation is very small when one of the frequencies is small, say, smaller than 10%. The correlation can be ignored in statements about totals over two or three patterns, when all of them are rare. The inflation of standard errors of the resulting estimators can also be ignored.

Table 6.5 displays the estimates and estimated standard errors for the EU-SILC panel data in 2007–2010 and the patterns listed in Table 6.3, with

TABLE 6.5
Estimated frequencies and MI standard errors for poverty patterns in 2007–2010; $\tau = 60\%$ and $\nu = 10\%$.

	0000	1111	→P	P→	3P	1P	3SW	2&2
AT	4.6 (0.9)	76.8 (1.7)	4.0 (0.8)	6.8 (1.1)	1.7 (0.6)	3.4 (0.8)	1.2 (0.5)	1.6 (0.6)
BE	6.5 (1.0)	71.6 (1.7)	4.1 (0.9)	8.8 (1.2)	2.4 (0.7)	4.1 (0.9)	1.2 (0.5)	1.2 (0.5)
BG	9.5 (1.5)	59.5 (2.5)	11.0 (1.7)	8.0 (1.4)	1.5 (0.7)	6.1 (1.3)	1.3 (0.7)	3.1 (0.9)
CZ	3.2 (0.5)	83.9 (1.0)	3.7 (0.6)	4.2 (0.6)	0.8 (0.3)	3.0 (0.5)	0.6 (0.3)	0.8 (0.3)
DK	6.3 (1.3)	76.7 (2.2)	4.7 (1.3)	6.6 (1.5)	1.6 (0.9)	1.6 (0.8)	0.9 (0.6)	1.6 (0.8)
EE	11.2 (1.5)	57.6 (2.3)	4.5 (1.1)	14.9 (2.0)	1.8 (0.7)	5.9 (1.3)	1.7 (0.7)	2.5 (0.9)
ES	8.4 (0.9)	64.5 (1.5)	7.1 (0.9)	8.0 (0.9)	3.0 (0.6)	5.1 (0.7)	1.9 (0.6)	2.0 (0.5)
FI	8.2 (1.3)	72.8 (2.1)	3.5 (1.0)	5.7 (1.3)	1.5 (0.7)	5.0 (1.3)	1.0 (0.6)	2.2 (0.9)
FR	4.4 (0.4)	77.6 (0.9)	4.3 (0.5)	7.1 (0.6)	1.4 (0.3)	3.2 (0.4)	1.0 (0.2)	1.1 (0.3)
GR	11.9 (1.3)	63.8 (1.9)	6.7 (1.1)	7.9 (1.2)	2.2 (0.7)	4.3 (0.9)	1.1 (0.5)	2.2 (0.7)
HU	3.0 (0.5)	81.9 (1.2)	2.6 (0.5)	6.2 (0.8)	1.3 (0.4)	3.0 (0.6)	0.9 (0.3)	1.1 (0.4)
IS	2.2 (1.4)	77.3 (3.3)	3.2 (1.5)	8.4 (2.6)	0.4 (0.6)	4.1 (1.8)	1.5 (1.2)	2.8 (1.3)
IT	9.9 (0.7)	69.1 (1.0)	3.9 (0.5)	8.5 (0.7)	2.4 (0.4)	3.3 (0.4)	1.2 (0.3)	1.8 (0.3)
LT	9.4 (1.3)	55.7 (2.2)	7.9 (1.2)	14.0 (1.7)	1.6 (0.6)	5.9 (1.1)	2.2 (0.7)	3.2 (0.9)
LU	3.6 (0.8)	84.4 (1.5)	1.9 (0.6)	5.7 (1.1)	0.9 (0.5)	1.6 (0.6)	0.9 (0.5)	1.0 (0.5)
LV	12.2 (1.4)	48.8 (2.1)	7.4 (1.2)	16.6 (1.8)	1.5 (0.6)	7.0 (1.2)	1.6 (0.6)	5.0 (1.0)
NL	1.8 (0.4)	83.5 (1.1)	1.2 (0.3)	8.9 (1.0)	0.7 (0.3)	2.3 (0.5)	0.8 (0.3)	0.8 (0.3)
NO	7.7 (0.8)	78.8 (1.2)	2.8 (0.6)	6.0 (0.8)	0.8 (0.4)	2.1 (0.5)	1.1 (0.4)	0.8 (0.4)
PL	6.8 (0.7)	71.8 (1.3)	8.5 (0.9)	5.8 (0.7)	1.3 (0.3)	3.8 (0.6)	0.7 (0.3)	1.5 (0.4)
PT	9.4 (1.3)	68.2 (2.0)	6.7 (1.2)	6.0 (1.1)	2.7 (0.8)	4.0 (0.9)	1.6 (0.6)	1.4 (0.6)
RO	10.3 (0.9)	67.2 (1.4)	6.2 (0.8)	8.9 (0.9)	2.7 (0.5)	2.5 (0.5)	1.0 (0.3)	1.2 (0.4)
SE	4.0 (0.7)	80.3 (1.4)	6.6 (1.0)	2.6 (0.6)	1.1 (0.4)	3.0 (0.7)	0.8 (0.4)	1.6 (0.5)
SI	10.7 (1.1)	76.1 (1.5)	2.0 (0.6)	4.8 (0.9)	1.4 (0.5)	2.7 (0.7)	1.1 (0.4)	1.2 (0.5)
UK	6.8 (0.8)	61.0 (1.6)	7.9 (0.9)	9.8 (1.1)	2.5 (0.6)	6.3 (0.9)	2.8 (0.6)	2.8 (0.6)

the setting $\tau = 60\%$ and $\nu = 10\%$. The table confirms that estimates are sufficiently precise for identifying the countries that have one of the highest or one of the lowest rates of the patterns 0000 (persistent poverty) and 1111 (no experience of poverty), but the countries with the highest and lowest values of the frequencies for any of the rarer patterns cannot be identified with any level of confidence.

We emphasise that the results in Table 6.5 are not definite in any sense, because they are conditioned on our representation of the uncertainty about the poverty status by the centre $\tau = 60\%$ and half-width of the transition zone $\nu = 10\%$. However, the setting of ν has only slight effect on the results, as does the definition of the probability of being poor for eHI% in the transition zone.

We draw a distinction between a probability and a partial score. A probability captures our (the analyst's) uncertainty about the poverty status, and it is reflected in the MI procedure by the binomial variation of the plausible poverty states generated by a random mechanism. In contrast, a (partial) score entails no uncertainty, even when it represents a compromise between the two extremes of (certain) poverty and its total absence.

As a consequence of the uncertainty about (some of) the individual-by-year poverty states, the MI estimators have two components of uncertainty: due to sampling and due to representing the poverty states by probabilities. In the application summarised in Table 6.5, the sampling variance dominates. It is unwise to rely on this in general. We can treat the probabilities as scores only when the estimators that are evaluated are linear functions of the imputed values. In our application they are not, because the estimators count the households within the patterns; thus, they are discontinuous. Even without such discontinuity, the sampling variances or standard errors would be estimated with bias because they are nonlinear functions of the imputed values. In an alternative viewpoint, the uncertainty is bound to cause an inflation of the sampling variance; if every poverty status were established without uncertainty, the additional information would be reflected by reduced sampling variation.

The presented analysis is based on the panel data for the cohort from 2007. Other cohorts as well as cross-sectional data could contribute with some information. For example, the annual standards could be estimated with greater precision. Some of the logistical problems can be resolved by combining data-specific estimates.

Multiple imputation could be applied also to account for our uncertainty about the poverty threshold (set to 60%) and about its uniformity. Suppose each household has a different centre of the transition zone, τ, and these centres are uniformly distributed in the range $(50, 70)\%$, but remain constant for a household over the years. Suppose further that the transition zone for each household has the same half-width ν. Then in a replication we draw a centre for every household, define the annual probabilities of the household being poor with respect to this centre and half-width in each year and form the completed-data table of the rates of transition patterns. Implementing this scheme is suitable for a substantial project.

Suggested Reading

Atkinson, Morrisson and Bourgignon (1992) is a widely quoted text on profiles of income over the years (mobility of earnings).

Fuzzy logic and fuzzy sets (Zadeh, 1965, and Klir and Yuan, 1996) are often used for modelling uncertainty. Lemmi and Betti (2006) apply these methods to poverty measurement, including multiple aspects and the dimensions of time and space. Rubin (2002) is an essential text on MI. Rubin (1996) gives a concise review and cogent arguments for its application. An elementary introduction to MI is in Schafer (1999); Schafer (1996) gives a much more detailed account, with associated software in R. Carpenter and Kenward (2013) and van Buuren (2012) are recent monographs on the subject, the latter with a practical (computational) orientation; see also Longford (2005, Part I).

Models for longitudinal data, both continuous and discrete, and algorithms for fitting such models and selecting among them are the subject of Zeger,

Liang and Albert (1998), Verbeke and Molenbergs (2000) and Diggle *et al.* (2002). They develop and implement the original framework of Laird and Ware (1982) based on linear models with random coefficients. Molenberghs and Verbeke (2006) give a comprehensive account of models for discrete longitudinal data.

6.A Appendix. Programming Notes

6.A.1 National Panel Databases

The database for a national panel has person-by-year records. For example, the database for Austria has 34 259 records, 5169, 8021, 11 146 and 9923 from the respective years 2007–2010. The records originate from 14 599 distinct individuals in 6133 households. These figures are established by data input,

```
ATlngDt <- read.dta("~/Splus/EUsilc12/PDATA/AT.dta")
```

followed by the expression

```
list(Recs=nrow(ATlngDt), Recs.by.year=table(ATlngDt[, "year"]),
    Subjects=apply(ATlngDt[, c("pid", "hid")], 2, LeUni) )
```

The records are from three rotational groups, with 14 957, 10 762 and 8540 records for the cohorts started in respective years 2007, 2008 and 2009.

```
table(ATlngDt[, "db075"])
```

More detail is obtained by tabulating the groups and years:

```
table(ATlngDt[, "db075"], ATlngDt[, "year"])
```

	2007	2008	2009	2010
1	0	0	4563	3977
3	5169	3737	3147	2904
4	0	4284	3436	3042

From this table we can identify the cohorts and assess the attrition in the panel. Similar tabulation for the households is not as useful because a household appears in the database once for each of its members. Therefore, we have to tabulate the unique combinations of households and years:

```
## The records that are not duplicates  (hhld within years)
hhe <- !duplicated(paste(ATlngDt[, "year"], ATlngDt[, "hid"]))

ATlngDc <- table(ATlngDt[hhe, "db075"], ATlngDt[hhe, "year"])
```

The result is the table

	2007	2008	2009	2010
1	0	0	2016	1718
3	2124	1468	1262	1171
4	0	1861	1453	1297

A person appears in the database twice in a year if he or she changes household. The table obtained by the expression

```
table(table(ATlngDt[, "pid"]))
```

1	2	3	4	5
3650	5037	3175	2675	62

shows that 62 persons appear five times in the database. Further exploration, to confirm that every one of these persons appears twice in one year and once each in the other three years, is left for an exercise. It is practical to convert this R code to a function, especially if you want to apply this exploratory analysis to several countries. This is done in function `EUreadC` in supplementary materials.

The function `EUreadQ` returns a table of counts within the rotational groups for a country. Part of this table is used in the country's row of Table 6.1. The only mandatory argument of the function is the country acronym (e.g., `AT` for Austria).

```
EUreadQ <- function(co, dire="~/Splus/EUsilc12/PDATA",ext="dta")
{
###    Input of the 4-year panel data and tabulation

## file    The filename  (country acronym)
## dire    The directory
## ext     The extension

## Data input and reduction to the relevant variables
dat <- read.dta(paste(dire, "/",co,".", ext, sep=""))[,
    c("hid","pid","year","hx040","hy020","hx050","db075","db090")]

## The missing records
MS <- apply(is.na(dat), 1, sum) > 0

## Missing records by rotational group
MSm <- table(dat[MS, "db075"])

## Listwise deletion
dat <- dat[!MS, ]
```

```
##   The years and rotational groups
RoYe <- table(dat[, "year"], dat[, "db075"])

##  Missing records within the rotational groups
MSn <- rep(0, ncol(RoYe))
names(MSn) <- colnames(RoYe)
MSn[names(MSm)] <- MSm

##  The number of years
nyr <- nrow(RoYe)
```

The (unique) patterns of presence are defined in their numerical form as

```
##   The pattern of presence  (zeros and ones)
ZRs <- t(RoYe > 0) %*% matrix(10^(rev(seq(nyr)-1)))
```

Next, summaries are evaluated within the patterns: the numbers of units with full participation, households that are also intact, and persons who appear more than once.

```
##   Initialisation for the table of counts
PHD <- c()

##   Patterns of presence
for (zr in unique(ZRs))
{
##   The relevant rotational groups
rtg <- as.numeric(colnames(RoYe)[ZRs==zr])

##   The records in the rotational groups
who <- dat[, "db075"] %in% rtg

##  Persons and households appearing in every year
prs <- apply(table(dat[who, "pid"], dat[who, "year"]), 1,min) > 0
hds <- apply(table(dat[who, "hid"], dat[who, "year"]), 1,min) > 0

##  Persons appearing more than once in a year
prt <- prs & apply(table(dat[who,"pid"], dat[who,"year"]), 1, max) > 1

##  The counts of persons and households
PHD <- rbind(PHD, c(LeUni(dat[who,"pid"]), LeUni(dat[who,"hid"]),

##   Full participation
   c(sum(prs), sum(hds)),

##   Intact households
   sum(hds & tapply(dat[who,"hx050"], dat[who,"hid"], sd)==0),

##   Multiples
      sum(prt)) )
} ## End of the loop over the patterns (zr)
```

The results to be returned by the function are set in a matrix with two columns for the individual-level counts and five columns for households.

```
##   Add the info about missing records
PHD <- cbind(PHD, tapply(MSn, ZRs, sum))

##   Labelling
ZRs <- unique(as.numeric(ZRs))

dimnames(PHD) <- list(ZRs, c(rep(c("Persons", "Households"), 2),
          "Intact", "Multiples", "Missing"))

##   Sort the rows by the rotational group
PHD[sort.list(ZRs), ]
}  ##  End·of function  EUreadQ
```

The function is applied for Austria by the expression

```
ATq <- EUreadQ("AT")
```

It yields the table

Rot. grp	Persons	H-holds	Persons	Households	Intact	Multiples	Missing
11	4680	2045	3827	1688	1497	32	1
111	4487	1901	2844	1260	993	45	7
1111	5430	2186	2683	1119	765	59	2

Note that 0011 and 0111 are the character versions of the labels for presence in 2009–2010 and 2008–2010, instead of 11 and 111. The list of tables for all the countries is obtained by the expression

```
EU710A <- lapply(as.list(EU710), EUreadQ)
names(EU710A) <- EU710
```

Table 6.1 is formed by the function HTable, which uses the auxiliary function EUshapQ.

```
EUshapQ <- function(mat)
apply(mat[, c(2,4,5)], 2,  rev)

HTable <- function(vec, spa)
{
###   Formatting a table of words/integers for LaTex

##  vec      Vector of table entries
##  spa      The spacing between entries

##  Initialisation
tbl <- c()

##  Place the spacing symbols intermittently with the values
```

```
for (i in seq(length(vec)))
tbl <- paste(tbl, spa[i], vec[i], sep="")

## Add the symbol for the end of the line, if available
if (length(spa) > i)
tbl <- paste(tbl, spa[i+1])

## Labelling
names(tbl) <- rep("", length(tbl))
tbl
} ## End of function HTable

EU710At <- apply(t(sapply(EU710A[-15], EUshapQ)), 1, HTable,
        c(rep(c(" && ", " & "," & "), 3), "\\[0.3mm]"))
```

The table does not have the row for Luxembourg (the 15th element of the list EU710A), because the country does not have the rotational group 0111. It has to be entered in Table 6.1 separately.

6.A.2   Rates of Transition

The rates of transition for a pair of (consecutive) years are estimated by the function EUtransA. The function reads a dataset (for a country), reduces it to the relevant years and, optionally, to the intact households.

```
EUtransA <- function(cou, dire="~/Splus/EUsilc12/PDATA", ext="dta",
        yrs=c(9,10), STD=60, intact=T)
{
### The rates of transition

## cou      The country
## dire     The data directory
## ext      The data-file extension
## yrs      The years (digits)
## STD      The treshold(s)
## intact   Whether to deal only with intact households

## Sort out the years
if (max(yrs) < 100) yrs <- 2000+yrs

## The number of years
nyr <- length(yrs)

## The thresholds
while (max(STD) > 1) STD <- STD/100

## Input of the dataset
dat <- read.dta(paste(dire, "/", cou, ".", ext, sep=""))[, c(
      "hid","year","hx040","hy020","hx050","db075","db090")]
```

```
##   Reduction to the years
dat <- dat[dat[, "year"] %in% yrs, ]

##   Listwise deletion
dat <- dat[apply(is.na(dat), 1, sum) == 0, ]

##   The households - initialisation
hds <- c()

##   The households present in every year (at least once)
for (yr in yrs)
hds <- c(hds, unique(dat[dat[, "year"]==yr, "hid"]))

##   The households that are present in every year
hds <- table(hds)
hds <- names(hds)[hds == nyr]

##   The records of these households
who <- dat[, "hid"] %in% hds

hdt <- hds

##   Intact households
if (intact)
hdt <- hds[tapply(dat[who,"hx050"], dat[who,"hid"], sd) == 0]

##   The net dataset of the intact households
dat <- dat[dat[, "hid"] %in% hdt, ]

##   Reduction to unique household x year records
dat <- dat[!duplicated(paste(dat[, "hid"], dat[, "year"])), ]
```

In the last expression, the first record for each combination of household Id.
and year is selected. The values of eHI are reformatted into an annual list and
the relative eHI (eHI%) evaluated for each year (element of the list).

```
##   Reshaping the dataset to annual list
dbt <- list()

for (yr in yrs)
{
cyr <- as.character(yr)

##   The dataset for the year
dbt[[cyr]] <- dat[dat[, "year"]==yr, ]

##   Alignment with hdt
dbt[[cyr]] <- dbt[[cyr]][sort.list(hdt), ]
}  ##   End of the loop over the years
```

```
##  Initialise for the scaled eHI%
seHI <- c()

##  The annual weighted medians and scaled eHI%
for (yr in yrs)
{
cyr <- as.character(yr)

##  The weighted median
QW <- QuantW(dbt[[cyr]][, "hy020"] / dbt[[cyr]][, "hx050"],
            dbt[[cyr]][, "db090"] * dbt[[cyr]][, "hx040"])

##  The scaled equi-HI%
seHI <- cbind(seHI, dbt[[cyr]][,"hy020"] / dbt[[cyr]][,"hx050"] / QW)
}  ##  End of the loop over the years (yr)
```

The patterns of poverty are tabulated and the tables converted to percentages.

```
##  Initialisation of the transition tables
TBL <- c()

##  The transition tables
for (st in STD)
TBL <- rbind(TBL, as.vector( tapply(dbt[[1]][, "db090"],
    list(sign(seHI[, 1] - st), sign(seHI[, 2] - st)), sum)) )

##  Labelling and conversion to percentages
dimnames(TBL) <- list(100*STD, c("PP", "PR", "RP", "RR"))

100 * TBL / sum(dbt[[1]][, "db090"])
}  ##  End of function  EUtransA

##   Application to Austria, 2009-2010
stds <- seq(40,80,1)
ATlngA <- list(Intact=EUtransA("AT", yrs=c(9,10), STD=stds),
        All=EUtransA("AT", yrs=c(9,10), STD=stds, intact=F))
```

The three sets of transition rates (2007–2008, 2008–2009 and 2009–2010) can be evaluated in a loop over the initial year, or a function can be compiled with the loop within it. The latter option is computationally more efficient because the country's dataset is read only once. However, the evaluation for all countries using a loop over the years and countries takes only about 40 seconds, so the programming effort is probably not worth it.

The function EUtransD is a copy of function EUtransA up to the point where the values of eHI% are evaluated (matrix seHI). It returns this matrix with the vectors of the weights and the indicator of intact households attached to its columns. This matrix is the input for the functions that summarise substantial and partial transition. For example,

```
ATlngT <- EUtransD("AT", yrs=c(9, 10))
```

returns the 4124×4 matrix with the values of eHI% for 2009, 2010, the sampling weights for 2009 and the indicator of being intact.

The function EUtransF estimates the rates of substantial transition. Its data-argument is the output of function EUtransD, such as ATlngT, the value of the centre and the values of half-widths. A single value of the centre should be submitted.

```
EUtransF <- function(dat, std, znt)
{
###    The rates of substantial transition

##  dat      The dataset, generated by function  EUtransD
##  std      The standard (centre of the transition zone)
##  znt      The half-width(s) of the transition zone

##  The total of the weights
Swe <- sum(dat[, 3])

##  Initialisation
Res <- c()

##  For each zone-width
for (zn in znt)
Res <- rbind(Res,
c(sum(dat[(dat[, 1] < std-zn) & (dat[, 2] > std+zn), "Weight"]),
  sum(dat[(dat[, 1] > std+zn) & (dat[, 2] < std-zn), "Weight"]),
  sum(dat[(dat[, 1] < std-zn) & (dat[, 2] < std-zn), "Weight"]),
  sum(dat[(dat[, 1] > std+zn) & (dat[, 2] > std+zn), "Weight"]) ) )

dimnames(Res) <- list(znt, c("Exit from poverty", "Entry to poverty",
  "Stay poor", "Stay prosperous"))

##  Scaling to percentages
100*Res/Swe
} ## End of function  EUtransF
```

The result is a matrix, in which each row is the estimated composition of the four patterns for a transition zone given by its half-width. For example, in

```
ATlngF <- round(EUtransF(ATlngT, 60, seq(0,10,0.1)), 2)
```

the row "7.5" is

	Exit pv	Enter pv	Stay poor	Stay prosp
7.5	2.29	1.73	4.54	74.78

Partial transitions are evaluated by the function EUtransK. It uses two auxiliary functions, Pos10 and PwTra.

```
Pos10 <- function(vec)
{
```

```
### Function to truncate a vector at zero and unity
vec[vec < 0] <- 0; vec[vec > 1] <- 1
vec
} ## End of vector Pos10


PwTra <- function(vec, quo)
{
### The power transformation of vector vec the exponent quo
whi <- vec < 0.5
vec[whi] <- vec[whi]^quo * 2^(quo-1)
vec[!whi] <- 1 - (1-vec[!whi])^quo * 2^(quo-1)


vec
} ## End of function PwTra
```

The latter function is for evaluating the scores. Linear scores correspond to
quo=1. In the function EUtransK, the scores for exit from and entry to poverty
are calculated, subjected to a power transformation, and the weighted means
are returned.

```
EUtransK <- function(dat, ctr=60, nu=5, rq=1)
{
### Estimation with partial transition (scores)

## dat   The country's dataset (with eHI% and weights)
## ctr   The centre of the transition zone
## nu    The half-width of the transition zone
## rq    The exponent of the transition function

## The limits of the transition zone
lw <- ctr-nu
up <- ctr+nu

## Linear scoring for the EXIT from poverty
s12X <- cbind(s1=Pos10((ctr - dat[, 1])/nu),
              s2=Pos10((dat[, 2] - ctr)/nu))

## Linear scoring for the ENTRY from poverty
s12N <- cbind(s1=Pos10((dat[, 1] - ctr)/nu),
              s2=Pos10((ctr - dat[, 2])/nu))

## Transformation
if (rq > 1)
{
s12X <- PwTra(s12X, rq)
s12N <- PwTra(s12N, rq)
}

## The total of the weights (denominator)
Swe <- sum(dat[,"Weight"])
```

```
##  Weighted totals/means of the products of partial scores
100*c(sum(apply(s12X, 1, prod) * dat[, "Weight"]),
         sum(apply(s12N, 1, prod) * dat[, "Weight"])) / Swe
}  ##  End of function EUtransK
```

The following expressions evaluate the rates of partial transitions for the default setting ($\tau = 60\%$ and $\nu = 5\%$) and exponents 1, 1.1, ..., 4.

```
ATlngK <- c()

for (rr in seq(1, 4, 0.1))
ATlngK <- rbind(ATlngK, EUtransK(ATlngT, rq=rr))

dimnames(ATlngK) <- list(seq(1,4,0.1), c("Exit", "Entry"))
```

Multiple imputation for the patterns of transition is implemented in the function EUtraMI. The probabilities of being poor are set in the object PRB using the function Pos10.

```
EUtraMI <- function(dat, tau, nu, nMI, intact=T)
{
###    Estimation with multiply imputed poverty states

##   dat      The dataset generated by  EUtransM
##   tau      The poverty threshold -- centre of the transition zone
##   nu       The half-width of the transition zone
##   nMI      The number of imputations/completions
##   intact   Whether restrict to intact households

##  The number of years
nyr <- ncol(dat) - 2

##  Keep only intact households
if (intact)  dat <- dat[dat[, nyr+2]==1, ]

##  The number of households
nsb <- nrow(dat)

##  The total of the weights and the effective sample size
Swe <- sum(dat[, nyr+1])
Sw2 <- Swe^2 / sum(dat[, nyr+1]^2)

##  The probabilities
PRB <- Pos10((tau + nu - dat[, seq(nyr)])/(2*nu))
```

The plausible sets of poverty patterns are generated in a loop, using the vector $\mathbf{g} = (\ldots, 100, 10, 1)^\top$ as the 'glue'. For example, $(1, 0, 0, 1)\mathbf{g} = 1001$.

```
##  Initialise for the sets of plausible values
PLAU <- list()
```

```
## To stick the indicators together to form the pattern
glue <- rev(10^seq(0, nyr-1))

## Replicates for multiple imputation
for (mi in seq(nMI))
{
## Plausible patterns
rmat <- matrix((runif(nsb * nyr) > PRB), nsb, nyr) %*% glue

## Store the plausible totals of weights within the patterns
PLAU[[mi]] <- tapply(dat[, nyr+1], rmat, sum)
}
```

The list PLAU contains as its elements the totals of the weights for the plausible sets of patterns of poverty. They are then summarised, inserting zeros for patterns that do not occur is some plausible sets.

```
## All the patterns
ptr <- unique(sort(unlist(sapply(PLAU, names))))

## Initialisation for the results
PLAM <- PLAN <- matrix(0, length(ptr),nMI,dimnames=list(ptr, seq(nMI)))

## Loop over the replicates
for (mi in seq(nMI))
{
## Frequency, mean and variance
frq <- PLAU[[mi]] / Swe
PLAM[names(PLAU[[mi]]), mi] <- frq
PLAN[names(PLAU[[mi]]), mi] <- frq*(1-frq) / Sw2
}

## The within- and between-imputation variances
Wrbn <- 100^2*apply(PLAN, 1, mean)
Brbn <- 100^2*apply(PLAM, 1, var)

rbind(Mean=100*apply(PLAM, 1, mean), Within.var=Wrbn, Between.var=Brbn,
      MI.ste=sqrt(Wrbn + (1+1/nMI)*Brbn))
} ## End of function EUtraMI
```

In the results, the standard errors are estimated by combining the within- and between-imputation variances. The following is an example for Austria:

```
## The sequence of values of eHI%
ATlngM <- EUtransM("AT", yrs=seq(7, 10))

## Application
ATlngP <- EUtraMI(ATlngM, 60, 5, nMI=100)
```

The function EUtransM generates the country's dataset with a column for each year's values of eHI%, the sampling weight for the earliest of the years and an indication as to whether the household is intact. For years 2007–2010 it has 6 columns. The function reads the panel dataset, forms a vector of Id.s for the households that appear on the panel every year, estimates the annual medians of eHI and compiles the dataset of eHI% for the selected households, with the sampling weights and the indication of being intact attached.

7

Multivariate Mixtures

This chapter studies the joint distribution of equivalised household income (eHI) in a country over a sequence of years. The relevant data is from the longitudinal part of the European Union Statistics on Income and Living Conditions (EU-SILC). The study complements Chapter 4, where we studied the univariate distributions of eHI for the separate years, and Chapter 6, where we studied transitions between poverty states in consecutive years. Among classes of multivariate distributions, the normal has no competitor with its tractability, completeness of covariance structures and our understanding of its properties. We consider the multivariate normal distributions as the basis of mixtures for describing the joint distribution of log-eHI over the four years of a panel. We describe the EM algorithm for fitting mixtures, devise methods for summarising the fit and compare the results for the countries in EU-SILC.

7.1   Multivariate Normal Distributions

The multivariate normal distribution is defined by the density

$$\phi(\mathbf{y}; \boldsymbol{\mu}, \boldsymbol{\Sigma}) = \frac{1}{\sqrt{(2\pi)^p \det(\boldsymbol{\Sigma})}} \exp\left\{ -\frac{1}{2}(\mathbf{y} - \boldsymbol{\mu})^\top \boldsymbol{\Sigma}^{-1}(\mathbf{y} - \boldsymbol{\mu}) \right\} \qquad (7.1)$$

for a p-dimensional vector $\mathbf{y}$, where $\boldsymbol{\mu}$ is the $p \times 1$ vector of expectations and $\boldsymbol{\Sigma}$ the $(p \times p)$ variance matrix. The matrix $\boldsymbol{\Sigma}$ has to be symmetric and positive definite. We denote the distribution given by Equation (7.1) as $\mathcal{N}_p(\boldsymbol{\mu}, \boldsymbol{\Sigma})$, and write $\mathbf{Y} \sim \mathcal{N}_p(\boldsymbol{\mu}, \boldsymbol{\Sigma})$ for a random vector $\mathbf{Y}$ with this distribution. The subscript p is omitted in general statements that apply for all dimensions or when the dimension is obvious from the context. An equivalent definition states that the distribution of a random vector $\mathbf{Y}$ is multivariate normal if and only if every nontrivial linear combination of its elements, $\mathbf{a}^\top \mathbf{Y}$, $\mathbf{a} \neq \mathbf{0}$, has univariate normal distribution. Here, $\mathbf{0} = \mathbf{0}_p$ is the $(p \times 1)$ vector of zeros.

The multivariate normal distribution has some very attractive properties that have no parallel among the known classes of multivariate distributions. First, any marginal distribution (of a subvector of $\mathbf{Y}$) of the multivariate normal is also multivariate normal. More generally, the class of multivariate normal distributions is closed with respect to nonsingular linear transformations. That is, if $\mathbf{Y} \sim \mathcal{N}_p(\boldsymbol{\mu}, \boldsymbol{\Sigma})$ and $\mathbf{A}$ is a $(r \times p)$ matrix of full rank $r \leq p$, then $\mathbf{AY} \sim \mathcal{N}_r(\mathbf{A}\boldsymbol{\mu}, \mathbf{A}\boldsymbol{\Sigma}\mathbf{A}^\top)$.

Multivariate normal distributions are also closed with respect to conditioning. Suppose $\mathbf{Y}$ is partitioned to two subvectors as $\mathbf{Y}^\top = (\mathbf{Y}_1^\top, \mathbf{Y}_2^\top)^\top$, and $\boldsymbol{\mu}$ and $\boldsymbol{\Sigma}$ are partitioned compatibly as

$$\boldsymbol{\mu} = \begin{pmatrix} \boldsymbol{\mu}_1 \\ \boldsymbol{\mu}_2 \end{pmatrix}, \quad \boldsymbol{\Sigma} = \begin{pmatrix} \boldsymbol{\Sigma}_1 & \boldsymbol{\Sigma}_{12} \\ \boldsymbol{\Sigma}_{21} & \boldsymbol{\Sigma}_2 \end{pmatrix},$$

so that $\mathbf{Y}_h \sim \mathcal{N}(\boldsymbol{\mu}_h, \boldsymbol{\Sigma}_h)$ for $h = 1$ and 2. Then the conditional distribution of $\mathbf{Y}_1$, given that $\mathbf{Y}_2$ is equal to a vector $\mathbf{y}_2$, is

$$(\mathbf{Y}_1 \mid \mathbf{Y}_2 = \mathbf{y}_2) \sim \mathcal{N}\left\{\boldsymbol{\mu}_1 + \boldsymbol{\Sigma}_{12}\boldsymbol{\Sigma}_2^{-1}(\mathbf{y}_2 - \boldsymbol{\mu}_2), \boldsymbol{\Sigma}_1 - \boldsymbol{\Sigma}_{12}\boldsymbol{\Sigma}_2^{-1}\boldsymbol{\Sigma}_{21}\right\}.$$

The multivariate standard normal distribution has zero means, unit variances and zero covariances: $\mathcal{N}(\mathbf{0}, \mathbf{I})$, where $\mathbf{I}$ is the identity matrix of the appropriate dimensions $(p \times p)$. A random vector with this distribution is formed by setting its elements to independent draws from the univariate standard normal distribution $\mathcal{N}(0, 1)$. From a p-variate random vector $\mathbf{Y}$ with $\mathcal{N}(\mathbf{0}, \mathbf{I})$, we can form any p-variate normal distribution by the following process. Let $\boldsymbol{\Sigma}$ be the variance matrix of the target distribution. It has the Cholesky decomposition $\boldsymbol{\Sigma} = \mathbf{L}\mathbf{L}^\top$, with a lower triangular matrix $\mathbf{L}$; see Golub and van Loan (1996) or Horn and Johnson (1985) for background. Then the random variable $\boldsymbol{\mu} + \mathbf{LY}$ has the target distribution $\mathcal{N}(\boldsymbol{\mu}, \boldsymbol{\Sigma})$. Thus, $\mathcal{N}(0, 1)$ serves as a building block for generating the entire class of multivariate normal distributions by replicate draws and linear transformations.

The expectations and variances of a multivariate normal distribution are essentially univariate characteristics, because they relate to a single element of the distribution, and their estimation entails a set of univariate analyses. In contrast, a covariance in $\boldsymbol{\Sigma}$ is an essentially bivariate characteristic, related to two elements. However, the entire multivariate distribution is fully characterised by these univariate and bivariate characteristics.

The multivariate normal distribution is defined only with positive definite variance matrices $\boldsymbol{\Sigma}$. Thus, $\mathbf{AY}$ for a $r \times p$ matrix $\mathbf{A}$ and a random vector $\mathbf{Y}$ with distribution $\mathcal{N}_p(\boldsymbol{\mu}, \boldsymbol{\Sigma})$ does not have a normal distribution if $\mathbf{A}$ has a rank smaller than r. A subvector of $\mathbf{AY}$, selected so that it would be a nonsingular transformation of $\mathbf{Y}$, has a multivariate normal distribution. For more background, see Mardia, Kent and Bibby (1979).

7.1.1 Finite Mixtures of Normal Distributions

The definition of mixtures for univariate normal distributions, formulated in Section 4.1, has an obvious extension to multivariate normals. The motivation for mixtures of univariate normals also carries over to multivariate data. Thus, we seek a description for a multivariate sample that is more detailed than what can be afforded by a single multivariate distribution. Formally, the components define subpopulations, and it is natural to inquire about each subject (unit in the sample) as to which component it belongs. However, a mixture component does not automatically warrant its interpretation as a subpopulation of some relevance. A case for such an interpretation is stronger when a component is well-separated from another component, from all components, or from their union. We revisit this issue is Section 7.7.

The theoretical support for using mixtures of normal distributions, adding to the tractability of the multivariate normal densities, is that any distribution can be approximated arbitrarily closely by a finite mixture of normals (Marron and Wand, 1992). In algebraic terminology, finite mixtures of normals are dense in the space of all distributions. The pragmatic argument that a single multivariate lognormal distribution is a good starting point for modelling income (eHI) profiles, and that we are unlikely to need many components for a good fit, is just as powerful.

In a univariate mixture, we can relate the (fitted) expectations and variances of the components to a vague concept of their distance. Our judgement can be informed by graphical representation of the components by the scaled densities. In multivariate mixtures, there are too many parameters to consider and relate, and so the confusion index (and matrix), introduced in Section 4.5, cannot be substituted by an informal assessment of the fitted parameters. Also, our capacity to represent the components graphically is reduced, especially when the correlation structure is of importance.

7.2   EM Algorithm

A finite mixture of multivariate normal distributions is constructed from a (finite) set of distributions by the multivariate version of the process described in Chapter 4. Suppose the component distributions are $\mathcal{N}(\boldsymbol{\mu}_k, \boldsymbol{\Sigma}_k)$, $k = 1, 2, \ldots, K$. First we draw an integer from the composition given by the probabilities $p_1, \ldots, p_K$ on the set $(1, 2, \ldots, K)$. If the draw is equal to h, then we draw from the distribution $\mathcal{N}(\boldsymbol{\mu}_h, \boldsymbol{\Sigma}_h)$.

The EM algorithm for fitting a finite mixture of multivariate normals proceeds by iterations comprising steps E and M. They are adaptations of their counterparts described in Section 4.2. In the E step, the conditional probabilities of belonging to component k are estimated provisionally by the Bayes theorem as

$$\hat{r}_{j,k} = \frac{\hat{p}_k \, \phi\left(\mathbf{y}_j; \hat{\boldsymbol{\mu}}_k, \hat{\boldsymbol{\Sigma}}_k\right)}{\hat{p}_1 \phi\left(\mathbf{y}_j; \hat{\boldsymbol{\mu}}_1, \hat{\boldsymbol{\Sigma}}_1\right) + \cdots + \hat{p}_K \phi\left(\mathbf{y}_j; \hat{\boldsymbol{\mu}}_K, \hat{\boldsymbol{\Sigma}}_K\right)} \tag{7.2}$$

for outcome vector $\mathbf{y}_j$ of unit j; the density ϕ is defined by Equation (7.1). The reference to the iteration is suppressed in Equation (7.2). As in the univariate version of the E step, we have to evaluate for each unit j only the K scaled densities $p_k \phi(\mathbf{y}_j; \hat{\boldsymbol{\mu}}_k, \hat{\boldsymbol{\Sigma}}_k)$ at the provisional (current-iteration) fit.

In the M step, we estimate the parameter objects $\boldsymbol{\mu}_k$ and $\boldsymbol{\Sigma}_k$ by adapting the standard procedure for maximum likelihood estimation to the uncertainty about the composition of each component:

$$\hat{\boldsymbol{\mu}}_k = \frac{1}{\hat{r}_{+k}} \sum_{j=1}^{n} \hat{r}_{j,k} \, \mathbf{y}_j$$

$$\hat{\boldsymbol{\Sigma}}_k = \frac{1}{\hat{r}_{+k}} \sum_{j=1}^{n} \hat{r}_{j,k} \left(\mathbf{y}_j - \hat{\boldsymbol{\mu}}_k \right) \left(\mathbf{y}_j - \hat{\boldsymbol{\mu}}_k \right)^{\top}, \tag{7.3}$$

where $\hat{r}_{+k} = \hat{r}_{1,k} + \hat{r}_{2,k} + \cdots + \hat{r}_{n,k}$ is the within-component total of the estimated conditional probabilities and n is the sample size. The marginal probabilities p_k are estimated by $\hat{r}_{+k}/n$. The sampling weights are incorporated in the analysis by replacing in Equation (7.3) each $\hat{r}_{j,k}$ with $w_j \hat{r}_{j,k}$ and each r_{+k} with their total $\sum_j w_j \hat{r}_{j,k}$. Equation (7.2) is used without any change because it refers to a single unit.

The iterations are terminated when two consecutive provisional fits differ by less than a set standard. We define the standard as follows. Let $\boldsymbol{\theta}$ be the vector of all the parameters that are estimated, and let q be the length of $\boldsymbol{\theta}$. The updating in iteration t is defined as $\Delta \hat{\boldsymbol{\theta}} = \hat{\boldsymbol{\theta}}^{(t)} - \hat{\boldsymbol{\theta}}^{(t-1)}$, and its norm as $\|\Delta \hat{\boldsymbol{\theta}}\| = \sqrt{(q^{-1} \Delta \hat{\boldsymbol{\theta}}^{\top} \Delta \hat{\boldsymbol{\theta}})}$. The iterations are stopped when this norm is smaller than 10^{-6}. Without the division by q the standard for models with many components and for data with many elements would be too strict.

7.3 Example

In this section, we analyse the data on eHI from the Austrian panel in 2007–2010. There are 1119 households with complete records for the four years; see Table 6.1. Of them, 765 are intact over the four years. The values of eHI are subjected to the log transformation with a token Euro added, $\log(1 + \text{eHI}_+)$, after truncating the values of eHI at zero. The estimated means and standard deviations for them are listed at the top of Table 7.1. They correspond to the fit by a single univariate lognormal distribution for each year. The fitted correlation matrix is

$$\frac{1}{1000} \begin{pmatrix} 1000 & 544 & 571 & 554 \\ 544 & 1000 & 747 & 733 \\ 571 & 747 & 1000 & 783 \\ 554 & 733 & 783 & 1000 \end{pmatrix}.$$

TABLE 7.1

Mixture model fits to the values of eHI with multivariate lognormal distributions as the basis; intact households in the Austrian EU-SILC panel 2007–2010.

Component		Year				$\hat{p}_k$ / corr.s
		2007	2008	2009	2010	
1 component						
	$\hat{\boldsymbol{\mu}}_1$	9.799	9.844	9.900	9.926	1.000
	$\hat{\mathbf{s}}_1$	0.616	0.513	0.489	0.460	0.544–0.782
2 components						
1(2)	$\hat{\boldsymbol{\mu}}_1$	9.835	9.866	9.903	9.933	0.753
	$\hat{\mathbf{s}}_1$	0.407	0.401	0.395	0.389	0.823–0.913
2(2)	$\hat{\boldsymbol{\mu}}_2$	9.689	9.777	9.894	9.906	0.247
	$\hat{\mathbf{s}}_2$	1.006	0.753	0.703	0.621	0.337–0.646
3 components						
1(3)	$\hat{\boldsymbol{\mu}}_1$	9.885	9.909	9.951	9.981	0.645
	$\hat{\mathbf{s}}_1$	0.438	0.415	0.409	0.405	0.758–0.883
2(3)	$\hat{\boldsymbol{\mu}}_2$	9.599	9.726	9.848	9.860	0.194
	$\hat{\mathbf{s}}_2$	1.070	0.795	0.737	0.639	0.278–0.611
3(3)	$\hat{\boldsymbol{\mu}}_3$	9.696	9.728	9.761	9.783	0.160
	$\hat{\mathbf{s}}_3$	0.342	0.364	0.360	0.354	0.983–0.987
4 components						
1(4)	$\hat{\boldsymbol{\mu}}_1$	9.933	9.987	10.015	10.044	0.441
	$\hat{\mathbf{s}}_1$	0.423	0.411	0.433	0.396	0.859–0.903
2(4)	$\hat{\boldsymbol{\mu}}_2$	9.775	9.737	9.815	9.842	0.245
	$\hat{\mathbf{s}}_2$	0.454	0.395	0.341	0.408	0.494–0.954
3(4)	$\hat{\boldsymbol{\mu}}_3$	9.598	9.752	9.866	9.883	0.176
	$\hat{\mathbf{s}}_3$	1.076	0.812	0.752	0.644	0.276–0.589
4(4)	$\hat{\boldsymbol{\mu}}_4$	9.667	9.694	9.730	9.752	0.137
	$\hat{\mathbf{s}}_4$	0.335	0.356	0.352	0.345	0.985–0.988
5 components						
1(5)	$\hat{\boldsymbol{\mu}}_1$	9.936	9.990	10.018	10.050	0.425
	$\hat{\mathbf{s}}_1$	0.428	0.417	0.440	0.400	0.861–0.907
2(5)	$\hat{\boldsymbol{\mu}}_2$	9.770	9.736	9.815	9.841	0.254
	$\hat{\mathbf{s}}_2$	0.457	0.395	0.341	0.407	0.499–0.953
3(5)	$\hat{\boldsymbol{\mu}}_3$	9.586	9.753	9.822	9.896	0.162
	$\hat{\mathbf{s}}_3$	1.149	0.842	0.759	0.663	0.275–0.624
4(5)	$\hat{\boldsymbol{\mu}}_4$	9.664	9.688	9.725	9.750	0.133
	$\hat{\mathbf{s}}_4$	0.337	0.357	0.353	0.347	0.986–0.989
5(5)	$\hat{\boldsymbol{\mu}}_5$	9.847	9.880	10.198	9.807	0.025
	$\hat{\mathbf{s}}_5$	0.181	0.140	0.329	0.185	−0.765–0.892

Note: $\hat{\mathbf{s}}$ denotes the vector of estimated standard deviations. The sets of six fitted correlations are summarised by their ranges.

This matrix is omitted from Table 7.1 because the rest of the table refers to several more 4×4 matrices that would not fit on a single page. Instead, the six estimated correlations are summarised by their range (the minimum and maximum values), given in the right-hand column. The correlations are lower for year 2007 with the other years, around 0.55 vs. around 0.75. Such a disparity in the correlations, or in any other characteristic of the distribution of log-eHI, is rather unexpected, given our understanding that the economic situation in the country was stable in the studied years. The estimated annual standard deviations have decreased over the years substantially, from 0.616 to 0.460. A similar trend in the studied population is highly unlikely.

For a model with K multivariate normal components, the fit is completely described by K vectors of means, K vectors of standard deviations, a K-variate composition (probabilities) and K correlation (or covariance) matrices. We use the term *majority* component for a mixture component with fitted marginal probability greater than 0.5, and *minority* component for a component with probability much smaller than any other component. For component k of the K component model (or fit), we use the notation $k(K)$, as we did in Chapter 4. The components of a model fit are listed in Table 7.1 in the descending order of the fitted probabilities.

The differences of the fitted means on the log scale are translated to the original scale by exponentiation. For a small difference $\Delta \doteq 0$, $\exp(\Delta) \doteq 1 + \Delta$. Thus, the additive difference of Δ on the log scale converts to the multiplicative change by $100\Delta\%$. However, this approximation ignores the sampling variation related to Δ. If $\widehat{\Delta}$ has sampling distribution $\mathcal{N}(\Delta, \sigma^2)$, then $\mathrm{E}\{\exp(\widehat{\Delta})\} = \exp(\Delta + \frac{1}{2}\sigma^2)$, so the sampling variation should be reflected by an inflation of the transformed expectation $\exp(\widehat{\Delta})$. Further, σ^2 is usually also estimated and the uncertainty about it should perhaps also be reflected in the transformation. See Longford (2009) for a study of this problem.

The fit of the model with two mixture components is listed near the top of Table 7.1. The first component, with large estimated probability (0.75) has the vector of means with steady annual increments, by 0.031, 0.037 and 0.030, that is, by $3\%-4\%$, small annual standard deviations that are in a narrow range $(0.39-0.41)$ and high correlations, in the range $0.82-0.91$. In contrast, the second component has erratic means, with large increases from year 2007 to 2009, by 0.088 and 0.117 on the log scale, followed by an increase by only 0.012. The fitted standard deviations are very large and in a wide range, and some of the correlations are quite small. The correlations for year 2007 with the other three years are all smaller than 0.40, and the other three correlations are greater than 0.59.

In the fit with three components, the first and third components have similar (small) annual increments of the means and annual standard deviations in narrow ranges $(0.40-0.44$ and $0.34-0.36)$. The third component has extremely high correlations. Component 2(3) is similar in all features to component 2(2). We assess later whether the two components have many households in common.

In the fit with four components, components 3(4) and 4(4) closely resemble the respective components 2(3) and 3(3), so the only important change seems to be the decomposition of the component 1(3) to 1(4) and 2(4). Component 1(4) has slightly higher means, lower standard deviations and higher correlations than 1(3), and component 2(4) contains less regular sequences, with smaller correlations and a decrease of the mean from 2007 to 2008 that is then more than offset in 2009.

In the fit with five components, $k(5)$ closely resembles $k(4)$ for $k \leq 4$, and 5(5) is a 'new' minority component. It has small fitted standard deviations, an unusual pattern of the fitted means, with an exceptionally large value for 2009, 10.20. All three correlations for year 2009 are negative; their estimates are -0.612, -0.765 and -0.401 for the respective years 2007, 2008 and 2010. The 2008–2010 correlation is 0.892. We conjecture that this component (2.5%) accounts for some aberrations in the data collection process in 2009 which were resolved in 2010.

The two- to five-component models took respectively 31, 142, 236 and 284 iterations to converge. By the standards of the EM algorithm this is quite rapid convergence. Mixture models with more components can be fitted, but their main features are the abberant patterns in very small subpopulations. This suggests that the five-component model is perhaps sufficiently complex.

The level of uncertainty about the assignment of households to the mixture components for the two-component fit can be judged by a histogram of the fitted probabilities or by tabulating their rounded values. The top part of Table 7.2 lists the counts of households assigned to the majority component with probabilities that have the leading decimal digits 0, 1, ..., 9. Only 121 households (15.8% out of 765) have probabilities of assignment to a component in the range (0.1, 0.9).

For the three-component fit, the leading decimal digits of the probabilities of the components 1(3) and 2(3) are tabulated in the main part of Table 7.2. Positive counts are only on the anti-diagonal, in cells (9,0), (8,1), ..., (0,9), accounting for 562 households out of 765 (73.5%), and in column 0 (203 households, not counting the anti-diagonal cell (9,0); 26.5%). The 50 households in cell (0,0) are very likely to belong to component 3(3), because their probabilities of belonging to 1(3) and 2(3) are both smaller than 0.1. In fact, their probabilities of belonging to component 2(3) are all smaller than 0.002, whereas the probabilities of belonging to 1(3) are all greater than 0.033. Thus, component 2(3), in which the correlations are lower, can be ruled out for quite a few households. But the assignments for these households to 1(3) or 3(3), components with higher correlations, are much more uncertain.

In the five-component fit, we can identify four households with probabilities $\hat{r}_{5,5} > 0.99$; they almost certainly belong to the minority component 5(5). All four households have much higher values of eHI in year 2009 than in the other three years, but even these values (30–50 thousand Euro) are not exceptional in the general population, and neither are the values in the other three years (13–22 thousand Euro). These households would not appear exceptional in

TABLE 7.2
Leading digits of the probabilities of assignment in two- and three-component model fits; Austria, 2007–2010.

Two components									
Digit	0	1	2	3	4	5	6	7	8	9
Count	121	11	9	11	10	9	9	15	47	523

Three components										
0	50	0	0	0	0	0	0	0	0	88
1	39	0	0	0	0	0	0	0	8	0
2	17	0	0	0	0	0	0	6	0	0
3	20	0	0	0	0	0	9	0	0	0
4	15	0	0	0	0	10	0	0	0	0
5	15	0	0	0	10	0	0	0	0	0
6	9	0	0	3	0	0	0	0	0	0
7	18	0	23	0	0	0	0	0	0	0
8	20	28	0	0	0	0	0	0	0	0
9	377	0	0	0	0	0	0	0	0	0

any univariate (annual) analysis. Only ten other households have fitted probabilities $\hat{r}_{5,5} > 0.5$, and all but 45 households have fitted probabilities smaller than 0.001.

We say that the component 5(5) is *exclusive* because most of the households are extremely unlikely to belong to it. A component is called inclusive if most households have a nontrivial probability of belonging to it, such as greater than 0.01. A minority component is not necessarily exclusive, because many households may have appreciable probabilities of belonging to it (say, in the range 0.01–0.05). Similarly, a majority component is not necessarily inclusive, because there may be a relatively large subset of households that almost certainly do not belong to it.

The fitted probabilities in a three-component solution can be presented in a ternary plot; for details, see Section 4.3.1. In the plot in Figure 7.1, the points for all the households are on or close to the sides (1,2) and (1,3). (A small amount of random noise is added to each point in both horizontal and vertical directions.) One of the components can almost be ruled out for every household. Further, when component 1 is ruled out another component can also be ruled out, because the side (2,3) is empty.

The profiles of the households in the components $k(4)$, $k = 1, \ldots, 4$, are explored in Figure 7.2. In each panel, the sequences of values of eHI are plotted for a random sample of size up to 20 from the analysed households that are very likely to belong to component $k(4)$. 'Very likely' is interpreted here as

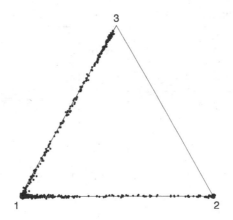

FIGURE 7.1
Ternary plot of the fitted probabilities in mixture model with three components; Austria 2007–2010.

$r_{i,k} > 0.9$ for unit i. For component 2(4), all 15 households that satisfy this condition are represented in the panel.

The diagram confirms what we can infer from the tabular summary of the fit in Table 7.1, associating high correlation with uniform patterns. Households in component 4(4) have extremely regular profiles and very small annual variances. Most households in component 2(4) have lower eHI in 2008 than in 2007. The annual changes in eHI in component 3(4) are not implausible, not even for intact households, but their sizeable probability, $\hat{p}_{3,4} = 0.18$, suggests that some of the households in the component have erroneous entries. For orientation, $e^8 \doteq 3000$ Euro. We note that the households represented in Figure 7.2 are not typical for the indicated components; they are samples of households that almost certainly belong to the indicated components. For example, a household with $\hat{r}_{i,2} = 0.6$ is ruled out from panel B, as well as from any other panel. Drawing samples that better represent the components is left as an exercise.

The model fits, that is, the estimated mean vectors, variance matrices and marginal probabilities, are represented graphically in Figure 7.3. Each panel corresponds to a mixture model, as indicated in the title. In a panel, each fitted component is represented by the line connecting the fitted means (mean profile) drawn by a solid line, and two dashed lines symmetrically placed around the mean profile, which combine the information about the fitted standard deviations and correlations for the neighbouring years. For year a, these lines connect the values $m_a + f s_a$, where m_a is the fitted mean and s_a the fitted standard deviation for the component and the year, and f is a positive constant, set to 0.125. Further, the dashed lines are 'cut in' (indented) at mid-year points, 7.5, 8.5 and 9.5, according to the fitted correlation of the

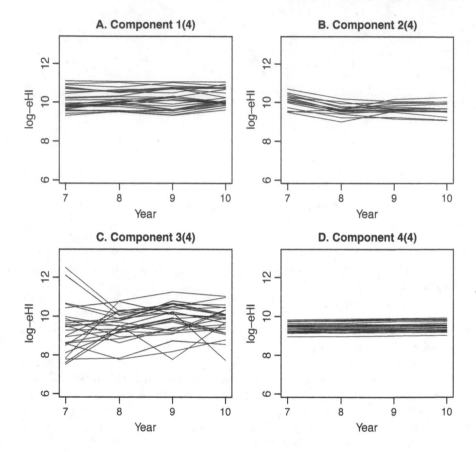

FIGURE 7.2
Random samples from subsets of the intact households that are very likely to belong to the four mixture components; Austria 2007–2010.

neighbouring years. Their values at these points are equal to the averages of the neighbours, e.g., of years 7 and 8 at 7.5, but they are shrunk toward the average of the two fitted means by the correlation as the factor. The thickness of each line is positively related to the marginal probability of the component it represents. This probability is also indicated at the right-hand margin by the size of the circle. The values $m_a - fs_a$ are also connected, and are symmetrically cut in.

Panel A, for the two-component model fit, indicates that there is a majority component with small annual increases of the fitted mean, small standard deviations (in relation to the other component), and high correlations, because the dashed lines are not cut in. Component 2(2) has much smaller fitted means in years 2007 and 2008, greater standard deviations, and the correlations for the consecutive years are positive and small, especially for 2007–2008. Panel

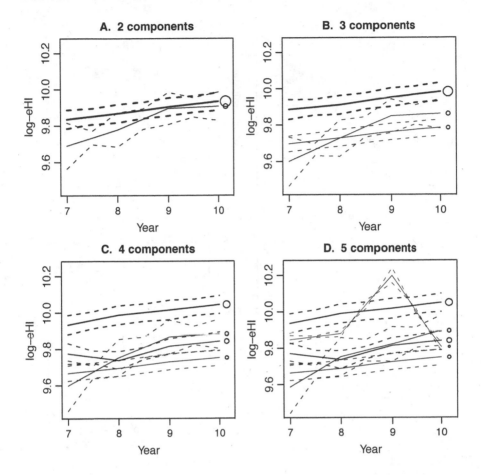

FIGURE 7.3
Fits of mixture models with two to five components; Austria 2007–2010.

B suggests that there are two components with regular profiles, one with substantially smaller means of eHI. These two components reappear in panel C with two other components for highly irregular patterns of income and low inter-year correlations. One 'irregular' component has large variances. The panels become cluttered as the number of components is increased, but the 'new' minority component 5(5), with a singularly high mean in 2009, is easy to discern in panel D.

It is customary to conclude an analysis using mixture models by a choice of the model that is most appropriate, so that all further discussion would use this model as a premise. This amounts to a 'denial of uncertainty'. We prefer to consider a small number of plausible models. We do not regard the number of components as an inferential target because we do not assume that any one

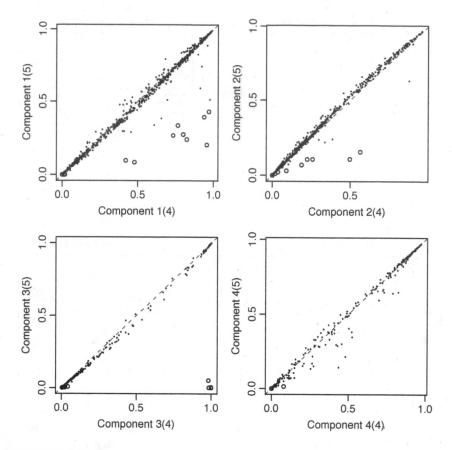

FIGURE 7.4
Fitted probabilities of assignment to the mixture components in the four- and
five-component models; Austria 2007–2010.
The households with $\hat{r}_{5,5} > 0.5$ are marked by circles.

of the mixture models considered is valid. They are merely approximations to
the population distribution, and we have no formal criterion for what amounts
to a satisfactory approximation. Criteria related to the likelihood, such as the
Akaike and Bayes information criteria (Akaike, 1973, and Schwartz, 1978,
respectively) and their successors, entail a denial of uncertainty and their
verdicts are affected by sample size.

A limiting factor in the assessment of a mixture model fit is our ability
to comprehend and interpret the long list of fitted components and their parameters. This can be aided by simple descriptions of the changes that take
place as the number of components is increased. Some of these changes can
be inferred from Figure 7.3, but finer detail may be desired. Figure 7.4 illustrates the changes of the fitted probabilities from the four- to five-component

model by plotting the fitted probabilities $\hat{r}_{k,4}$ and $\hat{r}_{k,5}$ for $k = 1,\ldots,4$. The diagram shows that the assignments to the pairs of components $k(4)$ and $k(5)$ are changed substantially only for a few households. The households with $\hat{r}_{5,5} > 0.5$ are marked by circles. Most of these 14 households have been recruited from components $1(4)$ and $3(4)$ where several of them have large fitted probabilities.

7.4 Improper Component

One of the mixture components can be declared as improper, with a constant (or another) density that has an infinite integral. Most details carry over directly from the univariate setting discussed in Section 4.4. A multivariate improper density is constructed as a product of univariate improper densities for the elements. This corresponds to independent densities of the elements. In practice, it suffices to consider constant improper densities. In Section 4.4, the purpose of an improper component was to account for univariate observations with outlying (unusual) values. With multivariate outcomes, it can account for observations with unusual profiles (patterns) of values. One might argue that the minority component in the five-component model considered in the previous section has dealt with such observations satisfactorily, although we have neither established the outlier status of any of the households nor formulated a definition of the term 'outlier'.

As in Section 4.4, the improper component has index 0, so that f_0 is its density and p_0 its probability. By D we denote the (constant) value of f_0, but use its negative decimal logarithm $d = -\log_{10}(D)$ in all discussion. The EM algorithm requires only a minor adjustment for an improper component. The initial solution for p_0 does not have to be set. The density f_0 is added to the other (proper normal) densities, and the marginal probability of the component is updated as the average of the fitted conditional probabilities.

We fit models with up to four proper components and an improper component with constant density set to 10^{-d}, $d = 1,\ldots,8$. This is far too many model fits to display in a systematic manner, and some selective presentation is necessary. First, we have to reduce the range of the values d that we would regard as plausible and assess how the estimates of the means and (co-)variances depend on them.

The changes of the fitted components in models with three proper and one improper component are recorded in Figure 7.5. The three columns of panels represent the three proper components, and the rows are for the means, standard deviations and the correlations, plotted as functions of the negative log-density $d = -\log_{10}(D)$. In the first and second row, there is a function for each of the four years (elements), drawn by lines of increasing thickness. In the bottom row, there are six functions, one for each pair of years. The pairs

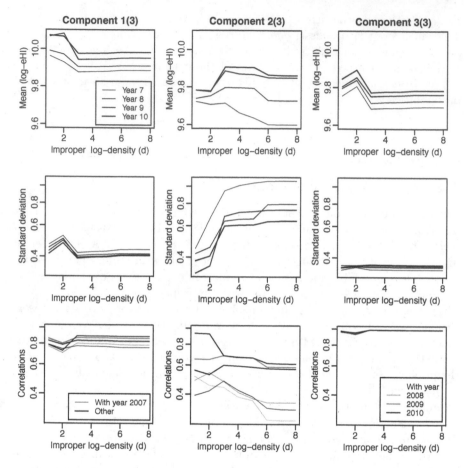

FIGURE 7.5

Fitted means, standard deviations and correlations as functions of the improper density for models with three proper components; Austria 2007–2010.

of years that each one involves are identified by line thickness and colour, as indicated in the two legends.

The parameter estimates for components 1(3) and 3(3) change only slightly for $d > 3$, whereas the estimates for component 2(3) involve much more substantial changes for all parameters, for $d < 6$. For $d \doteq 2$, when $\hat{p}_0$ is nearly 0.10, the standard deviations of 2(3) are of the magnitude similar to the other two components, but they increase with d, unlike the standard deviations of the other two components. This suggests that as d is reduced the improper component recruits additional households mainly from component 2(3). The plots of the fitted probabilities $\hat{p}_k$ of the proper components, as functions of d, in Figure 7.6 (middle panel) contradict this; $\hat{p}_2$ decreases with d for $d > 3$. As d is changed the proper components are realigned and they 'trade' households

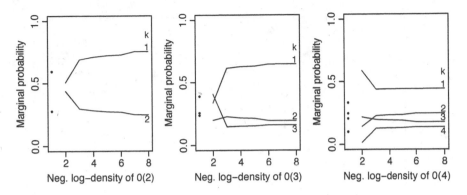

FIGURE 7.6
Fitted marginal probabilities of the proper components for the mixture models with two to four components; Austria 2007–2010.

among themselves, as they do with the improper component. The functions $\hat{p}_k(d)$ are drawn by lines only for $d \geq 2$ and the estimated probabilities for $d = 1$ are marked by dots. They are not connected with the lines because the link between the results for $d = 1$ and 2 is not clear. The vectors of means and the variance matrices are changed so much that it is not obvious which (proper) component of one fit corresponds to which component of the other fit. This is an example of the problem of labelling (label switching). In our case, it occurs for all three models.

Figure 7.7 displays the marginal probabilities of the improper component in the fits of the models with up to four proper components. The fitted marginal probabilities $\hat{p}_0$ decrease with d and also with the number of components, although the differences between $\hat{p}_0$ for three and four components are very small, except for d around 2.0. As more components are introduced, the recruitment by the improper density becomes more difficult for large d. For $d \leq 2$, the differences are much smaller, although the improper density with these values of d has perhaps too large probability. We would expect the sample to have not more than 1% or 2% outliers, marked in the diagram by a horizontal gray strip.

More detail can be extracted from the output of the function EUMmx2 which fits mixture models. For example, in the fit of the five-component model (with one component improper), there are only four households i with $\hat{r}_{i,0} > 0.99$ and one further household has $0.5 < \hat{r}_{i,0} < 0.99$. The near-certain members of the improper component have vectors of log-eHI values $(10.40, \underline{13.15}, 11.55, 10.70)$, $(10.64, 9.87, 10.21, \underline{7.75})$, $(10.09, \underline{7.31}, 10.00, 8.36)$ and $(9.62, \underline{6.80}, 9.53, 9.49)$; each household has an exceptional value of log-eHI in a year (underlined). These are not the four households identified as near-certain members of the minority component 5(5) in Section 7.3. The difference of 2.0 on the scale of log-eHI corresponds to $e^2 \doteq 7.4$ times greater

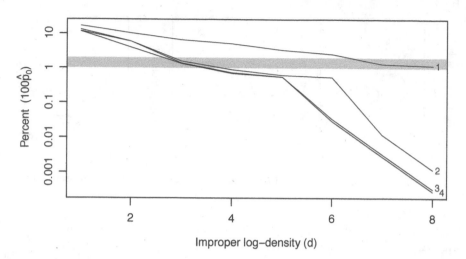

FIGURE 7.7
Fitted marginal percentages of the improper component for the mixture models with up to four components; Austria 2007–2010.
The horizontal gray strip marks the plausible range of $100p_0$. The numbers of proper components are indicated at the right-hand margin.

or smaller value of eHI, an exceptional change that is bound to be extremely rare.

We cannot argue with conviction that any highlighted observations are erroneous. It is the large estimated proportion of households with unusual profiles of eHI that raises doubt and deserves an investigation. Also, the uneven distribution of some of the features of the vectors of eHI across the years is implausible. These features become less prominent in the proper components when an improper component is included in the model. This suggests that the improper component fulfils its role of recruiting outliers or, more generally, households with exceptional profiles. It is preferable to a proper (lognormal) component because it involves far fewer parameters and less model specification.

7.5 Mixture Models for the Countries in EU-SILC

Figure 7.8 presents the summaries of the model fits with up to four proper and no improper components for the 24 countries with an EU-SILC panel started in 2007. Each panel has the same layout as the panels in Figure 7.3. The model for Slovenia (SI) has three components and the model for Iceland

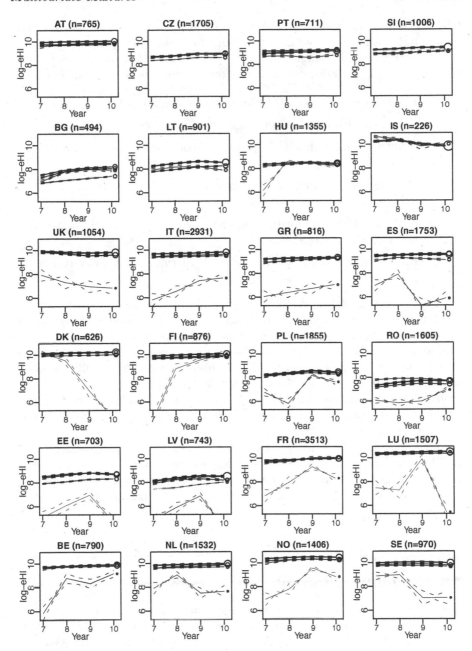

FIGURE 7.8
Summaries of the fits of the mixture models with four components for the
countries with panels in 2007−2010.
Sample sizes are listed in the titles. The fit for Iceland is based on two and
for Slovenia on three components.

(IS) only two; four components are fitted for the other countries. For Iceland, a singularity is encountered in fitting both four- and three-component models, and for Slovenia a singularity arises when fitting four components. The countries' panels are arranged in the following order, in sets of four in a row:

1. countries without a distinctive minority component;

2. countries with a component that (probably) accounts for outliers;

3. countries with a distinctive minority component with small fitted means;

4. countries with a minority component with small fitted means for some years;

5. countries with a minority component with a mean much higher in one year than in the others;

6. the remaining countries, each with a distinctive minority component.

This grouping of the model fits (countries) is informal, constrained by the layout of the diagram and the attempt to find a model suitable for all countries. For example, a distinctive minority component appears in the five-component fit for Austria. From the diagram we conclude that most countries have a minority component with some or all its means very small, large variances and small correlations. Such 'exceptional' components can be effectively replaced by an improper component, but its (constant) density has to be set individually for each country, matching it to the anticipated probability of exceptional patterns of eHI.

Setting aside the exceptional components, Belgium, France, Luxembourg and Hungary appear to have values of log-eHI with the smallest dispersion. The fitted components can be studied in much more detail by redrawing each panel with a narrower vertical axis set specifically for each country.

7.6   Stability of Income

One general conclusion from the mixture analysis in the previous section is that in every country many intact households have income profiles that do not deviate substantially from the national trend of small annual changes, increases by a few percent. In this section, we construct a summary for this feature, to which we refer as *stability*. We will work with the scaled equivalised household income, eHI%, related to the national median, evaluated separately for each year.

We say that an intact household has stable income over the years of the panel if the change in the value of eHI% from one year to the next is always smaller than a given threshold S. We estimate the percentage of intact households in the population that have income stable with respect to set values of S. As with the other summaries (indices) introduced in Chapters 1, 3 and 6, we consider a range of values of S, so stability is summarised by a function of S. For example, a household with the following values of eHI% in years $1-4$,

$$60, 66, 62, 57, \tag{7.4}$$

is regarded (and counted) as stable for $S \geq 0.06$, because the largest change in eHI%, increase by 6%, occurred between years 1 and 2. Note that the changes in eHI% are calculated as differences, not as ratios. The ratios are very volatile for very small values of eHI%. For example, the increase of eHI% from 15% to 30% would be by 200%, the same as from 55% to 110%, even though the former remains within extreme poverty. We treat the units of eHI% as a particular transformation of eHI that renders the comparisons we make additive. Small relative changes, e.g., by 10% in the example in Equation (7.4), are likely to make a lot of difference as regards the threat or proximity of poverty. In contrast, similar changes on the relative scale, even if numerically much greater, say from 150% to 160%, might reasonably be regarded as irrelevant. This suggests that we should simply truncate the values of eHI% from above at a value above which any differences are irrelevant. We set this upper limit to eHI% $= 125\%$. For example, the reduction of eHI% from 140% in one year to 110% in the next is treated on par with the reduction from 125% to 110%, that is, by 15%. Similarly, all negative values of eHI are truncated to zero.

The stability curves are plotted in Figure 7.9 for the 24 countries with EU-SILC panels in $2007-2010$. The two panels of the diagram split the range of stability thresholds S to achieve higher resolution especially for larger values of S in the right-hand panel. The countries' acronyms are printed on their curves and from left to right in the ascending order of their values at the left-hand extreme. The curves intersect a great deal, but most countries do not change their ranks substantially over the plotted range of S.

The diagram confirms that income stability is generally high in the Scandinavian countries and Luxembourg and low in some of the new members of EU, Latvia, Lithuania and Bulgaria in particular. Czech Republic and Slovenia are notable exceptions; their stability is nearly as high as in the Scandinavian countries for low thresholds ($S < 10\%$), and even exceeds some of them for higher thresholds. Among the established members of EU, the stability is uniformly lowest in Spain and the UK. We note that quality of the data may affect our estimation of stability, particularly when income is occasionally incorrectly or inappropriately recorded as negative or zero.

An analysis similar to the one in this chapter but applied to the data from the European Community Household Panel (ECHP), the predecessor of EU-SILC, is described by Longford and Pittau (2006). Its results on stability differ substantially from those in this section.

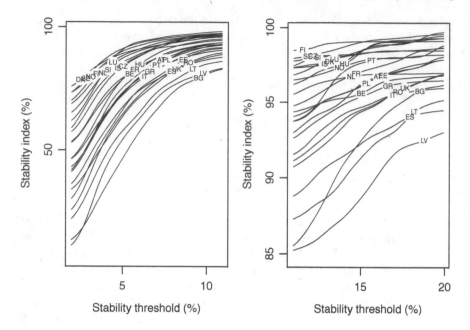

FIGURE 7.9
Estimated stability curves for the countries with panels since 2007.
The left-hand panel is for stability threshold up to 11% and the right-hand
panel from 11% on. In the evaluations, the values of eHI% are truncated at
125 eHI%.

7.7   Confusion and Separation

The confusion index and confusion matrix defined in Section 4.5 can be applied
to multivariate distributions, because the index depends only on densities
and marginal probabilities of components. The index is particularly useful
for panel data because a judgement as to the separation of two components is
more difficult to make in many dimensions, when many parameters have to be
inspected, than in a single dimension, where simple graphics can be effective.

The results of multivariate mixture analyses suggest that the differences
among the components are more pronounced in their covariance structures,
except perhaps for a minority component which for several countries has much
smaller means than the other components. The estimated confusion matrices
for the models with two to four components, listed in Table 7.1 (Austria's

panel 2007 – 2010), are

$$\begin{pmatrix} 500 & 133 \\ 44 & 500 \end{pmatrix} \quad \begin{pmatrix} 500 & 165 & 237 \\ 50 & 500 & 8 \\ 59 & 6 & 500 \end{pmatrix} \quad \begin{pmatrix} 500 & 86 & 6 & 42 \\ 48 & 500 & 98 & 199 \\ 2 & 71 & 500 & 67 \\ 13 & 111 & 52 & 500 \end{pmatrix},$$

with all entries multiplied by 1000. Suppose the plausible range for the threshold for confusion is 0.05 – 0.10. Then component 2(2) is a satellite of component 1(2) because their confusion indices are on either side of the range. In the three-component model fit, component 2(3) has large variances and small correlations and component 3(3) has small variances and large correlations. The two components are well-separated, and they are either satellites of component 1(3), or are confused with it. Recall that in the ternary plot in Figure 7.2 a household with substantial fitted probability of belonging to component 3(3) almost certainly does not belong to component 2(3), and vice versa.

In the fit with four components, 4(4) has small variances and extremely high correlations (0.99) and component 1(4) has larger variances (1.3 – 1.6 times) and somewhat smaller correlations (around 0.90). Component 3(4) has extremely large variances and small correlations. Component 4(4) is well-separated from 1(4) and is confused with 2(4). With component 3(4), it may be confused (if $T > 0.067$) as well as well-separated (if $T < 0.052$), and also its satellite (if $0.052 < T < 0.067$). Component 3(4) is well-separated from 1(4), and is well-separated from or confused with 2(4). Component 2(4) is either well-separated from 1(4) or is its satellite. These associations can be represented by the matrix

$$\begin{pmatrix} \bullet & CP & C & C \\ Cs & \bullet & CX & X \\ C & CX & \bullet & CX \\ C & X & CX & \bullet \end{pmatrix}$$

in which C, s, P and X indicate that the row and column components are clusters, the row-component is a satellite of the column-component, vice versa (parent or pivot), and the two components are confused. Cells with two symbols indicate that there is uncertainty about the status because the threshold is defined only by a plausible range. Note that the uncertainty due to estimation has been ignored in these descriptions. See Longford and Bartošová (2014) for more background to the confusion index and other examples.

7.A Appendix

7.A.1 What Can Go Wrong in Iterations of EM

The general EM algorithm has some very desirable properties. Convergence to a local minimum of the deviance (-2 log-likelihood) is assured and the value of the deviance decreases at every iteration. However, the convergence is often very slow, and in extreme settings it may be difficult to assess whether it has been reached. This occurs when the changes of the estimates and the deviance from one iteration to the next are small but they diminish so slowly that they would still accumulate nontrivially over a large number of iterations.

A whole host of problems are related to the convergence at the boundary of the parameter space. We refer to them collectively as *singularity*. An example of singularity is that one marginal probability $\hat{p}_k$ approaches zero. When it happens the corresponding values of $\hat{\mu}_k$ and $\hat{\Sigma}_k$ become indeterminate because they are based on no observations. The matrix $\hat{\Sigma}_k$ becomes indeterminate also when the component is reduced to a single observation. Note that this is a problem only with a proper component, not with an improper one.

The likelihood is defined only for positive definite variance matrices $\hat{\Sigma}_k$. When the provisional estimate of one of these matrices approaches singularity (deficient rank), the iterations fail because the deviance cannot be evaluated. Another form of singularity could arise in principle if the distributions of two components converged (became identical). In this case, the deviance can be evaluated, but we have to revise the number of components, because the two that cannot be distinguished should be merged.

These problems can be addressed by inserting in the programming code warnings to inform about various (potential) problems. This has to be done sparingly because it may be impractical to inspect a large number of warnings. A useful warning is to indicate that the iterations were stopped when the specified maximum number has been reached without satisfying the convergence criterion. When a marginal probability $\hat{p}_k$ converges to zero, the iterations are stopped. The model with one fewer component should be specified for the same dataset. When one of the fitted matrices is singular, a simpler model should be specified. The units with positive probabilities of belonging to this component could be identified. If there are only a few of them (or only one), then the model can be fitted after excluding these observations.

For the analyses of the datasets for the 24 countries we have foregone most of these arrangements because no problems with convergence were encountered, and the diagnosis of the problems if they did occur would be straightforward from the displayed error message. The R function `traceback` is useful in this context. An obvious threat to trouble-free running of the EM algorithm are zero values eHI, especially when many of them occur for a single year, because a component may converge toward zero mean and zero variance (with other variances positive), exclusively for households that declared negative or zero income for the year.

7.B Programming Notes

The EM algorithm for fitting multivariate normal mixtures is implemented in the function EUMmx1. It is developed from the function EUmix2 introduced in Section 4.A.1. The two functions have the same set of arguments. They are the matrix of the outcomes in EUmx1 (a vector in EUmix2), the vector of weights, optional initial solution (or the number of components to be fitted) and two scalars that control the iterations.

```
EUMmx1 <- function(dat, wei=1, sta=list(), K=0, maxit=20, tol=6)
{
###    Fitting a multivariate normal mixture model

## dat    The dataset (matrix)
## wei    The vector of weights  (or scalar for equal weights)
## sta    The initial solution (optional)
## K      The number of components (effective when  sta  is empty)
## maxit  The maximum number of iterations
## tol    The convergence criterion
```

First, compatibility of the dataset and the vector of weights is checked. If wei is set to a scalar (vector of length one), it is interpreted as equal weights.

```
## The number of units
ndt <- nrow(dat)

## Provision for equal weights
if (length(wei)==1) wei <- rep(1, ndt)

## Check the compatibility of the data and the weights
if (length(wei) != ndt)
    stop(paste("The number of subjects", ndt,
        "and the number of weights", length(wei), "disagree."))
```

The number of components is set as the length of the list sta when it is declared as a nontrivial argument; otherwise the value of the argument K is essential and it defines the number of components. When it is to be applied, the initial solution sta has to be a list, with one element per component. Each element itself has to be a list, comprising the vector of means, the variance matrix and the marginal probability of the component. When an initial solution is not supplied, then the function InitMx sets it by default. The function is described below.

```
## The number of components
```

```
if (is.list(sta) & (length(sta) > 0)) K <- length(sta)

##  The default initial solution
if (length(sta) == 0)    sta <- InitMx(dat, K)

estN <- sta
```

For the iterations, we set their counter (`iter`), the precision of the current update (`dist`) and initialise a matrix that records the marginal probabilities (the convergence monitor, `Conver`):

```
##  Initialisation for the iterations
Conver <- c()
dist <- 0
iter <- 0
```

The iteration loop has the same structure as its univariate counterpart in EUmix2, see Section 4.A.1. Only the evaluation of the scaled (multivariate) normal density is more involved, implemented in a separate function SNdns; see below. The values of these densities are stored in the matrix **dns**, and their within-row (within-unit) totals are the denominators for the probabilities of assignment (`prb`). The marginal probabilities are the weighted means of the columns of `prb`. In the remainder of the M step, the vector of means and the variance matrix are estimated for each component. The new solution is placed in the elements of the list `estN` and its distance from the previous solution, `estP`, is evaluated in `dist` as the negative logarithm of the scaled norm of the update for all the parameters. The logarithm has base 10, so that it can be interpreted as the number of decimal places of precision.

```
while (dist < tol & iter < maxit)
{
##  The next iteration
iter <- iter + 1

##  Store the current solution (N) as the previous (P)
estP <- estN

##  Initialisation for the scaled densities
dns <- c()

##  The scaled normal densities
for (est in estN)
dns <- cbind(dns, SNdns(dat, est))

##  The E step  --  conditional probabilities
prb <- apply(dns, 1, Prob)

##  The weighted totals of the probabilities
Prw <- prb %*% matrix(wei)
```

```
Srw <- Prob(Prw)

##   The M step
for (k in seq(K))
{
##   The weights that combine E step and sampling
wpr <- prb[k, ] * wei / Prw[k]

##   The estimates of the mean vector and variance matrix
Mm <- t(dat) %*% matrix(wpr)
Vv <- t(dat) %*% (dat * matrix(wpr, ndt, ncol(dat))) - Mm %*% t(Mm)

##   The updated (new) estimates
estN[[k]] <- list(M=t(Mm), V=Vv, p=Srw[k])
} ## End of the loop over the components (k)

dist <- -log(mean((unlist(estN) - unlist(estP))^2), base=10) / 2
```

At the end of each iteration, the values of `dist` and of the marginal probabilities are stored. If one of the probabilities is too small the iterations are stopped by altering the value of `dist`.

```
pp <- sapply(estN, ExtrL, 3)

##   Updating the convergence monitor
Conver <- cbind(Conver, c(dist, pp))

##   Check on singularity
if (log(min(pp)) < -tol) dist <- tol+1
} ## End of the iteration loop
```

After convergence, warnings are issued, as applicable, and the results are packaged into a list, one element of which is the set of fitted probabilities of assignment.

```
if (dist < tol)
  warning(paste("Convergence not achieved after", iter, "iterations."))

if (log(min(pp)) < -tol)
  warning(paste("Singularity reached after",iter,"iterations."))

list(Estimation=estN, Conver=Conver, Probs=t(prb))
} ## End of function  EUMmx1
```

The function `EUMmx1` can be expanded in several ways to deal with the various contingencies that might be encountered. For example, the current estimates of the variance matrices could be checked whether they are positive definite. If they are not, or one of the eigenvalues of a matrix is too small, the iterations would be stopped by altering the current value of `dist`; see the end of the iteration loop.

The initial solution for the EM algorithm is set by the function `InitMx`.

```
InitMx <- function(Dat, KK)
{
###    Setting the initial solution for fitting a mixture

## Dat      The dataset
## KK       The number of components
```

A single variable, Dbt, is defined for the units (households) by a simple form of elementwise standardisation. The units are then classified into KK groups according to this variable.

```
## The means and standard deviations
MSd <- apply(Dat, 2, MeVa)

## The standardized deviations
Dbt <- apply((Dat - matrix(MSd[1, ], nrow(Dat),ncol(Dat),byrow=T)) /
               matrix(MSd[2, ], nrow(Dat), ncol(Dat), byrow=T), 1, sum)

## The quantiles
qnt <- quantile(Dbt, seq(KK-1)/KK)
```

The initial solution is set to the fitted means and variance matrices for these groups, and the initial marginal probabilities are set to 1/KK.

```
## The assignment to the components
cmp <- 1

## Add a point for each exceedance
for (qn in qnt)
cmp <- cmp + (Dbt > qn)

## Initialisation
init <- list()

## Within-component estimation
for (kk in seq(KK))
{
## Who is in the component
whi <- (cmp==kk)

## The estimates
init[[kk]] <- list(M=apply(Dat[whi, ], 2, mean), V=var(Dat[whi, ]),
                   p=mean(whi))
} ## End of the loop over the components (kk)

init
} ## End of function  InitMx
```

A multivariate normal density is evaluated by the function SNdns. It evaluates first the residual vectors $\mathbf{e} = \mathbf{y} - \hat{\boldsymbol{\mu}}$, and then in separate statements the log-determinant $\log\{\det(\hat{\boldsymbol{\Sigma}})\}$ and the quadratic form $\mathbf{e}^{\top}\hat{\boldsymbol{\Sigma}}^{-1}\mathbf{e}$. A failure to evaluate the log-determinant indicates that the variance matrix provided in the second element of the list-argument **Est** is not positive definite. An error could be prevented by checking the current estimates of the variance matrices at the end of each iteration.

```
SNdns <- function(DT, Est)
{
###   Evaluation of the scaled multivariate normal density

##  DT   The net dataset  (matrix)
##  Est  The estimates (the current fit for a component)

##  The number of observations
ns <- nrow(DT)

##  The residuals as a matrix
Rsd <- DT - matrix(Est[[1]], ns, ncol(DT), byrow=T)

##  The log-determinant
LDT <- log(det(Est[[2]]))

##  The quadratic form
QDF <- apply(Rsd * t(solve(Est[[2]], t(Rsd))), 1, sum)

##  The scaled density
Est[[3]] * exp(- (LDT + QDF)/2)
}  ##  End of function  SNdns
```

Another auxiliary function converts a variance matrix to a list with the standard deviations and the correlation matrix as its two elements.

```
CVcor <- function(mat)
{
###   Convert a variance matrix  mat  to
###   its vector of standard deviations and the correlation matrix

if (diff(dim(mat)) !=0) stop("The matrix has to be square.")

##  The variances
DG <- diag(mat)

if (min(DG) <= 0)
 stop("All the variances in the matrix have to be positive.")

##  The divisors for the correlations
DH <- diag(1/sqrt(DG))
dimnames(DH) <- list(colnames(mat), colnames(mat))

list(Std=sqrt(DG), CorrM=DH %*% mat %*% DH)
}  ##  end of function  CVcor
```

The function EUMmxF rearranges a list of results for fitting several models to lists that contain the estimates by their type.

```
EUMmxF <- function(lst, dgt)
{
###   Extraction and rearrangements of the results (mixture model fit)

## lst   The result of a set of mixture model fits
## dgt   The number of digits in rounding
...
list(Means=MNS, Standard.deviations=SDE, Correlations=CRS,
     Probabilities=PRB, Iterations=ITS)
} ##  End of function  EUMmxF
```

The first element of the result-list is a list of matrices of the estimates of the means, the second a list with the same structure for the standard deviations, the third with matrices in which the rows contain the minima and maxima of the correlations, the fourth with vectors of the fitted marginal probabilities and the fifth with the numbers of iterations. The code is rather lengthy and contains only simple expressions, so it is not reproduced here.

The profiles of subsamples of households defined by their fitted probabilities are plotted by the function EUMgr.

```
EUMgr <- function(dat, res, ssz, prt=0.9)
{
###   Plots of the profiles for samples of households

## dat   The dataset
## res   The mixture model fit
## ssz   The sample size for the graphs
## prt   The threshold probability

## The years
yrs <- as.numeric(colnames(dat))

## The fitted probabilities and the number of components
res <- res[[3]]
ncm <- ncol(res)

## The number of units
ndt <- nrow(dat)

if (nrow(res) != ndt)
    stop("The two arguments, dat and res, are not compatible.")

## Permute the columns according to descending marginal probability
res <- res[, sort.list(-apply(res, 2, sum))]
```

The columns of the fitted probabilities are permuted to be in the descending order of the marginal probabilities, to agree with all the other displays. Only units with fitted probabilities $\hat{r}_{i,k}$ greater than `prt` are considered for the panel for component k. These units are referred to as candidates. If a component has more than `ssz` candidates a sample of size `ssz` is selected from them using the system-defined function `sample`.

```
##  Initialisation for the candidates for the components
smp <- list()


##  The candidates for the panels
for (k in seq(ncm))
smp[[k]] <- seq(nrow(res))[res[, k] > prt]


##  Select random samples where there are more than  ssz  candidates
for (k in seq(ncm)[sapply(smp, length) > ssz])
smp[[k]] <- sample(seq(ndt)[smp[[k]]], ssz)


##  Labelling (to be the titles of the panels)
names(smp) <- paste(LETTERS[seq(ncm)], ". Component ", seq(ncm),
                "(", ncm, ")", sep="")
```

Next, the profiles of the selected households are recovered and are plotted using a loop over the samples. Within the loop, an empty shell of the plot is drawn first, with the labelling of the horizontal axis controlled by the argument `xaxt="n"` of function `plot` and the application of the function `axis`. Finally, the profiles are drawn by repeated application of the function `lines`. The profiles of the selected samples are returned by the function `EUMgr` in the form or a named list.

```
##  Initialisation for the profiles
prf <- list()


##  The profiles of the selected units
for (nm in names(smp))
{
prf[[nm]] <- dat[smp[[nm]], ]
rownames(prf[[nm]]) <- smp[[nm]]
}


##  The range of the vertical axis
rng <- range(dat)


##  The panels/plots
for (i in seq(length(smp)))
{
##  Empty shell of the plot
plot(range(yrs), rng, type="n", xaxt="n", main=names(smp)[i],
    xlab="Year",ylab="log-eHI", cex.lab=1.1,cex.axis=1.1, cex.main=1.1)


##  Labelling the x-axis
axis(1, yrs, yrs%%/2000)
```

```
## The profiles
for (h in seq(length(smp[[i]])))
lines(yrs, prf[[i]][h, ], lwd=0.4)
} ## End of the loop over the plots (i)

prf
} ## End of function EUMgr
```

The graphical summary of a mixture model fit (see Figure 7.3) is drawn by the function EUMmxG.

```
EUMmxG <- function(res, fra=0.125, ylm=c(), ttl="")
{
### Graphical summary of mixture model fit(s)

## res   The fit of the mixture model
## fra   The fraction to be used for the standard deviations
## ylm   The range for the vertical axis
## ttl   The title of the plot
```

The vectors of means and standard deviations, the correlation matrices and the fitted probabilities are extracted into separate objects using the function ExtrL as the function-argument of lapply, sapply or their composition. These expressions are omitted.

The correlations for the neighbouring years are extracted by the expressions

```
## Indices for the neighbour-correlations
dcr <- 2 + seq(0, nnm-2) * (nnm+1)

## The neighbour-correlations
CRR <- t(sapply(lapply(VCR, ExtrL, 2), ExtrE, dcr))
```

Elements of a $p \times r$ matrix can be referred to by pairs of indices or by a single index which treats the matrix as a vector of length pr, with the columns stacked to the vector from left to right. The vector dcr holds the indices of the first sub-diagonal of the variance (or correlation) matrix. The set of these indices is the same for every mixture component; nnm is the dimension of the matrix.

Each mean profile (row of MNS) is accompanied by a lower and upper profile (LW and UP) which are cut in at the midpoint between two consecutive years (indices ind) to the values MID $\pm$ CID.

```
## The upper and lower limits
LW <- MNS - fra*SDE
UP <- MNS + fra*SDE

## The cut-ins to indicate correlations
MID <- 0.5*(MNS[,-1] + MNS[, -nnm])
CID <-  fra*0.5*(SDE[,-1] + SDE[, -nnm])*CRR
```

```
##  The indices that correspond to the years and mid-years
ind <- sort.list(c(seq(nnm)*2-1, seq(nnm-1)*2))

##  The values of the x-axis for these indices
nmf <- c(nme, 0.5*(nme[-1]+nme[-nnm]))[ind]

##  The cut-in
LW <- cbind(LW, MID-CID)[, ind]
UP <- cbind(UP, MID+CID)[, ind]
```

In the plot, the mean profiles are drawn by solid lines and the upper and lower profiles by dashes with thickness lwe specified as a function of the marginal probability PRB. The vector lwe also serves as the sizes (argument of cex) of the circles drawn at the right-hand margin. The default range of the vertical axis covers all the lines. By defining the argument ylm (a vector of length 2), we can zoom in on a narrower range of values of eHI.

7.B.1   Improper Component

Mixture models with an improper component with constant density are fitted by the function EUMmx2. It is compiled by a few simple adaptations of the function EUMmx1. EUMmx2 has one additional argument, the value of the improper density:

```
EUMmx2 <- function(dat, wei, H, sta=list(), K, maxit=20, tol=6)
{
###   EM algorithm for mixture model with an improper component

##  dat     The dataset
##  wei     The vector of weights
##  H       The value of the improper component
##  sta     The initial solution (proper components)
##  K       The number of (proper components)
##  maxit   The maximum number of iterations
##  tol     The convergence criterion
```

In the initial solution the values of the improper density and its probability are stored in an additional element of the list estN:

```
sta[[K+1]] <- c(Level=H, Probability=0)
estN <- sta
```

In the E step, the values of the proper densities are evaluated with the same expressions as in the function EUMmx1, but the constant density is added to them, and the probabilities of assignment are then evaluated also for the improper component.

```
##  The improper component added in column K+1
dns <- cbind(dns, H)
```

```
##  E step -- conditional probabilities
prb <- apply(dns, 1, Prob)

##  The weighted totals of the probabilities
Prw <- prb %*% matrix(wei)
Srw <- Prw / sum(Prw)
```

After each M step, the same as in EUMmx1, the updated estimate of p_0 is added to the list estN as a separate element.

```
##  The improper component updated
estN[[K+1]] <- c(Level=H, Probability=Srw[K+1])
```

The models with four components are fitted for all the countries expect Iceland and Slovenia by a loop in which the panel data is read and the intact households extracted by the function EUreadMx.

```
for (eu in EU710[-c(12, 23)])
{
##  Read the data
Dat <- EUreadMx(eu)

warning(paste("Country", eu, "Hssz =", nrow(Dat)))

EUMmx4[[eu]] <- EUMmx1(Dat[, seq(4)], Dat[, 5], K=4, maxit=1000)
}  ##  End of the loop over the countries (eu)
```

The function EUreadMx is a simple adaptation of EUreadQ to extract the minimal dataset required for the mixture analysis (the four annual values of eHI and the weight for the last year). The models for Iceland (country 12), with two components, and Slovenia (country 23), with three components, are fitted using separate expressions. As an alternative, the argument KK in the application of EUMmx1 could be set to

```
KK = 4 - (eu == "SI") - 2*(eu == "IS")
```

Figure 7.8 is drawn by the following code:

```
##  The countries' sample sizes
EU710n <- c(765,   790,   494, 1705,   626, 703, 1753,   876,
           3513,   816, 1355,   226, 2931, 901, 1507,   743,
           1532, 1406, 1855,   711, 1605, 970, 1006, 1054)
names(EU710n) <- EU710

Figure("MmxH",
for (eu in c("AT", "CZ", "PT", "SI", "BG", "LT", "HU", "IS",
             "UK", "IT", "GR", "ES", "DK", "FI", "PL", "RO",
             "EE", "LV", "FR", "LU", "BE", "NL", "NO", "SE"))
   EUMmxG(EUMmx4[[eu]], ttl=paste(eu, " (n=", EU710n[eu],")",
     sep=""), ylm=c(5.5, 10.75)),
     7.75, c(6,4), c(2.6, 2.6, 1.2, 0.5), 0.4)
```

The countries' sample sizes are pasted in from another output. Since the vector EU710n has names, the countries in the loop (**eu**) can be listed in arbitrary order.

7.B.2 Stability of Income

Income stability is estimated by the function EUsta.

```
EUsta <- function(eu, Ulm=125, S=seq(10,20), plt=T)
{
###   Stability of income

##  eu   The country
##  Ulm  The upper limit (for truncation of eHI%)
##  S    The set of max differences
##  plt  Whether to plot the stability curves
```

The data, read by the function EUreadMx, is reorganised to the matrix of values of eHI and the vector of weights.

```
##  Data input
Dat <- EUreadMx(eu)

##  Number of units and years
ncp <- nrow(Dat)
nyr <- ncol(Dat)-1

##  The weights and the eHI profiles
wei <- Dat[, nyr+1]
Dat <- Dat[, seq(nyr)]
```

The values of eHI are converted to eHI% and truncated at the specified upper limit Ulm.

```
##  The median eHI
MDS <- apply(Dat, 2, QuantW, wei)

##  The values of eHI%
for (yr in seq(nyr))
Dat[, yr] <- 100*Dat[, yr] / MDS[yr]

##  Truncation at the lower and upper limits
Dat[Dat < 0] <- 0
Dat[Dat > Ulm] <- Ulm
```

The within-household maximum difference from one year to the next is evaluated by the expression

```
##  The maximum inter-year difference
DF <- apply(abs(apply(Dat, 1, diff)), 2, max)
```

and the (weighted) percentages of stability for the thresholds in S are evaluated
by the loop

```
Spr <- c()

##  The stability totals
for (ss in S)
Spr <- c(Spr, sum(wei[DF < ss]))

##  Labelling
names(Spr) <- S

##  The stability index in percentages
100*Spr / sum(wei)
}  ##  End of function  EUsta
```

If plt is true, then a histogram of the values of DF is drawn.

The percentages of stability are estimated for all the countries and a fine
grid of thresholds, 2.0, 2.1, ..., 20.0, by the expression

```
EUstaR <- apply(matrix(EU710), 1, EUsta, S=seq(2, 20, 0.1), plt=F)
```

The result is a 181×24 matrix labelled by the grid values $(2, 2.1, \ldots, 20)$ and
the countries' names.

A panel of Figure 7.9 is drawn by the function EUMmxU. Its sole argument
is the matrix of the stability percentages (EUstaR or its submatrix) obtained
by application of function EUsta.

```
EUMmxU <- function(res)
{
###   Plot of the stability curves
##  res   The stability matrix  (Threshold x country)

##  The stability thresholds
pts <- as.numeric(rownames(res))
npt <- length(pts)

##  Empty shell of the plot
plot(range(pts), range(res), type="n", cex.lab=0.9, cex.axis=0.9,
   xlab="Stability threshold (%)", ylab="Stability index (%)")

##  Kernel smoothing
res <- apply(res, 2, KernSq, 5)

##  The countries' curves smoothed
for (eu in colnames(res))
lines(pts, res[, eu], lwd=0.5)

##  The order of the countries' markers
srt <- sort.list(-res[1, ])
```

```
##  Placement of the markers
EUpla(res[, srt], pts, floor(npt/25)+seq(ncol(res))*3.2,
     colnames(res)[srt], cx=c(2.1, 0.5))
}  ##  End of function  EUMmxU
```

The horizontal axis of the plot drawn by EUMmxU is restricted by submitting a subset of columns of the output EUstaR of EUsta.

Figure 7.9 is generated by the expression

```
Figure("MmxU",
list(EUMmxU(EUstaR[seq(90),]), EUMmxU(EUstaR[90+seq(90),])),
     3.75, c(1,2), 0.8, 0.75)
```

The first 90 columns correspond to the values of S smaller than 11.0; this is checked by inspecting rownames(EUstaR)[89:91]. Details of kernel smoothing are given in Sections 8.4 and 8.A.1. Some fine-tuning may be required in the application of the function EUpla to place the countries' acronyms in the diagram.

7.B.3   Confusion Index

The confusion index for a pair of multivariate normally distributed components is evaluated by the function EUconM. It is an adaptation of the function EUconf for the univariate version of the index; see Section 4.A.3. The arguments of EUconM are the two components, given as lists of three elements each, comprising the vector of means, the variance matrix and the marginal probability. The two lists have to refer to identical dimensions. Note that they are not interchangeable ($r_{A|B} \neq r_{B|A}$, in the notation of Chapter 4). The sample size for the empirical evaluation of the confusion index is another (optional) argument.

```
EUconM <- function(CA, CB, nre=50000)
{
###   The confusion index for two multinormal distributions

##  CA    The component  A
##  CB    The component  B
##  nre   The number of replications

##  The variance matrix for B  --  Cholesky decomposition
SigB <- chol(CB[[2]])

##  The number of elements
nlm <- nrow(SigB)

if (nlm != length(CA[[1]]))
    stop("The two components have different dimensions.")

##  Generate a random sample from the B distribution
```

```
SMP <- matrix(CB[[1]], nre, nlm, byrow=T) +
       matrix(rnorm(nre * nlm), nre, nlm) %*% SigB

##   Evaluate the scaled densities
sdA <- SNdns(SMP, CA)
sdB <- SNdns(SMP, CB)

##  Numerical evaluation of the expectation
mean(sdA/(sdA + sdB))
}   ##   End of function  EUconM
```

A sample from the distribution of component B is generated using the Cholesky decomposition of the variance matrix of B. The values of the two densities are evaluated on the sample and the confusion index is estimated by the average fraction of the scaled density of A with respect to the total of the two scaled densities. Bear in mind that the components of a model fit are not in the same order as reported in this section, and have to be permuted.

8

Social Transfers

Poverty reduction is a key goal of the social policy of most European national governments. Apart from generating employment opportunities and promoting their generation, direct payments to qualifying individuals and households are the most important element of such policies. In this chapter, we study how and to what extent this expectation is satisfied by social transfers as a source of income. Our principal device is the comparison of poverty, as assessed by various indices, in the 'real world' (R-world), in which social transfers are regarded as one of the sources of income, and in the 'alternative world' (A-world), in which all the social transfers are disregarded (discounted). The ideal system of social transfers would eradicate all poverty. This would entail raising the value of the equivalised household income (eHI) of each household classified as poor in the alternative world (A-poor) to the poverty standard. This ideal is clearly not attainable, even if the dispensers of social transfers had the intention to achieve it. First, future income from (self-)employment and other sources is not known, and so the level of social transfers has to be based on past income. Next, the real-world threshold (R-threshold) differs from the alternative-world threshold (A-threshold) if eHI of some households is raised from below the standard to above it. Further, social transfers include some payments not related to income directly, such as child support and disability entitlements. And finally, there is no single measure of poverty that could be used by the social transfer system for all its purposes.

8.1 Capacity of Social Transfers

In this section, we define several characteristics of a national social transfer system. Having assumed that combatting poverty is its main purpose, we relate how much funding is required for it to the funds that were disbursed by the system. We identify poverty with the poverty status, so this is a comparison of the overall poverty gap with the budget of the social security system. A hypothetical country with no poverty would require no funds for social transfers, whereas a country with a lot of poverty is doubly handicapped; first, more funds are required, and second, assuming that unemployment and low income from employment are the main causes of poverty, the source of such funds, principally the income tax, is eroded. Note that old-age and survivor's benefits (pensions) are not regarded as social transfers.

The total amount of social transfers, or their average (per capita or per household), is informative only when related to the level of eHI and the poverty gap in the country. Beyond the averages, the distribution of social transfers is also of interest, whether the system tends to make generous payments to a few, or spreads the funds more evenly among the individuals or households. The *capacity* of social transfers is defined as the comparison of their total amount with the poverty gap in the A-world.

Figure 8.1 displays the plot of the annual national means of the social transfers and the annual national means of disposable income of the countries with cross-sectional data in the European Union Statistics on Income and Living Conditions (EU-SILC) 2004–2010. Each country is represented by a set of connected segments that join the coordinates for the two summaries across the years; the country's symbol is printed near its point for 2010. The means refer to households, and no equivalisation is applied. Both axes are on the multiplicative scale. The thin dashes mark the percentages indicated at the left-hand margin. The R function for drawing this diagram has provisions for zooming in on part of the plotting range. With them, the small differences among the cluster of countries that include Austria, France, Germany, the Netherlands and the UK would be suitably magnified.

The countries can be classified according to the mean disposable income by their location on the horizontal axis as the former communist countries, except for Slovenia, the Mediterranean countries (including Slovenia, but excluding France), and the remaining countries of western and northern Europe, among which Ireland, Norway and Luxembourg might be singled out as having the highest mean income. The same grouping is not appropriate for the mean social transfers. For example, the mean for Greece was lower than for Czech Republic in the recent years up to 2010.

The equi-percent lines, drawn by dashes, help us to compare the fractions of mean income that are covered by social transfers. They have been consistently the lowest in Greece, despite a steady increase in the recent years (from 2.9% in 2004 to 3.8% in 2010). The social transfers in 2010 were most generous in Ireland, both by the mean amount (8850 Euro) and the percentage of all income (21.3%), followed, according to the mean amount, by Norway, Denmark and Luxembourg. The percentage was above 15% in all seven years only in Denmark. Among the countries with lower mean disposable income, the estimated share of the social transfers in overall income is the highest in Hungary (15.0% in 2008 and 13.3% in 2010). In 2010, the estimated percentages in Lithuania and Latvia exceeded 10%, after substantial increases in the recent years. An extreme example is Bulgaria, even though we have data from only four years, since 2007. Both mean income and mean social transfers have increased steeply over the four years; in 2007, both values were by far the lowest. The changes have been much smaller in Romania.

Countries with higher mean income tend to have smaller annual changes of

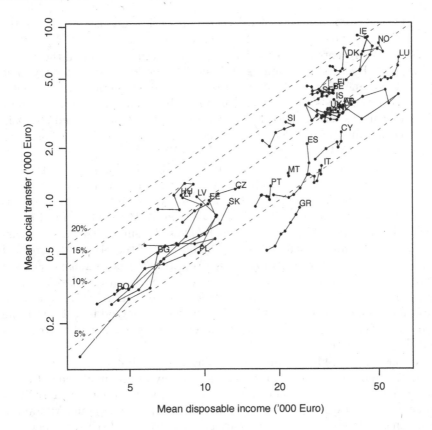

FIGURE 8.1
Annual national social transfers and disposable income in 2004–2010.
The country's symbol is printed near its point for 2010.

the mean social transfers. A notable exception is Iceland, which experienced a sudden drop in both mean income and level of social transfers between 2008 and 2009, with a recovery of the latter in 2010 accompanied by further drop in mean income. Note that the exchange rates for countries that use currencies other than the Euro may affect these comparisons.

Next we explore how the social transfers are distributed. One extreme is the uniform distribution, by which every household (or individual) receives the same amount. In the other, a select few households receive social transfers and the rest are ignored. The Lorenz curve and Gini coefficient, introduced in Section 3.3, could be applied to the social transfers, but we want to study not their distribution (of amounts received by households) but the contribution they make to the total income. Practical summaries that assess the distance from the two extremes are the percentage of households that receive no social transfers, and the percentages that receive more than given fractions of their overall income from social transfers. These summaries can be presented in the form of a function of the fraction for a country and year.

Figure 8.2 presents these functions for all the countries and all their years in EU-SILC. Each panel contains the annual functions for a country. The left-hand extreme of each function is the estimated percentage of households that receive some (positive) social transfers. This is in the range 40%–60% for most countries, but is much higher for Malta (84.6% in 2009 and 85.0% in 2010) and Ireland (78.6%–83.1% over the seven years), and much lower for Greece (21.9%–35.4%) and Spain (23.2%–36.1%).

The right-hand extreme of each function is the percentage of households that receive all their income from social transfers. This is lowest for Malta (0.15% in 2009 and 0.4% in 2010), followed by Cyprus in 2009 (0.6%), and is highest in Ireland (7.9%–15.5%) after a rise from 8.0% in 2008 and 11.5% in 2009, and Denmark (5.9%–11.1%), which followed a similar trend. The years for which data is available are listed at the bottom of the plot in the order of the percentage, with 0 printed for 2010. The estimated percentage decreased every year in Czech Republic and, with one exception, also in Belgium and Poland. Owing to the multiplicative scale for the vertical axis, the details are clearer for small percentages. By applying the odds or log-odds scale, with the extent of the vertical axis reduced, some other features can be highlighted; this is left for an exercise.

By construction, all the functions are decreasing, but they differ in their shapes. For most countries and years, they closely approximate smooth log-concave functions, but several countries have a downward hook near 100% of income (Austria, Belgium, Finland, France and Luxembourg), and Malta, the Netherlands and Norway in particular. Further, we can distinguish between countries with substantial changes across the years (Bulgaria, Estonia, Spain, Ireland, Iceland, Lithuania and Poland), and very small differences (Belgium, Denmark, Finland, Italy, Slovenia and the UK).

8.2   Impact of Social Transfers

So far, we have described the amounts of social transfers disbursed and their contribution to the overall income. To assess their impact, they have to be related to the need for them. Assuming that their sole purpose is poverty reduction, we construct the poverty rate that would apply if all the social transfers were withdrawn and compare it with the poverty rate evaluated for the income in which the social transfers are included. Thus, we consider the A-world, in which no social transfers are distributed, and the R-world, in which the social transfers are distributed. There is bound to be some poverty, as defined by the indices introduced in Chapters 1 and 3, in the A-world, but in our ideal there would be none in the R-world.

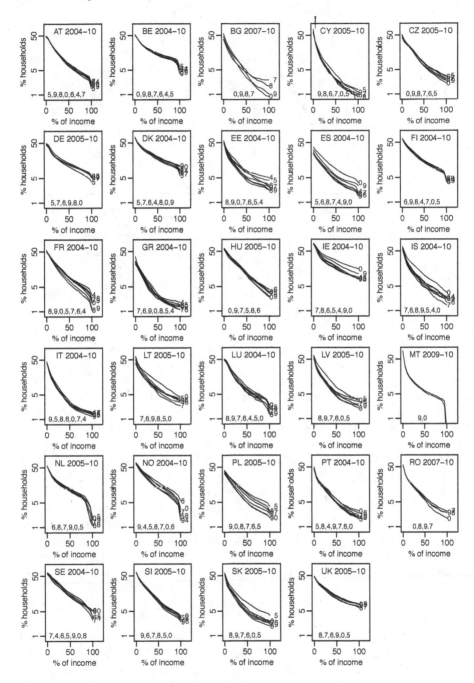

FIGURE 8.2

Estimated percentages of households that received more than a given percentage of their disposable income from social transfers; 2004–2010.

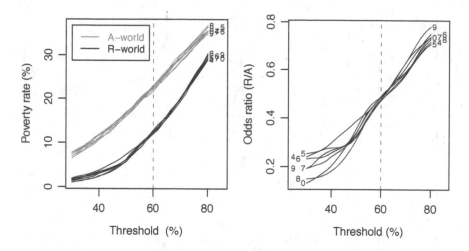

FIGURE 8.3
Estimated poverty rates in the A- and R-worlds and their odds ratios for Austria, 2004–2010.

Figure 8.3 presents the poverty rate functions in the two worlds for Austria for years 2004–2010 in its left-hand panel. The poverty rates in the A-world are much higher than in the R-world, so the social transfers perform largely as intended. Let p_A be the probability of being in A-poverty in a particular context, and p_R the probability of R-poverty in the same context. The odds for a probability p are defined as $p/(1-p)$, and the odds ratio of p_A and p_R, or of the corresponding events, is defined as $p_R/p_A \times (1-p_A)/(1-p_R)$; see Section 1.2.1 for another application of the odds ratio, and Section 6.2 for a use of the log-odds.

The estimates of the odds ratios of being in poverty, as functions of the poverty threshold τ, are plotted in the right-hand panel. Apart from a few narrow ranges, the estimates are increasing functions of τ. This indicates that the social transfers are focusing on those in acute poverty, who are poor even by the standards of a low threshold, while they are less successful in reducing poverty with respect to a higher threshold, which may be rightly regarded as a lower priority. This can be inferred also from the left-hand panel, where the R-world rates increase with τ more steeply than the A-world rates for $\tau > 50\%$.

The inter-year differences of the odds ratios are more pronounced for the small thresholds. For $\tau = 30\%$, the odds ratios for years 2004–2006 were 0.23–0.25, and have dropped to 0.15 in 2008 and 0.12 in 2010, with a reversal in 2009 (0.19). In contrast, they are in a narrow range, 0.47–0.49, for $\tau = 60\%$ throughout 2004–2010. Their range for $\tau = 80\%$ is 0.72–0.80. At around $\tau = 50\%$ the estimated odds ratio for year 2004 is the highest, whereas for $\tau > 75\%$ it is highest for 2009.

The percentage of households that receive social transfers used to be the lowest by far in Greece and Spain (22%–26% in 2004–2008), but has risen sharply since (to 36.1% in Spain and 35.4% in Greece in 2010). In contrast, it has decreased in Poland steadily from 40% in 2005 to 32% in 2010. The highest percentages are in Malta (85% in 2009 and 2010) and Ireland (78.5%–83% in 2004–2010), followed by Denmark (67%–72%) and Norway (62%–68%). In general, these percentages are not strongly related to the fractions of the social transfers in the overall or disposable income. For example, 57.2% of households in Cyprus received some social transfers in 2010, but these payments are very low on average—they constitute only 6.8% of the overall income. In Sweden, a similar percentage of households, 57.5%, received some social transfers, but they constituted 13.5% of overall income, twice as much as in Cyprus. These comparisons do not reflect the level of need for social transfers. In principle, where there is more poverty, as measured by the relative poverty gap or a similar index, the social transfers should on average be a greater fraction of the income.

The impact of social transfers can be assessed also by counting the individuals (or households) that have different poverty states in the two worlds. By success we refer to those who would be (classified as) poor in the A-world but are not poor in the R-world, and by failure to those who would not be poor in the A-world but are poor in the R-world. The latter cases do occur because some social transfers are disbursed to households with eHI above the poverty standard in the A-world, and therefore the R-world poverty standard is higher. Households that have the same poverty status in both worlds are called unequivocally poor or not poor, as appropriate.

We estimate the success and failure rates for each country and year and, as they are estimated for a continuum of poverty thresholds, we have a pair of functions (curves). In fact, these rates have their absolute and relative versions. The absolute rates are defined (and used in Chapter 6) as percentages with the entire population as the denominator (the basis). Only those who are poor in the A-world could possibly be a success, and so it is meaningful to use this subpopulation as the basis. The resulting percentage is the relative success rate. The relative failure rate is based on the subpopulation of those who would not be poor in the A-world.

In Austria in 2010, the estimated distribution of individuals by poverty status in A- and R-worlds using the threshold of 60% was as follows: 79.7% of households were classified as unequivocally not poor, 11.0% as unequivocally poor, 8.2% were poor only in A-world and 1.1% were poor only in R-world. If we measure the success of the social transfers by those whom they relieved of poverty, then 8.2% and $100 \times 8.2/(8.2 + 11.0) = 42.7\%$ are the absolute and relative rates of success. The absolute and relative rates of failure are 1.1% and 1.3%. Note that the difference of the absolute rates (success – failure) is equal to the difference of the poverty rates in A- and R-worlds, so the success and failure rates provide more detailed information. Further, the poverty rate in the R-world is equal to the total of the percentages of failures

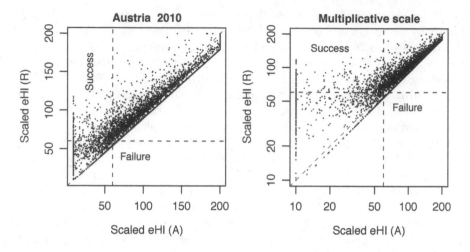

FIGURE 8.4
Scaled eHI in the A- and R-worlds; Austria 2010.

and unequivocally poor. For $\tau = 60\%$, the estimated poverty rate in 2010 is
$1.1 + 11.0 = 12.1\%$.

If we evaluate the rates of success and failure for several thresholds it is
useful to compute the scaled eHI (eHI%), defined as the ratio of eHI and the
national median eHI. For example, having scaled eHI of 55% in A-world and
65% in R-world is classified as a success with respect to the threshold $\tau = 60\%$.
For orientation, Figure 8.4 displays the plot of eHI% in the two worlds for
Austria in 2010. To increase the resolution of the plot, the values are truncated
from below at 10% and from above at 200%. The horizontal and vertical
lines drawn at the threshold of 60% split the plotting area to four parts. The
top right-hand part contains the households that are unequivocally not poor
(79.7%), and the bottom left-hand part the households that are unequivocally
poor (11.0%). The top left-hand part contains the successes (8.2%) and the
bottom right-hand part the failures (1.1%). The failures are concentrated in
the small triangle delimited by the horizontal and vertical threshold lines
(60%, drawn by dashes) and the line that corresponds to receiving no social
transfers, that is, the same value of eHI in the two worlds. Of course, a given
value of eHI corresponds to different values of scaled eHI in the A- and R-
worlds.

We can attach a meaning to the rates of success and failure only by comparing them with their counterparts for other countries. The comparisons
become more complete when the single values are replaced by functions of the
threshold. Figure 8.5 displays the estimates of these functions for the countries
in EU-SILC in 2010. Scaling, from absolute to relative rates, makes a lot of
difference for the success rates (compare the left-hand panels), but very little
for the failure rates (the right-hand panels). The reason for this is that there

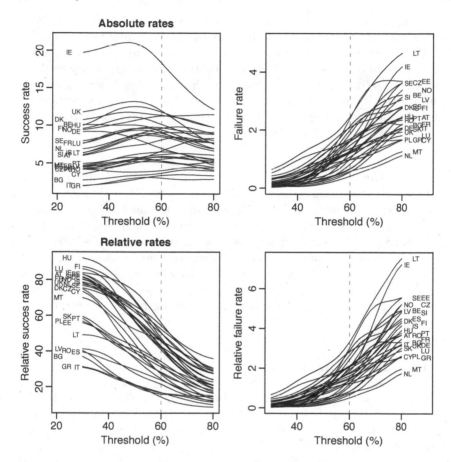

FIGURE 8.5
Absolute and relative rates of success and failure of the social transfers in the
EU-SILC countries in 2010.

are relatively few A-poor individuals who are counted in the denominator of
the expression for the relative success rate, whereas the percentage of individuals who are not A-poor is large in all settings. Therefore, the denominator of
the expression for the relative failure rate is close to unity (100%) and makes
little difference in scaling. Recall a similar discussion of Figure 6.7 in Section
6.3.

The absolute rates of success for most countries are relatively flat functions, with segments of increases and decreases, whereas the relative rates
are decreasing functions and they converge with increasing threshold. Ireland
has by far the highest absolute rate of success for nearly the entire range of
thresholds, but its relative success rate, although among the highest, is far
from exceptional. For lower thresholds, Greece and Italy have the lowest rates

of success, followed by several east European countries and Spain and Portugal. Hungary has the highest relative success rate for small thresholds. Note that if social transfers are received by many, then the relative success rate at a low threshold is bound to be high. For illustration, suppose every household in a country receives (positive) social transfers. Then the poverty rate at 0% is eliminated, so the success rate at $\tau = 0$ is 100%. A high success rate is achieved at a low threshold if most households with very low income in the A-world receive some social transfers.

The relative success rates for $\tau = 30\%$ are around 80% for all the west European and Scandinavian countries and Slovenia, Czech Republic and Cyprus. At $\tau = 60\%$, three sets of countries can be distinguished: all west European and Scandinavian countries, except for Germany and including Czech Republic and Slovenia, have relative success rates in the range 41%–56%, several east European and Mediterranean countries have relative success rates in the range 22%–34%, and the rates are lowest for Greece, Italy, Bulgaria and Romania, in the range 14%–17%. Such a classification of countries gains credence only when it applies for several years. Establishing whether this is the case is left as an exercise.

The failure rates, both absolute and relative, attain much smaller values and are mostly increasing with the threshold. Their estimates, and probably also the underlying values, are highly volatile; in the diagram they are smoothed. The estimated curves intersect a great deal, and their ranking changes even with small changes of τ. Ireland and Lithuania have the highest rates for high thresholds and the Netherlands and Malta the lowest. A crude way of summarising the failures of the social transfer system is by the increase of the median eHI brought about by the transfers. In 2010, these increases range from 5% (Greece 5.2% and Poland 5.6%) to 17% (Lithuania, 17.2%), with Ireland (31.6%) a distant outlier.

A country may have high success rates because its treasury is generous and makes social transfers quite liberally. Such a country would also have high failure rates if the transfers are made with little discrimination. Ireland is an obvious example. Its system of social transfers is least selective, with awards least strongly related to the shortfall (and poverty) in the A-world. The estimated relative rates of success and failure in Denmark at $\tau = 60\%$, 2.5% and 49.9%, respectively, are both among the highest. The social transfers in Hungary discriminate quite well at $\tau = 60\%$; its relative success rate, 50.6%, is the fourth highest, and its relative failure rate, 0.8%, the fourth lowest. Latvia and Lithuania have the highest estimated failure rates, 3.2%, but Latvia's success rate is only 22.3%, the fifth lowest, compared to Lithuania's 32.9%, close to the average.

TABLE 8.1
Linear poverty gap in the A- and R-worlds and the potential and effectiveness
of the social transfers in Austria in 2010.

World	Threshold (τ, in %)						
	30	40	50	60	70	80	90
A	214.00	357.90	558.20	851.60	1274.80	1831.50	2528.20
R	19.70	52.90	134.80	329.65	655.10	1160.30	1875.70
Potential	11.21	6.70	4.30	2.82	1.88	1.31	0.95
Effectiveness	8.10	12.71	17.64	21.75	25.83	27.97	27.19

8.3 Potential and Effectiveness

A country would have no need for social transfers if it had no poverty in the A-
world. Given that any complex system is imperfect and poverty is ubiquitous
in every country, the extent of poverty, as measured by the mean poverty
gap or a similar index, has to be a factor in the assessment of adequacy of
a social transfer system. The resources, whether meagre or generous, can be
disbursed with purpose, to deserving cases, indiscriminately, to some cases
that are deserving and others that are not, or even perversely, mainly to those
who do not need them, or to whom they were not intended originally.

The capacity of the social transfers was defined in Section 8.1 as their
average amount per individual. It can be scaled further by expressing it as the
fraction of the median eHI; see Figure 8.1. The potential of social transfers
is defined by relating the (linear) poverty gap in the A-world to the poverty
gap in the hypothetical R-world in which the social transfers are distributed
optimally—only to those who would be poor in the A-world, and only to
increase their eHI to the poverty standard. Thus, the potential is defined as
the ratio of the linear poverty gap and the capacity. By effectiveness we mean
the proximity of the actual distribution of the social transfers to the ideal.
Effectiveness is defined as the ratio of the reduction of the poverty gap and
the mean value of social transfers. Even in the ideal setting, the effectiveness
is smaller than unity when the capacity exceeds the poverty gap.

As an example, Table 8.1 presents the estimated poverty gaps in the A-
and R-worlds with respect to thresholds from 30% to 90% in Austria in 2010.
The estimated amount of social transfers per capita is 2399.40 Euro. This
amount is $2399.40/214.00 = 11.2$ times greater than the amount necessary to
eliminate the poverty gap at $\tau = 30\%$. But it is only 1.31 times greater than
the linear poverty gap at 80%, and could cover only 95% of the poverty gap
at $\tau = 90\%$. The potential of the social transfers can be characterised by the
so-called *equilibrium* threshold τ for which the potential is equal to unity. It is

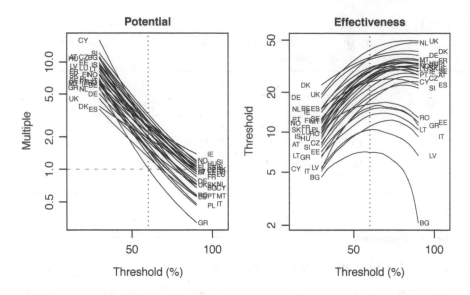

FIGURE 8.6
Potential and effectiveness of the social transfers in EU-SILC countries in
2010.

the highest threshold τ for which poverty could be eliminated by purposeful
allocation of social transfers. For Austria in 2010, it is estimated by 88.3%.

The social transfers can be credited with the reduction of the poverty gap
at $\tau = 30\%$ from 214.00 to 19.70 Euro, that is, by 194.30 Euro. However,
that has been achieved at the 'expense' of 2399.40 Euro; that is, each Euro
of social transfers is associated with the reduction by $194.30/2399.40 = 8.10$
cents. Thus, the effectiveness of the social transfers at $\tau = 30\%$ is estimated
by 8.1%. However, the largest possible effectiveness at $\tau = 30\%$ is only $100 \times$
$214.00/2399.40 = 8.9\%$. At $\tau = 60\%$, the effectiveness is estimated by 21.7%,
and its upper limit by 35.5%.

The potential and effectiveness as functions of the threshold are plotted
in Figure 8.6 for the countries in EU-SILC in 2010. The potential is decreasing with τ for every country and has very little curvature on the multiplicative scale, but the slopes vary substantially. We can summarise the functions
crudely by the ratio of their values at $\tau = 30\%$ and $\tau = 90\%$. The potential
decreases most rapidly for Cyprus (ratio equal to 23.7), followed by Greece
(18.0) and Bulgaria (16.4), and is flattest for Denmark (3.85) and Ireland
(4.5). Thirteen of the 29 countries have their estimated ratios in the range
9–12.

The equilibrium is the threshold for which the social transfers would just be
sufficient to eliminate the poverty gap, that is, every household's shortfall. It
quantifies how the needs for assistance are matched by the available resources.

For Greece, the equilibrium occurs for $\tau = 60\%$, where the potential intersects the horizontal dashed line in the left-hand panel of Figure 8.6 at the height of 1.0. The equilibrium is much higher in Poland (67.5%), Spain (68.3%) and Italy (69.2%), which are next in the ascending order. At the other extreme, the equilibrium threshold in Ireland exceeds 100%, and in Norway, Slovenia, Hungary and several other countries it exceeds $\tau = 90\%$.

The effectiveness curves in the right-hand panel are concave. They are either increasing throughout, or have a maximum within the range of thresholds (30, 90)%. In general, countries with lower effectiveness tend to have the maxima for smaller values of τ. Bulgaria is an extreme case. The effectiveness of its social transfers is very low for τ around 90%. A group of six countries (Latvia, Italy, Lithuania, Greece, Estonia and Romania) are next in the order of effectiveness for high thresholds. They do not stand out for small thresholds; for example, Romania's effectiveness exceeds that of several other countries, including Cyprus by a wide margin. The effectiveness is highest in the UK, the Netherlands, Denmark and Germany. Their curves are flat at $\tau \in (80, 90)\%$, although the curves for the UK and the Netherlands rise steeply for small values of τ.

The comparisons of effectiveness of the countries' social transfer systems at a given threshold τ are iniquitous in one important aspect. In one country, the A-poverty rate at τ may be quite small and the social transfers may deal with most of it, whereas the other country's system may have difficulties with dealing with the much higher A-poverty gap. It is not obvious how to adjust for such differences. One approach is to compare the effectiveness at the equilibrium threshold of each country, and another is to refrain from any comparisons of the systems in countries with substantially different equilibria because, in effect, the systems have incomparable tasks.

We reiterate that we have taken the liberty of assuming that poverty reduction is the sole purpose of social transfers. Since they have other functions, success, failure, effectiveness and other characteristics introduced in this section cannot be regarded as a means of uncritical assessment of the national social transfer systems. The systems fulfil several other roles and in several aspects are handicapped by rules that are based not on current circumstances, but the circumstances in the recent past. A typical system has long-term continuity. Social transfers may be paid to households with attributes that were strongly associated with poverty in the past, but are no longer today. Also, we conjecture that a typical respondent in EU-SILC may not distinguish income and social transfers when he or she receives a single (monthly) payment that combines these two components of income.

8.4   Nonparametric Regression

The conclusion of the previous section suggests that we should study the *conditional* distribution of the social transfers given the value of eHI or eHI% in the A-world. Both quantities should be scaled, so that median eHI corresponds to unity, or 100%. We use the notation A-eHI% and R-eHI% for the scaled eHI in the two worlds. In the ideal case, the conditional density of social transfers decreases for values of A-eHI% up to the threshold τ, and attains small values (or zero) for eHI% $> \tau$. Small values of the conditional expectation for small τ (e.g., at around 30%) and large values for eHI% $> \tau$% can be interpreted as iniquity.

The regression of a variable Y on another variable X is defined as the conditional expectation $\mathrm{E}(Y \mid X = x)$, regarded as a function of x. We are familiar with the ordinary regression, in which this function is linear and the conditional variance $\mathrm{var}(Y \mid X = x)$, called the residual variance, is constant. The observations are independent. This assumption maybe raises no problems for our data, even though the sampling design involves clustering, but the other two assumptions, linearity and homoscedasticity, are untenable— we need much more flexibility to describe how eHI and social transfers are related. The only assumption we adopt is that the regression, as a function, is smooth, and we estimate it by a smooth function. Such regression (and the method for its fitting) is called nonparametric.

In textbook applications, X is often a variable that can be manipulated; its values can be set at our will, by design, or are external (exogeneous) to the process we intend to study. The outcome variable Y (the amount of social transfers) is interpreted as the response to X (eHI% in the A-world). In our case, eHI and social transfers are interrelated, so we have to break this rule.

With sufficiently many observations, nonparametric regression is fitted at a value x_0 by evaluating the average of the outcomes for all the observations with $X = x_0$. In practice, with one approach we define a 'window' $(x_0 - \Delta, x_0 + \Delta)$ and estimate $\mathrm{E}(Y \mid X = x_0)$ by the average of the observations in this window. We may adjust the window by discarding some observations or arranging by some other means (e.g., by assigning weights) that the observations in the window are (approximately) balanced around x_0. It is rather undesirable that an observation in the middle of the window, with $X = x_0$, is regarded as no more important than an observation at the edge of the window, say, with $X = x_0 + 0.98\Delta$. And an observation just outside the window, with $X = x_0 + 1.01\Delta$, is disregarded altogether. We say that the window is *flat*.

Kernel smoothing resolves this apparent disparity. With it, the conditional expectation at $X = x_0$ is estimated as the weighted mean

$$\hat{y}_{x_0} = \frac{\sum\limits_{i=1}^{n} y_i \, \rho(x_i - x_0)}{\sum\limits_{i=1}^{n} \rho(x_i - x_0)}, \tag{8.1}$$

where ρ is a suitable function (Cleveland and Devlin, 1988; Wand and Jones, 1994; Simonoff, 1996). It is called the (smoothing) kernel. Meaningful choices for $\rho(z)$ are nonnegative functions that are increasing for $z < 0$, decreasing for $z > 0$, and have zero limits at both $+\infty$ and $-\infty$. The function ρ can be motivated as the strength of the influence on the regression. Thus, the influence on $\hat{y}_{x_0}$, the fit at x_0, diminishes with the distance from x_0. In fact, $\rho(z)$ may be equal to zero for very large and very small values of z, which we want to disregard altogether in the estimation of $E(Y \,|\, X = x_0)$. The kernel function does not have to be symmetric, although a rationale for asymmetry is difficult to sustain in most applications. The flat window corresponds to the indicator function: $\rho(z) = 1$ for $-\Delta < z < \Delta$ and $\rho(z) = 0$ otherwise. In general, smooth (differentiable) functions ρ are preferred.

A practical choice for the kernel ρ is the density of a centred normal distribution, $\rho(z) = \phi(z/\eta)/\eta$. The standard deviation η has to be chosen with care, to control the relative influence of the observations on $\hat{y}_{x_0}$ as a function of the distance $|x - x_0|$. For very small η, the density is highly concentrated around $z = 0$, and observations even moderately distant from x_0 have almost no influence on the fit $\hat{y}_{x_0}$. For large η, the density is flat and the influence depends on the distance $|x - x_0|$ only weakly; even observations at x distant from x_0 have a say in the fit $\hat{y}_{x_0}$. Although $\phi(z)$ is positive throughout $(-\infty, +\infty)$, its values are negligible for $|z| > 4\eta$, so η can be interpreted as a scale for the width of a window on which $\hat{y}_{x_0}$ is based. The window is not flat; it magnifies 'images' (the influence of the observations) in the middle and reduces those at its margins or edges. This description motivates an extension of Equation (8.1) with several covariates, $\mathbf{X}$, which requires a multivariate function ρ. Simple choices for ρ include products of univariate kernels for the elements of $\mathbf{X}$ and functions of the norm $\|\mathbf{x}_i - \mathbf{x}_0\|$ after suitable scaling of each dimension of $\mathbf{X}$ or some other form of standardisation. The Mahalanobis distance (Mahalanobis, 1936) is often adopted for this purpose.

Throughout, we use the version of Equation (8.1) with sampling weights w_i:

$$\hat{y}_{x_0} = \frac{\sum\limits_{i=1}^{n} w_i \, y_i \, \rho(x_i - x_0)}{\sum\limits_{i=1}^{n} w_i \, \rho(x_i - x_0)}. \tag{8.2}$$

After suitable scaling, the sum in the denominator can be interpreted as the effective number of observations (sample size) on which $\hat{y}_{x_0}$ is based. The actual number is not of any interest, but the fluctuation of its values (on the multiplicative scale) as a function of x_0 informs us about the extent of smoothing. If the joint density of X and Y is known to be smooth, then this function should be smooth to a similar extent.

The conditional probability of receiving no social transfers given the value of A-eHI% is estimated similarly, with the relevant indicator (0/1) substituted for y_i. The conditional variance is estimated by replacing y_i with $(y_i - \hat{y}_{x_0})^2$.

Panel A of Figure 8.7 presents the fits of the regressions of the scaled social transfers on A-eHI% for Austria in 2010. Normal kernels with standard deviations η set to $0.04j$ for $j = \frac{1}{2}, 1, 2, \ldots, 5$ are used. The scaling of social transfers is by dividing them by their median of eHI, pro-rated for the composition of the household. The fit for smaller η, drawn by thinner lines, has sharp peaks and troughs that are more likely to be an accident of the sample drawn than a feature of the population (or the system of social transfers). The fit for $\eta = 0.20$, drawn by the thickest line, is perhaps too smooth for the purpose of identifying some idiosyncracies of the system. The dashes mark the ideal amount of social transfers, which eliminates poverty without raising the poverty standard.

For values of A-eHI% close to zero, the regression decreases. In an application of smoothing for its value x_0 close to zero, the values of y that contribute to the summation in Equation (8.2) are all smaller than the value of y at x_0. As a result, greater smoothing leads to smaller value of the fit and a flatter curve. This drawback of the method could be addressed by using gradually smaller standard deviation (less smoothing) as we approach either limit of the curve. Such flexible smoothing is difficult to specify, and we do not pursue it here. An alternative is to draw the smoothed curves only over the interval in which the contributions made to the fit from the left and the right are not too lop-sided. That is an incentive to use smaller standard deviation η.

From the diagram we learn that the conditional expectation of social transfers declines with A-eHI%, as does the standard deviation (panel B), defined as the square root of the fitted residual variance. Panel C shows that the estimated percentage of the households that receive no social transfers increases with A-eHI%. An ideal system would supplement the income of households with zero A-eHI% up to 60% of the median eHI. Many households with very small A-eHI receive no social transfers, and the average social transfers are not sufficient for their R-eHI to reach the level of the R-poverty standard. Further, many households with A-eHI well above the standard receive some social transfers.

Panel D presents the plot of the scaled effective sample sizes, the denominators in Equation (8.2). Since the distribution of eHI is assumed to be smooth, the effective sample size should also be a smooth function of eHI. This criterion eliminates from our consideration the normal kernels with $\eta < 0.12$.

The standard deviation curves, drawn in panel B, would in an ideal system be zero throughout, because the level of social transfers would be determined

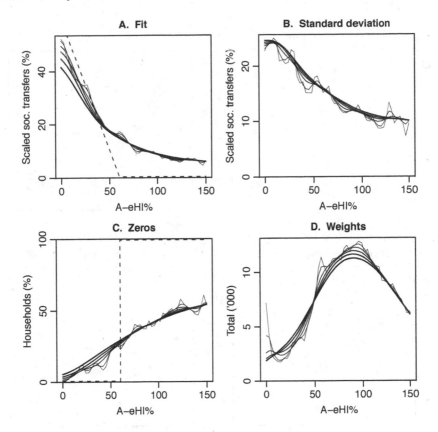

FIGURE 8.7

Nonparametric regression fits to equivalised social transfers for Austria in 2010.

The curves are the fits with the normal kernel with standard deviations 0.02, 0.04, 0.08, 0.12, 0.16 and 0.20, in ascending order of line thickness.

by A-eHI. The fraction of households receiving no social transfers would be zero for eHI% in A-world up to 60%, and unity for greater values of eHI% (dashes in panel C). This ideal is a tall order in general, and probably not the aim of any system. However, regarding it as a comparator is meaningful.

Figure 8.8 displays the fitted curves for the ten former communist countries in EU-SILC in 2010 in two columns, on the left with the standard deviation of the normal kernel $\eta = 0.12$ and on the right with $\eta = 0.16$. Each horizontal axis is for A-eHI%. The two sets of plots differ only in some details. For example, the regression fit for Slovenia is slightly above the dashed line that marks the threshold $\tau = 60\%$ for $\eta = 0.12$ and slightly below for $\eta = 0.16$. Also, the differences between Czech Republic and Lithuania are reduced by greater smoothing. The vertical axes for the standard deviation are cut off at 0.45

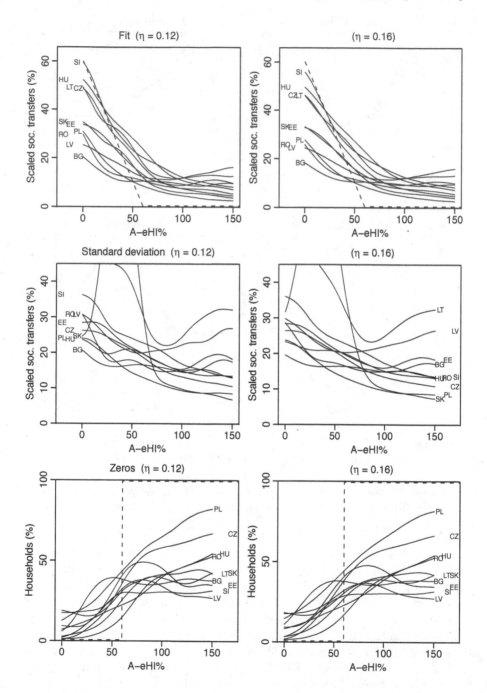

FIGURE 8.8
Nonparametric regression fits for countries in eastern Europe in 2010.

to maintain high resolution of the plots. The standard deviations for Slovakia have peaks for $\eta = 0.12$ at eHI% $= 36\%$, equal to 0.81, and for $\eta = 0.16$ at 33%, equal to 0.71. Lithuania has peaks at 0%, equal to 0.75 and 0.71 for respective values of $\eta = 0.12$ and 0.16.

The fitted regression is a decreasing function for all countries in the range $0 <$ A-eHI% < 50. For greater values of A-eHI%, the regression flattens out and increases for a few countries, notably Latvia, reaching 0.16 at eHI% $= 150$, followed by Lithuania (0.13). For small values of A-eHI%, Slovenia's regression fit is very close to the ideal of raising eHI% to the A-world threshold of 60% on average. This ideal is marked by dashes in the plots at the top. Hungary, followed by Lithuania and Czech Republic, are also quite close to this ideal, but their social transfers are less discriminating on average, not raising the income sufficiently for the poorest, and raising it above the A-world standard for those with A-eHI% > 30 (Hungary) and 40% (Lithuania and Czech Republic). By our standard of combatting poverty, Bulgaria's social transfer system is the poorest, even though it is quite generous to those who are above the A-world standard. For example, its fitted value at eHI% $= 150\%$ is 9.5%. Poland, Slovakia and Czech Republic are least generous to those with A-income at 150%; their respective fitted values are 2.7%, 4.0% and 4.7% for $\eta = 0.16$ and slightly less for $\eta = 0.12$. A majority of households with eHI% $\doteq 150\%$ in Poland and Czech Republic receive no social transfers, but this percentage for Slovakia is not exceptional (about the same as for Lithuania). The percentage is lowest for Latvia, followed by Slovenia. We can deduce from the diagram that payments of some standard amounts are made in Poland and Slovakia, because both the fitted means and standard deviations of the social transfers for households with A-eHI% $\doteq 150$ are small. Lithuania's social transfers have large standard deviations throughout the range of A-eHI%; we interpret it that the system is very selective with respect to a criterion not related to A-eHI%. The system in Slovakia is also highly selective, but the differentiation affects mainly households with low income, in the range $(10, 60)$ of A-eHI%.

Even if the *average* value of social transfers raises the value of eHI% to 60% for a group of A-poor households with a given (low) value of A-eHI%, it may leave many of these households in poverty. Therefore the standard deviation (as a curve) is an important additional summary. For example, the social transfer system in Slovenia would be exemplary if, in addition to the near-ideal fitted values (means) its fit had small standard deviations. The panels in the middle row of Figure 8.8 show that the standard deviations are large for small values of A-eHI%. Thus, some households that are extremely poor in the A-world are assisted by social transfers quite generously and others are not assisted sufficiently (or not at all). In Bulgaria, with small fitted means and small standard deviations, the social transfers raise a nontrivial fraction of households from A-poverty to R-prosperity, even though the fitted curve is well below the ideal for A-eHI% < 45.

One might expect that greater values of the fit would be accompanied by greater standard deviations. This is not the case for the studied countries.

The fitted curves of standard deviation have pronounced modes (mounds) for Slovakia and Lithuania for small values of A-eHI%, and Lithuania has another mode at A-eHI% near 150. Large standard deviation indicates that factors other than A-eHI% have a strong influence of the value of social transfers. The fit for the proportion of households receiving no social transfers would be an increasing function of eHI% in an equitable system. Poland and Czech Republic are good examples. The fitted curves for Latvia and Estonia have large mounds, and those for Bulgaria, Lithuania, Slovakia and Slovenia have smaller mounds that indicate some deviation from equitability.

Figure 8.9 compares the regression fits for the Mediterranean countries in 2009 and 2010 using the normal kernel with $\eta = 0.16$. In the left-hand panels, the relevant functions are drawn by distinct colours for the two years. The two functions for a country differ because they are independent estimates and there may be differences even between the underlying (population-related) functions. The conditional expectations of the relative social transfers have increased for all the countries and values of A-eHI%, except for a few short segments, for Cyprus around the median eHI (100) and for Malta at eHI% below 40. This is much easier to see in the plots of the ratios of the fitted conditional expectations in the top right-hand panel.

The countries' year-to-year differences of the fitted values are much smaller than the differences among the countries, especially for small values of A-eHI%. The plot of the standard errors is dominated by the large mound for Cyprus in 2009 around A-eHI% $= 100$. There is a smaller mound also for the fit in 2009 at eHI% $= 150$, which is clearer in the plot of the ratios because a similar mound is not present in 2010. There is a mound for Malta in 2010 at A-eHI% $= 120$ for both estimates of the mean and standard deviation.

We conclude that the social transfers have nearly uniformly risen from 2009 to 2010 on average throughout the Mediterranean countries, as have the percentages of recipients (see bottom panels), except for those with low eHI% in Malta and those with high eHI% in Cyprus. Compared to the east European countries, the social transfer systems of the large Mediterranean countries are much less comprehensive for those with small A-eHI%; compare the proportions of zeros in Figures 8.8 and 8.9. None of the Mediterranean countries disburses social transfers to households with low A-eHI% to the extent of Slovenia, Hungary, Lithuania or Czech Republic, and Greece and Italy fall even below the level of Bulgaria.

8.4.1　Smoothing Sequences

A sequence $y_1, y_2, \ldots, y_n$ can be treated as a collection of n pairs $(1, y_1)$, $(2, y_2)$, $\ldots$, (n, y_n), and therefore as a representation of a function $y(x)$ by its values at $x = 1, \ldots, n$. If the values of y are estimated at these points and we know that the target function y is smooth, then we may apply smoothing to the estimated sequence $y_1, y_2, \ldots, y_n$. There may be other pragmatic reasons

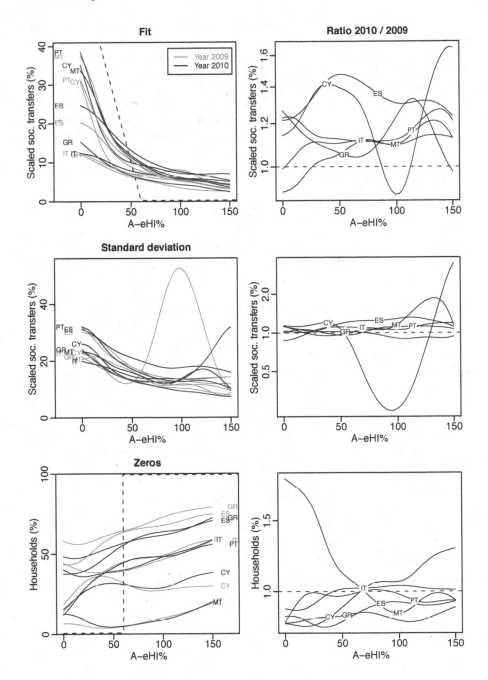

FIGURE 8.9
Nonparametric regression fits for the Mediterranean countries in 2009 and 2010.

for smoothing a sequence; the smoothed function helps us focus on its main features.

The normal (or another) smoothing kernel can be applied to a sequence by Equation (8.1), with the grid of values of x set to the integers $1, \ldots, n$. The smoothing can be applied also with noninteger values of x, but that is not necessary when the grid $1, \ldots, n$ is sufficiently fine (dense) or the sequence sufficiently long (large n). If smoothing is applied only at integer values of x and the kernel ρ in Equation (8.1) is symmetric, then the absolute differences in the argument of ρ are integers, and the function ρ can be replaced by a decreasing sequence of (nonnegative) weights $\rho_0, \rho_1, \ldots, \rho_{n-1}$. The smoothing of a sequence can then be defined as

$$
\hat{y}_k = \frac{\displaystyle\sum_{i=1}^{n} y_i \rho_{|k-i|}}{\displaystyle\sum_{i=1}^{n} \rho_{|k-i|}}.
\tag{8.3}
$$

This scheme is used throughout the book, with a normal kernel ρ. Multiplying each weight ρ_k by a positive scalar does not alter the smoothed value $\hat{y}_k$. Hence no generality is lost by setting $\rho_0 = 1$ or $\rho_0 + \rho_1 + \cdots + \rho_{n-1} = 1$.

In a special case of Equation (8.3), the smoothed value of y_k is defined as a composition (weighted mean) of y_k and its neighbours y_{k-1} and y_{k+1}. The two neighbours are assigned the same weight ρ_1, and all other elements of the sequence are ignored; $\rho_i = 0$ for $i > 2$. This method is called neighbourhood smoothing. Since the three weights add up to unity, $\rho_1 < \frac{1}{3}$ is a reasonable constraint that ensures that $\rho_0 > \rho_1$. Neighbourhood smoothing can be applied several times. Denote $\rho = \rho_1$ (assuming that $\rho_0 = 1 - 2\rho_1$). Then repeated application of neighbourhood smoothing yields

$$
\hat{y} = \left\{ (1 - 2\rho)^2 + 2\rho^2 \right\} y_k + 2\rho(1 - 2\rho)\left(y_{k-1} + y_{k+1}\right) + \rho^2\left(y_{k-2} + y_{k+2}\right)
$$

for $2 < k < n - 2$. This smoothing method involves the neighbours of the target and their neighbours (neighbours once removed). Extensions by repeated applications of Equation (8.3) are obvious. Also, different values of ρ can be applied in each repeat.

8.5   Perils of Indices

Having defined poverty indices and evaluated them for several countries and years, it might invite a policy-maker to adopt their changes in specific directions as a plan or a target for the future. The terms that we defined, potential and effectiveness, and the expressions we used, such as purposeful allocation of social transfers (to reduce poverty), have connotations of quality and responsible management of funds.

If we pursued the reduction of the poverty rate with no regard for any other obligations or priorities, the most effective strategy would focus first on single-member households with small shortfall, because their poverty status can be altered at least expense. Large households with small or no income would not be supported unless funds were left over after dealing with easier cases. Such a strategy is iniquitous because it dwells on a definition of poverty status and its summary by an index that are not universally appropriate, and policies can be devised to (further) subvert their validity. We want to point out that any index should be judged by the details of its definition, not its label (name).

If we use the poverty gap as an index, the social transfers would be allocated optimally only to those classified as poor in the A-world, and only to eliminate their shortfalls. If funds are not sufficient for this, it is immaterial whose shortfall is not eliminated, so other criteria can be applied to the allocation. If some funds are left over after eliminating the shortfalls, they should be distributed in a way that generates no new poverty.

When we define an index, we are aware of its deficiencies, which can be described as being a surrogate (a proxy) for an ideal quantity that would measure our target perfectly. So long as the deviation of the index from the target is inexplicable, the index can be used with confidence. However, a policy may be devised in such a way that the value of the index is improved (changed in the desirable direction) substantially, without a corresponding change in the value of the target it represents. The original association of the index with the target is then altered, and the value of the index as a means of assessment eroded. Such a policy is much more harmful than a policy that attends to the undesirable features of the reality without altering the value of the index.

These comments are not related to the general enterprise of estimation. That is, the error of estimation is a different and unrelated element of the imperfect inference about the target. When we declare our interest in the value of an index for a particular setting and estimate this index, the properties of the estimator can be assessed by standard approaches, such as the bootstrap. If we search through many indices for a given setting, and report the one that serves a particular cause best, the assessment of the properties of such an estimator is much more complex, because the process of selecting an index based on the values of its estimate and other sample quantities cannot be ignored and should be reflected in the assessment. In particular, a more thorough search can discover more remarkable and 'newsworthy facts'. However, their exceptionality should be downgraded by the thoroughness of the effort invested in the search. This problem can be resolved by declaring upfront (preparing a protocol for) which indices will be evaluated and how their values will be interpreted.

Suggested Reading

See Bibi and Duclos (2010) and Caminada and Goudswaard (2012) for an
analyses with a purpose similar to this chapter.

8.A   Appendix. Programming Notes

As the first inspection of the contribution of social transfers to the overall disposable income, function EUSTra0 evaluates the means and medians of the relevant variables: total household income, social transfers and their equivalised
versions. Treat this function as a template for evaluating other summaries of
these and other variables.

```
EUSTra0 <- function(Cou, Yea, dire="~/Splus/EUsilc12/DATA/", ext="dta")
{
###   Means and medians of variables related to social transfers

##  Cou   The country acronym
##  Yea   The year     (digit)
##  dire  The data directory
##  ext   The extension of the dataset

##  Data input (households)
dat <- EUread(dire, Cou, Yea, ext,
        VRS=c("hx040", "hy020", "hy022", "db090"), IH=2)

##  Truncate negative income
dat[dat[, "hy020"] < 1, "hy020"] <- 1

##  The amount of social transfers
STR <- dat[, "hy020"] - dat[, "hy022"]

##  The weights and their total (household level)
weiP <- dat[, "db090"]
Swe <- sum(weiP)

##  The values of x and y
DT <- cbind(dat[, "hy020"], STR)

##  Equivalisation
DT <- cbind(DT, DT / matrix(dat[, "hx040"], nrow(dat), 2))
colnames(DT) <- c("All", "Soc.Tra", "All.Eq", "Soc.Tra.Eq")

##  The sample summaries
cbind(Mean=t(DT) %*% matrix(weiP)/sum(weiP),
    Median=apply(DT, 2, QuantW, weiP))
```

```
}  ##  End of function  EUSTra0
```

Application to Greece in 2010, by the expression

```
GRtr10 <- round(EUSTra0("GR", 10), 2)
```

yields the table (array)

```
                Mean    Median
All          24230.02 19483.95
Soc.Tra        911.08     0.00
All.Eq        9760.67  8270.58
Soc.Tra.Eq     386.38     0.00
```

with summaries of the entire disposable income, social transfers and the equivalised versions of the two; the zero medians imply that more than half the households receive no social transfers.

The results for the countries can be organised in a list and the values of a summary compared by a plot, e.g., for two years. The values relevant for the plot are extracted by an expression list-applying (lapply or sapply) the function ExtrR or ExtrE.

The function EUSTra1 estimates the fractions of the households whose social transfers form more than a given percentage of the overall income. This function is developed from EUSTra0. The only additional argument is the vector of percentages, pcs. Apart from some 'housekeeping code' (e.g., adding zero to pcs as another element and sorting the vector in the ascending order), the essential addition are the expressions

```
##  Those who receive no soc-tra
Recs  <-  (STR==0)

##  Those who receive more than 100pc%
for (pc in pcs[-1])
Recs <- cbind(Recs, STR/dat[, "hy020"] >= pc)

##  Labelling
colnames(Recs) <- 100*pcs

##  The weighted averages and proportions
t(cbind(dat[, "hy020"], STR, Recs)) %*% matrix(weiP) /Swe
}  ##  End of function  EUSTra1
```

The first two elements of the result are the estimates of the mean disposable income and the mean value of social transfers. They are followed by the estimated fractions, labelled (named) by the corresponding percentages. The third element of the result is the estimated fraction of households that receive nonzero social transfers.

The following expressions yield these summaries for all the countries with data in years 2004–2010, and format them in a list of the annual matrices.

```
##  Initialisation
EUTra1R <- list()

##  Application to all EU-SILC countries, years 4-10
for (yr in seq(4, 10))
{
cy <- as.character(yr)

##  Evaluation
EUTra1R[[cy]] <- apply(matrix(EU410a[[cy]]), 1, EUSTra1, yr
    pcs=seq(5, 95, 5))

##  Labelling
dimnames(EUTra1R[[cy]]) <- list(c("hy020", "hy022", seq(0, 100, 5)),
    EU410a[[cy]])
}
```

Figures 8.1 and 8.2 are based on this list. The diagrams are drawn by the respective functions EUSTra1G and EUSTra2G. The functions have a common outline. First the relevant data is extracted and transformed, then the empty shell of the plot is drawn, with data- (or argument-) dependent extents of the axes, and finally the segments or curves are drawn by the function lines. The function EUloc is used for labelling the curves. The function HTable, introduced in Section 6.A.1, is used to print the years at the bottom of each plot in Figure 8.2.

The poverty gaps in the A- and R-worlds are evaluated by the function EUSTpf. It is an adaptation of the function EUpovA used in Chapter 3; see Section 3.A.3.

```
EUSTpf <- function(Cou, Yea, dire="~/Splus/EUsilc12/DATA/", ext="dta",
        STD=seq(30, 80, 10))
{
###   The poverty gaps in the A- and R-worlds

##  Cou    The country acronym
##  Yea    The year
##  dire   The data directory
##  ext    The extention of the data file
##  STD    The set of thresholds
```

The output of the function is a list with two elements: matrix of two rows (A- and R-worlds) of poverty gaps (at the individual level) for the thresholds specified by the argument STD and the mean value of the social transfers. The function EUpoef postprocesses this output to obtain the values of the potential and effectiveness:

```
EUpoef <- function(lst)
{
###   The potential and effectiveness of social transfers
##  lst    The result of EUSTpf
```

```
rbind(Potential=lst[[2]] / lst[[1]][1, ],
      Effectiveness=-100*(apply(lst[[1]], 2, diff) / lst[[2]]) ) )
} ## End of function EUpoef
```

Figure 8.6 is drawn by the function EUpoefG.

```
EUpoefG <- function(lst, shr=12, offs=c(1,5.5,10))
{
### Graphics -- potential and effectiveness curves (countries)

## lst  Output of function EUpoef (potential and effect)
## shr  The expansion of the horizontal axis
## offs The locations of the country's acronyms
```

Its first argument, lst, set to EUpf10, is a list of country-specific elements that have been preprocessed by the function EUpoef:

```
EUpf10 <- list()

## Country-by-country analysis
for (eu in EU10a)
EUpf10[[eu]] <- EUpoef(EUSTpf(eu, 10, STD=seq(30,90)))
```

8.A.1 Nonparametric Regression

The regression of the amount of social transfers received on eHI% using the normal kernel is implemented in the function EUSReg.

```
EUSReg <- function(Cou, Yea, grd=c(0,1.5), std,
       dire="~/Splus/EUsilc12/DATA/", ext="dta")
{
### Nonparametric regression by (normal) kernel smoothing

## Cou  The country acronym
## Yea  The year
## grd  The grid of values of x  for regression
## std  The standard deviation of the kernel
## dire The data directory
## ext  The extension of the datafile name
```

The regression is fitted to the amounts of social transfers scaled to the median of A-eHI. As the default, the fit is over the interval $(0, 150)$. The values of the standard deviation of the normal kernel are given in vector std. If grd is given as a vector of two elements, they are interpreted as the limits of an equidistant grid with 51 points. For example, the default corresponds to the grid 0, 3, 6, ..., 150.

After data input, the weights are established, the median eHI in A-world evaluated and the scaled social transfers calculated:

```
## The grid for evaluating the kernel estimates
if (length(grd) == 2)  grd <- seq(grd[1], grd[2], length=51)
```

```
##   Data input
dat <- EUread(dire, Cou, Yea, ext,
      VRS=c("hy020", "hy022", "hx050", "hx090", "db090"), IH=2)

##   The weights  (individuals)
weiP <- dat[, "db090"] * dat[, "hx050"]
Swe <- sum(weiP)

##   The median (A-world)
MDW <- QuantW(dat[, "hy022"]/dat[, "hx050"], weiP)

##   The amount of social transfers, equivalised, eST%
STR <- (dat[, "hy020"] - dat[, "hy022"]) / dat[, "hx050"] / MDW

##   Scale eHI
eHI <- dat[, "hy022"] / dat[, "hx050"] / MDW
eHI[eHI < 0] <- 0
```

In the computational core of the function, the following evaluations take place for each point on the grid **grd**: the weights that combine sampling and the normal kernel are evaluated for each standard deviation in **std**, and the fitted regression, fitted variance, estimated number of zeros and the total weight are evaluated. They are the output of the function, in the form of a list of four matrices.

```
##   Initialisation
TW <- FT <- VR <- ZE <- c()

for (x0 in grd)
{
W <- c()

##   The product of the sampling and kernel weights
for (st in std)
W <- cbind(W, weiP * dnorm(eHI, x0, st))

##   The within-bandwidth totals (denominators)
SW <- apply(W, 2, sum)

FT <- cbind(FT, t(W) %*% matrix(STR) / SW)
VR <- cbind(VR, t(W) %*% matrix(STR^2) / SW)
ZE <- cbind(ZE, t(W) %*% matrix(STR<=0) / SW)
TW <- cbind(TW, apply(W, 2, sum) )
}  ##   End of the loop over the gridpoints  (x0)
```

In the remainder of the function, the former three matrices are converted to percentages and suitably labelled. The totals of the weights TW are standardised.

The regression for the countries in year 2010 is fitted by the expression

```
EUreg10 <- lapply(EU10a, EUSReg, 10, std=c(2, seq(4, 20, 4)/100))
```

```
names(EUreg10) <- EU10a
```

An enterprising analyst may find good use for a function that implements kernel smoothing without the context of EUSReg. An example of such a function is Kern:

```
Kern <- function(XYW, sde, npt=101, plt=T)
{
###   Smoothing a function by normal kernel

## XYW   The values of x and y  and weights in an  n x 3 matrix
##                   n x 2  for data with no weights
## sde   The standard deviation of the kernel
## npt   The number of evaluation points for x

## The evaluation points (the grid)
pts <- seq(min(XYW[, 1]), max(XYW[, 1]), length=npt)

## If weights are not given  (equal weights)
if (ncol(XYW) == 2)  XYW <- cbind(XYW, 1)
```

The first argument of the function is the data, as an $n \times 2$ or $n \times 3$ matrix. The first two columns contain the respective vectors of values of the covariate X and the outcome variable Y, and the optional third column the sampling weights. If the latter column is not included, then equal weights are assumed. Further arguments of Kern are the standard deviation of the kernel, the number of grid-points and whether the result should be plotted. The grid-points are regularly spaced over the range of values of X. The default number of grid points, 101, is sufficient for most purposes. It can be set to a smaller value when a lot of computing is involved, as when many functions are smoothed. The computational core of the function is the evaluation of Equation (8.2) for each grid-point.

```
## Initialisation for the fitted values
FV <- c()

## Loop over the grid-points
for (pt in pts)
{
## The weights that combine sampling and smoothing
wei <- XYW[, 3] * dnorm((XYW[, 1] - pt), 0, sde)
swe <- sum(wei)

FV <- rbind(FV, c(sum(wei * XYW[, 2]) / swe, swe))
} ## End of the loop over the grid-points

## Labelling by the grid-points
rownames(FV) <- pts

## The plotting option
if (plt)
```

```
    plot(pts, FV[, 1], xlab="X", ylab="Y", type="l", lwd=0.75)

FV
} ## End of function  Kern
```

The result includes the column of totals of the weights. A simple instructional application of **Kern** is given by the following expression:

```
## A simulated application
KrnR <- Kern(cbind(seq(150), seq(150)*0.25+rnorm(150), 3+2*runif(150)),
             sde=3)

## Draw the 'true' line for comparison
abline(0, 0.25, lty=2, lwd=0.4)
```

The values of X are set to the sequence $1, 2, \ldots, 150$, and the values of Y and the weights are generated by a simple random processes. The reader working in R may relate the setting of **sde** (change it from **sde=3**) to the coarseness of the fitted regression. Consider how your choice would be influenced by information about the level of smoothness of the underlying regression. The bias of the fit at the limits for x can also be explored.

The function **KernSq** is for smoothing a sequence using a normal kernel.

```
KernSq <- function(ysq, sde)
{
### Kernel smoothing for a sequence

## ysq     The sequence (a vector)
## sde     The standard deviation of the normal kernel

## The length of the sequence
n <- length(ysq)

## The weights
smw <- dnorm(seq(n)-1, 0, sde)

## Initialisation for the smoothed values
ysm <- c()

for (k in seq(n))
{
wei <- smw[1 + abs(k - seq(n))]
ysm <- c(ysm, sum(ysq * wei) / sum(wei))
}
} ## End of function  KernSq
```

It is a reduced version of **Kern**, with X set to the sequence $1, 2, \ldots$, and these points being used as the grid. No sampling weights are assumed. This function can also be used for the exercise to study the smoothing of a randomly generated sequence.

8.A.2   Graphics for Nonparametric Regression

Figure 8.8 is generated by the function EUSReF2, applied to the sublist of EUreg10 for the selected countries.

```
EUSreFr <- list(EUSReF2(EUreg10[cos], sde=0.12, offs=2+8*seq(3),
        ylm=c(66, 45, 100), LEFT=c(T,T,F)),
        EUSReF2(EUreg10[cos], sde=0.16, titl=F, offs=2+8*seq(3),
        ylm=c(66, 45, 100), LEFT=c(T,F,F)))
```

The function EUSReF2 has the form

```
EUSReF2 <- function(lst, STD=60, sde="0.12", offs=c(), titl=T, ylm=c(),
        LEFT=T)
{
###   Graphics -- nonparametric regression for a set of countries

##  lst    The result list
##  STD    The threshold
##  sde    The smoothing st. dev.
##  titl   Whether the plot is to have a title
##  ylm    The extent of the vertical axes (maxima)
##  LEFT   Whether to mark the countries at the left-hand margin
##         (F  --  right-hand margin)
```

After some housekeeping, the data to be plotted is gathered in the list RES.

```
##  Placement of the country labels  -  a choice for each plot
if (length(LEFT) == 1)  LEFT <- rep(LEFT, 3)

##  Sort out the type of the argument
if (sde < 1)  sde <- as.character(sde)

##  Extract the three arrays
RES <- list()

for (i in seq(3))
RES[[i]] <- sapply(lapply(lst, ExtrL, i), ExtrR, sde)

##  Where to place the countries' acronyms
names(LEFT) <- names(RES) <- names(lst[[1]])[-4]
```

The grid-points are recovered from the labels of RES[[1]] (or any other element of RES), the offsets for marking the countries at the left- or right-hand margin are set and the three panels in a column of Figure 8.8 are drawn by a loop.

```
##  Shortcut for labelling
nms <- names(lst[[1]])[-4]

##  Use labels of RES for referencing
if (length(ylm)==3)  names(ylm) <- nms
```

```
##   The grid points
pts <- as.numeric(rownames(RES[[1]]))

##   Placement of the markers (left or right)
leri <- c(1, length(pts))

##   The offset
ofs <- 0
if (length(offs) > 0)    ofs <- offs[1]

##   The label of the y-axis  (later changed for plotting %Zeros)
ylb <- paste(c(rep("Scaled soc. transfers", 2), "Households"), "(%)")

##   The extents of the y-axes
mxt <- sapply(RES, max)
if (length(ylm) ==3)  mxt <- ylm

##   The loop for plotting
for (j in seq(3))
{
##   The data
rs <- RES[[j]]

ttl <- ""
if (titl)  ttl <- nms[j]

##    Switch the placement of the country labels
lr <- leri[2-LEFT[nm]]

gap <- c(2*ofs, 0)
if (lr > 1)  gap <- rev(-gap)

##   Change of the label for y axis
if (nm == "Zeros")  ylb <- "Households (%)"
```

The Greek letter η is printed by the function `substitute`. The countries'
curves are drawn by the function `lines` and features specific to a panel are
drawn in conditional expressions.

```
##   Empty shell of the plot
plot(range(pts)-gap, c(0, mxr), type="n", yaxs="i",
   main=substitute(paste(ttl, "  (", eta, " = ", sde,")", sep="")),
   xlab="eHI% (A)", ylab=ylb, cex.lab=1.2, cex.axis=1.2, cex.main=1.2)

for (eu in colnames(rs))
lines(pts, rs[, eu], lwd=0.6)

##   The dashes in the 'Fit' panel
if (j == 1)  lines(c(0, STD, 1.5*max(pts)), c(STD, rep(mxt[j]/150, 2)),
            lty=2, lwd=0.75)
```

```
## The dashes in the 'Zero' panel
else if (j == 3)
lines(c(0, STD, STD, 1.25*max(pts)),
      c(0,0,100,100)+c(1, 1, -1, -1)*mxt[j]/150, lty=2, lwd=0.75)

## Mark the countries
if (length(offs) > 0)
{
lcs <- EUloc(rs[lr,], offs)
srt <- sort.list(rs[lr,])

text(pts[lr]+lcs*(-1)^(1+(lr>1)), rs[lr,srt], colnames(rs)[srt],
      cex=0.8, adj=1-LEFT[nm])
}
}  ## End of the loop over the plots (j)
RES
}  ## End of function EUSReF2
```

To generate a postcript file of the diagram, the function **Figure** is not used because the panels are drawn in columns, with the setting

```
par(mfcol=c(3,2), mar=c(4,4,2.5,1)*0.75, mgp=0.5*c(3,1,0),lab=c(3,3,1))
```

The function **Figure** could be made more general to take care of this, but it is hardly worth the effort, since **mfcol** is used only once.

9

Causes and Effects. Education and Income

Many analyses in statistics are concerned with what would happen if the units in a population changed their values of a categorical variable (their status), while keeping all other circumstances (their background) unaltered. If we could control the process by which the values of this categorical variable are assigned, the problem would be resolved by experimental design—by assigning the values of the variable to the studied units completely at random. Numerous problems in social sciences cannot be studied by this approach because they relate to variables that cannot be controlled externally. Often they are recorded a long time after their values are set and would in normal circumstances remain unchanged. In any case, the units to be observed (individuals, households, schools, businesses, or the like) would reject such control for any purpose. Their right to do so is protected by legal code and ethical standards. Neither could we contact a unit twice; first to arrange the assignment, and then, some time later (e.g., in several years' time), to observe the outcome.

This chapter explores inferences of this nature from observational studies in which none of the variables are subject to any (experimental) control. The proposed solution uses the *potential outcomes framework* (Rubin, 1974, 2005 and 2008; Holland 1986), also known as the Rubin's causal model. It is imperfect, in that it cannot make up for all that has been 'lost' by failure (or impossibility) to implement experimental design. But in a well-defined sense it is the best that can be done with data from an observational study. The framework is applied to estimate the value of higher education, in terms of increased income, in several countries of the European Union (EU).

9.1 Background and Motivation

Our everyday lives are made interesting and challenging by numerous tasks of having to choose among a few options or alternatives, each time without being certain of the outcome that would follow. In this setting, the available options are the causes and the outcomes are the effects. Only the differences of these outcomes (or their contrasts on some other scale) matter, because we want to achieve the highest (or the lowest) values of these outcomes. We can realise only one of the options, yet we want to compare the outcomes following every one of the options. That is why the collection of the values of the outcome variable is called a set of *potential outcomes*. The task of comparing them seems to be impossible since out of any two outcomes we might want to compare at least one is not observed.

Should we encourage young people to pursue higher education because they would be rewarded for it by higher income? The unambiguous answer to this question would be obtained by having two identical individuals, one of whom would pursue studies and the other who would not. After years of adulthood we would compare their earnings. However, the comparison (the *effect*) would refer to the particular individual and could not be generalised to the relevant population. This issue could be addressed in principle by having a sample of twins (or clones) that represent well the studied population. We are likely to find that the effect is not constant, and that maybe neither is its sign (positive or negative). The effect is unit-specific—it is a variable defined (although not observed) in a population.

In real life, we do not have twins, or the twins that we might recruit for a study are not identical in all the relevant aspects, we cannot direct them to study or not to do so, and we do not wish to wait for a few decades while their adulthood and labour-market careers unfold. We postpone for later the discussion of some more subtle barriers to establishing a specified effect. They include the influence of one unit on the fortunes of another (contagion), the fact that the effect might be specific to time, so that the effect of education on (lifetime) earnings of people retiring nowadays may not be closely related to the effect defined for the 18-year-olds of today, even if the earnings were adjusted for inflation and the labour market and economy in general were not subjected to any structural changes over the decades that would elapse.

In the example motivated by an appeal to pursue higher education, one inferential target may be the value attributable to the education. This is meaningful to estimate for the subpopulation of those who have enrolled or will elect to enroll. Another target may be the value lost by those who do not enroll (but are qualified to do so). The two targets are likely to differ if the effect of education is not constant, because they refer to different subpopulations. In a setting of 'good advice', young people enroll when they have good prospects of gaining benefits from education, and abstain when they rightly believe or are correctly advised that they would not benefit. In this setting, the comparison of the outcomes of two groups would be meaningless, because the realised outcomes of those who did not study are a poor replacement for the would-be outcomes of those who did study, had they not done so.

Tertiary education is not a single treatment because there is a wide variety of universities and colleges, and they provide a variety of courses, learning environments and other elements of education, training and experience. Once committed to enroll, there is a choice between institutions and courses, and they should be regarded as separate alternative treatments. We set aside this issue altogether.

9.2 Definitions and Notation

In the example outlined in the previous section, we recognise three types of variables. The *outcome* variable is defined on an ordinal scale that corresponds to what we regard as superior, more desirable, more beneficial, or the like. The *treatment* variable indicates the cause. The target of a typical inferential problem is the effect (the change of the outcome variable) associated with a change of the treatment variable. The change is meant to be *autonomous*, not accompanied by any change of a set of variables called *background* and with no influence on the behaviour (and values of variables) of any other units (subjects).

The treatment variable, assumed to be categorical, is denoted by T. To simplify the discourse, we assume that it is binary, with values 0 and 1. The units with $T = 1$ are called *focal* and those with $T = 0$ are called *reference*. They form the respective focal and reference (treatment) groups. Note that the treatment labels 0 and 1 can be interchanged. We are interested in the changes that would take place if the focal units switched from $T = 1$ to $T = 0$. We define a separate outcome variable for each treatment, denoted by $Y^{(0)}$ and $Y^{(1)}$. These two variables are called potential outcomes. We assume that they are well-defined; that is, the value of $Y_i^{(t)}$ could, in principle, be realised for every unit i in the population by prescribing (manipulating) the value of T for the unit. The *realised outcome* variable is denoted by Y. It is a mixture (a combination) of the potential outcomes,

$$Y = (1 - T)Y^{(0)} + TY^{(1)}. \tag{9.1}$$

We can write $Y = Y^{(T)}$ or $Y_i = Y_i^{(T_i)}$. That is, the value of Y for unit i is established by selecting from the values $Y^{(0)}$ and $Y^{(1)}$, depending on the treatment T applied to the unit.

The *unit-level effect* of treatment T on the outcome Y is defined as the difference $\Delta Y_i = Y_i^{(1)} - Y_i^{(0)}$ for unit i. We make no assumptions about this variable; it is not assumed to be constant, nor to have a fixed sign (positive or negative), nor to be related to any other variables. Instead of the difference in the definition of ΔY we can use another contrast, such as the ratio or, indeed, the difference after a monotone transformation applied to both $Y^{(0)}$ and $Y^{(1)}$. Sometimes, the contrast can be back-transformed to the original scale. The *average effect* is defined as the mean of ΔY_i in a specified subpopulation (or subset) of units. We assume throughout that the target is the average effect for the focal group, although the averaging could be applied after a monotone transformation. Also, median and other measures of location could be contrasted in place of the mean.

We can define the potential (treatment-related) versions of any variable. A variable is called *background* if its two potential versions coincide for every unit. Examples of background variables are units' attributes that have been defined prior to contemplating the selection of a treatment. Thus, sex is a background variable for the education as the treatment, if it is unalterable after birth. We assume that each background variable is either observed completely, for every unit, or not observed at all. In summary, we assume that the values of the outcome and background variables are fixed, and the values of the treatment variable *could be* altered (or could have been, at least in principle). The populations we study are large, so we can refer to (joint) distributions of the outcome and background variables, and assume that they are continuous when they attain many ordinal values.

9.2.1   Treatment-Assignment Mechanism

The treatment-assignment mechanism is defined as the conditional distribution of T given the potential outcomes and the observed background variables,

$$\left(T \mid Y^{(0)}, Y^{(1)}, \mathbf{X} \right).$$

A treatment-assignment mechanism is called *unconfounded* if the conditioning on the potential outcome variables is redundant, when

$$\left(T \mid Y^{(0)}, Y^{(1)}, \mathbf{X} \right) = (T \mid \mathbf{X}).$$

(This is an identity of conditional distributions.) This is equivalent to conditional independence of T and the potential outcomes given the background variables $\mathbf{X}$. A simple example of an unconfounded mechanism arises when T is independent of both the potential outcomes and the background, when $P(T = 1 \mid Y^{(0)}, Y^{(1)}, \mathbf{X}) = P(T = 1) = p, 0 < p < 1$. This mechanism can be implemented by *randomisation*, assigning each unit to treatment T according to the realisation of a random draw from the binary distribution with probability p. The draws are mutually independent. It is applied in experiments. In experimental design, the two treatment groups have identical distributions of the background variables. More precisely, the frequencies of the background variables in the two groups differ only due to chance. Averaged over many replications of the design, the background frequencies would differ across the groups only slightly.

With randomisation, any information about the background is redundant and not useful in a comparison of the two treatment groups. In practice, even if experimental design is planned, some unwanted departures from it may occur, such as missing values and imperfect application and switches of the treatment (using treatment $1 - t$ despite having been assigned t). Then the background variables become instrumental in 'repairing' the deficiencies that arose in implementing the design, so recording their values is useful.

Establishing that a treatment is unconfounded is not trivial. Treatments are confounded in many situations because the treated units possess (not necessarily perfect) information about the potential outcomes (in addition to the background $\mathbf{X}$) and have the privilege of choosing the treatment they believe will yield the better outcome. An obvious example of a confounded treatment-assignment mechanism arises when the units know their values of $Y^{(0)}$ and $Y^{(1)}$, and they choose $T = 1$ when $Y^{(1)} > Y^{(0)}$ and $T = 0$ otherwise, and the values of $Y^{(0)}$ and $Y^{(1)}$ could not be predicted perfectly from $\mathbf{X}$.

The key property of an unconfounded treatment-assignment mechanism is that the contrast of the average outcomes within groups matched on the values of $\mathbf{X}$ (an $\mathbf{X}$ group) is an unbiased estimator of the average treatment effect in the $\mathbf{X}$ group. Let n_0 and n_1 be the numbers of units subjected to the respective treatments 0 and 1. Suppose they and $n = n_0 + n_1$ are constant; that is, every realisation of the treatment-assignment mechanism (a treatment assignment) is such that $T_1 + \cdots + T_n = n_1$. The contrast of the average outcomes is

$$\Delta Y = \frac{1}{n_1} \sum_{i=1}^{n} T_i Y_i - \frac{1}{n_0} \sum_{i=1}^{n} (1 - T_i) Y_i \,, \tag{9.2}$$

where Y is defined by Equation (9.1). This is a biased estimator of the average treatment effect (in a treatment group), because it is a contrast of two different variables ($Y^{(0)}$ and $Y^{(1)}$) on two different subsets of units (the two treatment groups). In any meaningful comparison, either two populations (or samples) are compared on the same (potential outcome) variable, or two outcome variables are compared within a population.

We express the conditional independence of the potential outcomes and the treatment symbolically as

$$f(T \,|\, \mathbf{X}, \mathbf{Y}) = f(T \,|\, \mathbf{X}) \,, \tag{9.3}$$

where $\mathbf{Y} = (Y^{(0)}, Y^{(1)})$ and f denotes the (conditional) density or probability of the random vector given in its argument, with the conditioning behind the vertical bar '$|$'. Thus, a treatment being unconfounded corresponds to the identity

$$\frac{f(T, \mathbf{X}, \mathbf{Y})}{f(\mathbf{X}, \mathbf{Y})} = \frac{f(T, \mathbf{X})}{f(\mathbf{X})} \,.$$

For the conditional distribution of $\mathbf{Y}$ given T and $\mathbf{X}$, we have the identity

$$\frac{f(\mathbf{Y} \,|\, T, \mathbf{X}) \, f(T, \mathbf{X})}{f(\mathbf{X}, \mathbf{Y})} = \frac{f(T, \mathbf{X})}{f(\mathbf{X})} \,,$$

which, assuming that $f(T, \mathbf{X}) > 0$ for $\mathbf{X}$ that are plausible, implies that

$$f(\mathbf{Y} \,|\, T, \mathbf{X}) = f(\mathbf{Y} \,|\, \mathbf{X}) \,, \tag{9.4}$$

so long as $f(\mathbf{X}, \mathbf{Y}) > 0$. The derivation is redundant if we take for granted that conditional independence is a symmetric property. Then the conditional independence of T and $\mathbf{Y}$ in Equation (9.3) is equivalent to the same property in Equation (9.4).

The expectation of the contrast ΔY in Equation (9.2) for an unconfounded treatment-assignment mechanism is

$$
\begin{aligned}
& \mathrm{E}\left\{Y^{(1)}\,\middle|\,T=1\right\} - \mathrm{E}\left\{Y^{(0)}\,\middle|\,T=0\right\} \\
=\ & \mathrm{E}_{\mathbf{X}}\left[\mathrm{E}\left\{Y^{(1)}\,\middle|\,T=1,\mathbf{X}\right\} - \mathrm{E}\left\{Y^{(0)}\,\middle|\,T=0,\mathbf{X}\right\}\right] \\
=\ & \mathrm{E}_{\mathbf{X}}\left[\mathrm{E}\left\{Y^{(1)}\,\middle|\,\mathbf{X}\right\} - \mathrm{E}\left\{Y^{(0)}\,\middle|\,\mathbf{X}\right\}\right] \\
=\ & \mathrm{E}\left(Y^{(1)}\right) - \mathrm{E}\left(Y^{(0)}\right)\ ;
\end{aligned}
$$

in brief, the conditioning on the treatment T can be ignored. In the second line, we introduced $\mathbf{X}$ and averaging over its values (the outer expectation $\mathrm{E}_{\mathbf{X}}$), in the third line we dropped T in the conditions, and in the ultimate line we applied the identity for conditional expectation.

In an observational study, such as a survey in the European Union Statistics on Income and Living Conditions (EU-SILC), we cannot assume that the treatment-assignment mechanism is unconfounded for any treatment, outcome and background variables. Unconfoundedness can be related to conditional independence. Just as independence is a special (limiting) case of dependence, with a wide range of types and strengths of the latter, so unconfoundedness is a special case of confoundedness which encompasses a much wider variety of treatment-assignment mechanisms.

If the outcome and treatment variables are conditionally independent for a given set of background variables $\mathbf{X}$, then adding further background variables to $\mathbf{X}$ will not undermine the conditional independence (unconfoundedness). This is an incentive to collect a rich set of background variables in any observational study of a treatment effect. No such set can be confirmed as sufficient, but its greater extent and variety make the assumption more plausible.

9.3 Missing-Data Perspective

If the potential outcome variables were observed completely the analysis would be straightforward, comparing their means. This provides a link to the methods for incomplete data. For n units, the $n \times 2$ matrix of values of the two potential outcomes is the complete dataset, and the values that were realised form the incomplete dataset. The missing data is its complement in the complete dataset. It comprises the values of $Y^{(1)}$ for units with $T = 0$ and the values of $Y^{(0)}$ for units with $T = 1$. In multiple imputation, we define several completions (completed datasets) in a manner that reflects our uncertainty about the missing values. The completions are based on a model that relates the missing values to the observed (recorded) values. Unconfoundedness (a sufficiently rich set of background variables) substantially simplifies the specification of such a model. The analysis intended for the complete data, called the complete-data analysis, is particularly simple—the difference of the means of the variables $Y^{(1)}$ and $Y^{(0)}$ in the focal group. It is also easy to apply on each completed dataset. The result of each completed-data analysis is a plausible value (an estimate) of the result of the complete-data analysis. The variation of these results reflects the uncertainty due to the missing values. The complete-data analysis entails no uncertainty when it is concerned with a fixed focal group, because the values of the potential outcomes are fixed.

If we had a single background variable X and it was categorical, the obvious method would impute for a missing value $Y^{(0)}$ (in treatment group 1) the value of $Y^{(0)}$ of a unit in treatment group 0 that is in the same category of X. This method is called *hot deck*. A unit i with background $X_i = x$ in treatment group t is associated with a set of *donors* for the missing value $Y_i^{(1-t)}$. The donors are the units with $X = x$ (matched on X with i) in the opposing treatment group. If we imputed the mean or a similar summary of the values of $Y^{(1-t)}$ of these units, we would end up with the problems of single (deterministic) imputation discussed in Section 6.6. We draw a donor at random and use its value of $Y^{(1-t)}$ as the imputed value of $Y_i^{(1-t_i)}$. It is a plausible value of $Y_i^{(1-t_i)}$. See Figure 9.1 for an illustration for a variable X with three categories, 1, 2 and 3. Black discs indicate the recorded values ($Y^{(t)}$ for treatment group $t = 0, 1$) and white circles the 'missing' values ($Y^{(1-t)}$ for treatment group $t = 0, 1$). In the diagram, the third unit in treatment group 0 and X-group 1 is (drawn at random to be) the donor of its value $Y^{(0)}$ to the seventh unit in treatment group 1 and the same X-group 1. In the other example, the fifth unit in the group ($T = 1, X = 3$) is the donor of $Y^{(1)}$ to the first unit in the group ($T = 0, X = 3$). For estimating the average treatment effect for the group $T = 1$, we do not need the latter imputation, but it may be useful for other inferences.

Hot deck is problematic when the group of donors is small because it does not represent well the variety of possible values of $Y^{(1-t)}$ that would be realised in replications of the treatment-assignment mechanism. The imputation, or donorship, is without replacement—a unit can be a donor at most once. When in an X-group there are more focal units than reference, some units will end up without a donor. In that case, we reduce the analysis to the units that received a donated value. In Figure 9.1, such an analysis would be reduced to $6+6+4 = 18$ focal units. Their count is fixed across replications, but different units may be involved in the replicate samples.

When we have more than one categorical background variable, a single variable can be constructed from them by defining a separate category of a new variable for each configuration of the values of the original variables. This is useful only when we have only a few variables or, more precisely, the product of the numbers of their categories is small. Hot deck is not practical with many categories, when several categories contain only a few (or no) donor

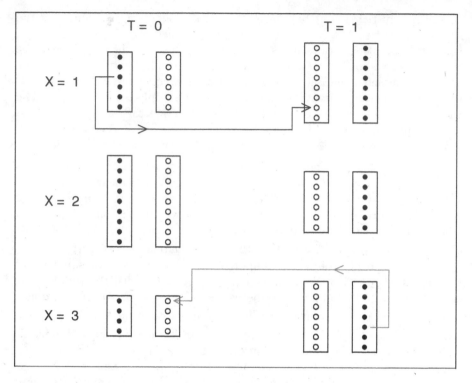

FIGURE 9.1
Hot deck for potential outcomes.
Arrows head from donors to recipients.

units each. However, the sparse categories can be aggregated or attached to categories that contain many units. A continuous variable X can be coarsened to a categorical one, making hot deck feasible, by defining a set of cutpoints c_k, $k = 1, \ldots, K - 1$, and assigning X_i to category 1 if $X_i < c_1$, to category 2 if $c_1 \leq X_i < c_2$, $\ldots$, and to category K if $X_i > c_{K-1}$.

A random draw from the donor group is a draw from a multinomial distribution. If the values of Y are categorical it may be more appropriate to consider the underlying (multinomial) probabilities which are estimated by their realised counterparts. Therefore the set of estimated probabilities should be replaced by a *plausible* set of probabilities, drawn from their estimated sampling distribution. If the values of Y are continuous, the underlying parameters for the donor group should be estimated and a draw made from a plausible distribution. Each set of imputed values is based on a different (independently drawn) plausible distribution.

9.4 Propensity and Matched Pairs

Usually there are several background variables, and hot deck or its adaptation are not feasible. This problem is relieved by a fundamental result (Rosenbaum and Rubin, 1983), which states that the best match on the background variables can also be obtained by matching on the *propensity* score. The propensity is defined as the conditional probability of receiving the focal treatment given the values of the background variables. This function can be approximated (modelled) by logistic regression of the treatment on the background variables, and the approximation estimated. Any monotone transformation of the propensity scores is equally useful because a match (equality) of values is maintained by such a transformation.

The propensity score, defined on a scale, is a continuous variable, and therefore we would usually find very few or no exact matches. The scores are therefore coarsened, classified into groups (ranges of values, or bins), and each range defines a donor group. It is also referred to as a propensity group. A donor can be used at most once (without replacement), and so some recipients may end up without a donor. The analysis is then reduced to the recipients that have donors. Having formed a set of matched (recipient-donor) pairs, we regard them as a completed dataset and apply the complete-data analysis on it. The processes of matching and complete-data analysis are replicated several times. The variation of these replicate results is a component of the overall uncertainty (variation associated with estimation). If inferences are desired for a fixed set of units, then this is the sole source of variation, because then the complete-data analysis is deterministic. In general, the two sources of variation, when both are nontrivial, can be regarded as independent.

Theory is short on recommending any particular way of forming propensity groups or constructing the model for propensities, other than preferring more complex models because of the imperative of attaining an unconfounded treatment-assignment mechanism. In particular, interactions and powers of continuous variables can be included in such models, especially when the focal group comprises so many units that little efficiency is lost by including a few possibly redundant covariates in the propensity model. The purpose of the propensity model is not a good fit, but a balance of the two groups on all the background variables, as it would be attained (approximately) in an experiment. The balance can be checked straightforwardly, by evaluating the contrasts of the proportions of each category of a discrete variable and by comparing the means and standard deviations of continuous variables across the treatment groups.

The propensity score can only be estimated, but the estimated scores can be used as if they were genuine, ignoring their sampling variation (Rubin and Thomas, 1996). For any given model, they entail no uncertainty because the set of (focal) units is fixed, as are their values of the background variables.

Forming matched pairs (data completion) entails some randomness. It reflects the uncertainty about the values we have declared as missing.

9.4.1   Regression as an Alternative

The arguments presented in this section are intended to sway any doubting reader toward using propensity score matching for comparing two groups. Regression of the outcome Y on the treatment and the background variables is an established alternative to the potential outcomes framework. Unlike the potential outcomes framework, regression discards no observations, so it may at first appear to be more efficient. However, the selection of the units into the matched-pairs analysis is purposeful.

The outcome variable may have a distribution other than normal, for which generalised least squares could be applied. A suitable distribution and link function have to be identified. That is a nontrivial matter in many settings. Such concerns do not arise with potential outcomes.

Model uncertainty is addressed in the regression by comparing the fits of alternative models. If a model used for estimating a target is set prior to data inspection, unconditionally or a priori, the resulting estimator is called *single-model based.* In every replication of data generation and estimation, the same estimator would be applied. If a model is selected by a data-dependent procedure, such as an information criterion or an hypothesis test, the estimator based on the selected model is called *selected-model based.* Such an estimator is a *mixture* of the candidate single-model based estimators. The model-selection process is usually ignored in the subsequent inferential statements. That is, if model $\mathcal{M}$ is selected (but in a replication another model might be), then we claim that the estimator has the same properties as if $\mathcal{M}$ were selected a priori. This is clearly problematic. Putting this right is difficult because establishing the properties of the selected-model based estimators, which are mixtures of single-model based estimators, is an unsolved problem. Maximising the probability of selecting an appropriate model is a dubious strategy because the consequences of the errors made in its application are ignored. For more discussion of this issue, see Longford (2012b).

When no treatment-by-background interactions are used as covariates, regression assumes a universal (constant) treatment effect. Even with some interactions included in the model, the assumption is a that the treatment effects follow a particular pattern. With matched pairs, the unit-level treatment effects are not constrained to any pattern.

Model selection takes place also in propensity modelling. However, this does not involve the outcome variable (log-eHI), and the estimation of the average treatment effect is not conditioned on the correctness of the selected model, but on the equality (proximity) of the distributions of the covariates in the matched pairs. Therefore the related model uncertainty does not affect the sampling distribution of the estimator of the average treatment effect. Rubin (2008) interpets propensity modelling or, more generally, the process of forming matched pairs, as the selection of a design, because its purpose is to configure the values of the background variables, **X**, so as to best serve the inferential agenda. By not involving the outcome variable in this process we introduce no selection bias (confounding). In contrast, the selection bias in model selection for regression is ignored, even though it is completely out of control. For example, models that failed to win the model-selection contest by a narrow margin are usually ignored. When they are not ignored, as in model averaging (Bayes factors; Kass and Raftery 1995, and Hoeting *et al.*, 1999), their properties are considered without conditioning on the outcome of the selection process.

Influential observations are a distinct weakness of the regression model. For illustration, suppose there is a single continuous background variable X and its distributions within the treatment groups differ. Say, X has smaller mean in the reference group than in the focal group. Then units in the reference group with the smallest values of X and units in the focal group with the largest values of X have a strong influence on the estimate of the average treatment effect, yet they are least relevant for the comparison we want to make because such values are rare in the opposing group. Influential observations may distort the fit of the propensity model, but the check on the balance is an effective gate-keeper for forming a suitable set of matched pairs.

Figure 9.2 illustrates this point. The values of X are collected at the bottom (gray colour) and top (black) for the respective reference and focal groups, with their means indicated by vertical segments both at the top and bottom, to highlight the difference in their distributions. A small amount of vertical noise is added to the points to better indicate the density of X. There are several obvious influential observations (outlying values of X), marked by circles. If they were removed, the regressions within the treatment groups would be altered substantially. The choice of the value of X at which the difference of the two regressions would be evaluated is both crucial, because the two slopes differ, and unclear, because the two groups have different distributions of X.

If the supports of the two distributions did not overlap the appropriate conclusion would be that the two groups cannot be compared, unless some very strong assumptions are made about the conditional distributions $(Y|T, X)$ for configurations of (T, X) that do not occur in the data. When the two distributions partly overlap, the analysis should focus on the units in the overlap. In this context, the influential observations have a potential to do much more harm then good, and should therefore be discarded. In any case, they are least relevant to the comparisons of interest. Such observations are excluded as a matter of course by propensity matching.

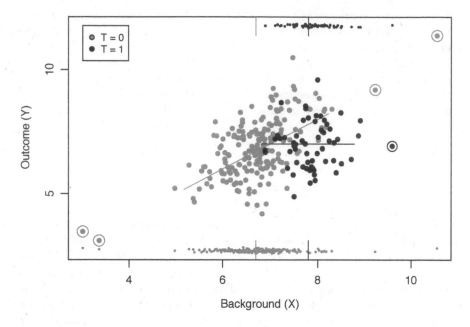

FIGURE 9.2
Influential values of a background variable.
The points (x, y) are printed in gray for the reference group and in black for
the focal group. The influential observations are marked by circles.

9.5   Application

We estimate the effect of postsecondary education on income in Austria from
the cross-sectional survey in 2011. We consider the income aggregated at the
household level, and so we have to reconcile it with education which is defined for individuals. We select a member of the household whose educational
level will represent the household. Priority in this selection is given to those
in full-time employment, followed by being under the age of 65. If these criteria do not identify a single individual, then the oldest of the candidates is
selected. The outcome variable is the log-transformation of the equivalised
household income, log-eHI, defined by the transformation $\log(1 + x)$ of the
variable "hy020" after its truncation at zero.

As the focal (treatment) group, we declare households whose representatives are in categories 4 or 5 of the variable "pe040" (highest level attained on the scale of the International Standard Classification of Education;
4: postsecondary nontertiary education; 5: tertiary education). In the cross-
sectional survey for Austria in 2011, there are 1838 such households (29.7%);
the reference group comprises 4349 households.

The estimated means of log-eHI within the treatment groups are 10.200 and 10.536; their difference is $100 \times \exp(0.336) - 100 = 40.0\%$ in favour of the focal group. The standard error of the difference is about 2.1%. The medians of the two groups are 10.287 and 10.640, leading to essentially the same conclusion.

For the propensity analysis, we use the background variables and their transformations and interactions listed in Table 9.1. Their choice is discussed below by assessing the resulting balance of the sets of matched pairs. Note that the sampling weight is one of the covariates. The propensity scores are the fitted probabilities in the logistic regression of the treatment T on these variables $\mathbf{X}$; $p(\mathbf{x}) = P(T = 1 \mid \mathbf{X} = \mathbf{x})$. The model fit is displayed in Table 9.2. The traditional discussion of significance of the regression parameters is irrelevant in our context because a balance on $\mathbf{X}$ of the two subgroups defined by the matched pairs is our sole objective. The propensity scores, that is, the fitted probabilities $\hat{p}(\mathbf{x})$ for the units in the sample, are plotted in Figure 9.3.

A majority of the households in treatment group 1 (the focal group) have propensities higher than the average, whereas many households in treatment group 0 (the reference group) have propensities much smaller. The propensities are grouped into 50 categories (propensity groups), defined by the cutpoints equal to the percentiles 2, 4, ..., 98, so that each category has approximately the same number of units (123 or 124). The split to focal and reference units varies across these groups; see Table 9.3 and Figure 9.4. In the groups with the lowest propensity scores the reference group is in a substantial majority (e.g., 110 vs. 14 in propensity group 1), and the representation in the highest propensity groups is close to even (e.g., 59 vs. 65 households in propensity group 48). In fact, the focal group is in a majority only in propensity groups 44, 48 and 50, by a total of $10 + 6 + 28 = 44$ households. Thus, a set of 44 focal households (out of 1838, 2.4%) will not have a donor from the reference group and will not be matched in a replication. Their count is constant, but the sets differ across replications. We present the information about the propensity groups in both tabular and graphical forms to show that the latter requires less effort to digest.

We construct 50 sets of matched pairs by hot deck. In the realised sets, the means of log-eHI are in the range $10.309 - 10.347$ for the replicate reference groups of households and $10.570 - 10.577$ for the replicate focal groups in the matched pairs. The range is so narrow for the focal group because the contribution to its mean varies across the replications only for the three propensity groups, 44, 48 and 50, in which a subset of $57 + 59 + 64 = 164$ focal households is matched and the remaining 44 households are not matched. The contrasts of the means for the matched pairs are in the range $0.226 - 0.266$; their mean is 0.243 and standard deviation is 0.010. We conclude that there is strong evidence of an income-related advantage associated with higher education, but the advantage is not as great as the difference of the subpopulation means, which we estimated by $10.536 - 10.200 = 0.336$. The matched-pairs estimate corresponds to $100 \times \{\exp(0.243) - 1\} = 27.5\%$ advantage in the focal group.

TABLE 9.1
Background variables in the propensity model.

Variable	EU-SILC Code	Recoding Transformation	Definition/Notes
Household type	hx060		Categories 5, 6, ..., 13
Sex of RP	pb150		1—Man; 2—Woman
Age of RP	pb140	$2011 - X$	Age (years) derived from the year of birth
Weight	pb040	$W/1000$	The sampling weight
General health	ph010	$(3, 4, 5) \to 3$	1—very good; 2—good; 3—fair to very bad
Economic status	p1030	$(2, 3, 4) \to 3$ $(6, 7, 8, 9) \to 3$	1—employed full-time; 5—retired; 3—other
Change of job in the last 12 months	p1160		1—Yes; 2—No
Interactions and transformations			
Age² of RP			Age squared
Sex.GHealth			Sex × General health
Sex.EconSt			Sex × Economic status
EconSt.GHealth			Economic status × General health
Sex.Age			Sex × Age
Sex.ChJob			Sex × Change of job
GHealth.Age			GHealth × Age (two variables)
EconSt.Age			Econ. status × Age (two variables)

Notes: RP: person representing the household.

Note that the estimated standard deviation of the replicate contrasts, 0.010, is the standard error of the contrast for the fixed set of 1838 focal households, with a compromise related to 44 unmatched households in a replication. A replication of the survey would yield a different set of focal (and reference) households, and therefore the matched-pairs analysis yields only one component of the overall (sampling) variation.

The sampling-design component of variation is estimated from the variation of log-eHI in the focal group. The standard error of the mean of log-eHI for the focal group that is related to the sampling process is 0.018. Therefore the standard error of the contrast of the two (educational 'treatment') groups is at most $0.018 \times \sqrt{2} = 0.025$, and by combining with the uncertainty of the matching process we obtain the overall standard error of $\sqrt{0.010^2 + 0.025^2} = 0.027$. It is a small fraction of the estimated mean treatment effect, 0.243, so we

TABLE 9.2
Propensity model fit (logistic regression); Austria 2011.

Covariate	Estimate	St. error	Covariate	Estimate	St. error
Intercept	−1.679	0.359	Econ.St.-5	−0.423	0.831
HHtype-6	−0.128	0.091	ChangeJob-2	−0.267	0.150
HHtype-7	−0.072	0.112	Age^2	−0.057	0.017
HHtype-8	−0.320	0.137	Sex.GHealth-2	0.127	0.143
HHtype-9	−0.115	0.154	Sex.GHealth-3	0.359	0.176
HHtype-10	0.086	0.116	Sex.EconSt.-3	−0.891	0.147
HHtype-11	0.223	0.114	Sex.EconSt.-5	−0.423	0.415
HHtype-12	0.049	0.161	EcoSt.GHlth-1	−0.032	0.177
HHtype-13	−0.631	0.172	Sex.Age	−0.022	0.005
Sex-2	0.996	0.304	Sex.ChJob-2	0.253	0.257
Age	0.067	0.014	GHealth2.Age	0.003	0.005
Weight/1000	−0.162	0.123	GHealth3.Age	0.004	0.007
GenHealth-2	−0.639	0.247	EconSt3.Age	−0.014	0.007
GenHealth-3	−1.397	0.348	EconSt5.Age	0.020	0.018
Econ.St.-3	1.023	0.321			

Note: The following reference groups are used for the categorical covariates: HHtype-5, Sex-1, GenHealth-1, Econ.St.-1, ChangeJob-1 and the corresponding groups in the interactions.

conclude with evidence that education confers a clear advantage on household income (eHI) on average for those who elect to pursue postsecondary education.

The comparison of the matched pairs is without any distributional assumptions. We merely have to check that the matching has been adequate, producing sets of pairs that have nearly identical profiles on all the background variables. The profiles would be nearly identical if the assignment to the treatments were randomised. Figure 9.5 plots the contrasts of the proportions for the categories of each background variable within the two treatment groups. For each category of a background variable, the horizontal segment connects the contrast of its proportions within the two treatment groups (marked by a thick long tick) with its value multiplied by −1 (thin long tick), which represents an imbalance of the same magnitude. The contrasts for 50 replicate sets of matched pairs are marked by thin short ticks, and their mean by a gray disc. The diagram shows that the match is nearly perfect for each category of a background variable. A large part of the deviation of the mean contrasts from zero can be attributed to the randomness of the matching procedure.

The match for the continuous background variables is summarised by the contrasts of their means and the ratio of their standard deviations for the

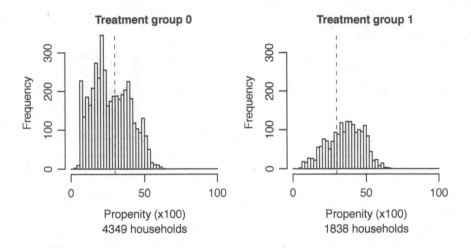

FIGURE 9.3
Histograms of the fitted propensity scores within the treatment groups. The vertical dashes indicate the average propensity.

treatment groups. The relative contrast is defined by dividing the (absolute) contrast by the standard deviation of the outcomes pooled across the two sets of matched pairs. Its version for the entire sample has respective means −0.267 and −0.105 for age and sampling weight. The corresponding figures for the replicate matched pairs are −0.010 for both variables, close to the ideal value of zero. The relative standard deviation of the contrasts is defined as the ratio of the standard deviations in the two treatment groups. Good match corresponds to its values close to unity. The values for the entire sample are 0.871 and 0.916 for age and weight, respectively. For the replicate sets of matched pairs, we evaluate the geometric mean of the 50 relative standard deviations. Its respective values for age and weight are 0.992 and 0.986, close to the ideal of unity.

Thus, both categorical and continuous background variables are well-balanced across the replicate matched pairs. They would be closely, but not perfectly, matched also in an hypothetical experiment in which the treatment (postsecondary education) would be assigned completely at random. The checks on the balance do not furnish a proof of validity of our analysis, because we cannot rule out imbalance with respect to some other background variables that were not recorded (and not included in the propensity analysis). The logically correct conclusion is that the inspection of the matched pairs revealed no contradictions with an important feature of the experimental design. We cannot undo the deficiencies of the observational studies that relate to the absence of control over the assignment of the treatment, but can obviate its impact, conditionally on clearly formulated assumptions. As an aside, we note that the covariates should have identical joint (multivariate)

TABLE 9.3
Numbers of households in the propensity groups; Austria 2011.

								Propensity group									
	1	2	3	4	5	6	7	8	9	10	11	12	13	14	15	16	17
$T=0$	110	119	111	108	108	103	113	113	103	103	105	98	101	102	107	96	98
$T=1$	14	5	13	15	16	21	11	10	21	21	18	26	23	22	16	28	26
	18	19	20	21	22	23	24	25	26	27	28	29	30	31	32	33	34
$T=0$	97	95	102	89	94	87	82	84	82	77	85	74	76	89	76	80	81
$T=1$	26	29	22	35	29	37	42	40	41	47	39	49	48	35	48	43	43
	35	36	37	38	39	40	41	42	43	*44*	45	46	47	*48*	49	*50*	
$T=0$	76	79	77	81	80	78	66	71	66	*57*	62	67	66	*59*	68	*48*	
$T=1$	48	44	47	43	44	45	58	53	57	*67*	62	57	57	*65*	56	*76*	

Note: The counts for the propensity groups in which the focal group is in a majority are printed in italics.

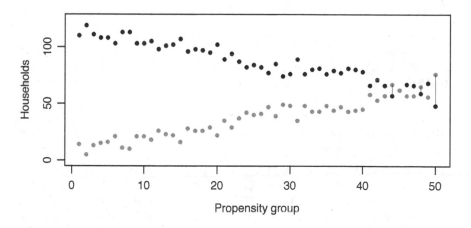

FIGURE 9.4
Numbers of households in the propensity groups; Austria 2011.
The vertical segments connect the counts for propensity groups in which the reference group has fewer households than the focal group.

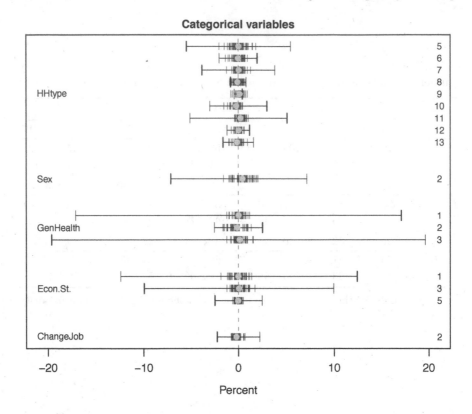

FIGURE 9.5
Balance plot for the categorical background variables used in the propensity model; Austria, 2011.
The category labels are printed at the right-hand margin. The gray discs indicate the average contrasts over 50 replications.

distributions within the treatment groups, and this is difficult to check when there are several covariates.

We summarise the results for Austria in the previous years. The respective estimates for years 2004–2008 are 0.216, 0.214, 0.232, 0.237, 0.226, with standard errors in the range 0.009–0.012 for the sets of focal units in the annual samples. Educational level is not recorded in the survey in 2009. The estimate for 2010 is 0.226, with standard error 0.011. The estimates of the differences of the population means of the two groups (without matching) are much higher: 0.301, 0.271, 0.310, 0.298, 0.300, 0.331 and 0.336 for the respective years 2004–2011, with 2009 omitted. The issue is not which set of estimates is correct, because they refer to different targets. The substantial differences between the two sets indicate that the details of how the target is defined are important. If there were a universal (constant) treatment effect, it would be the same

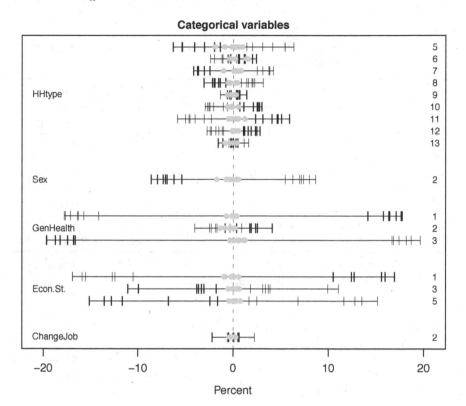

FIGURE 9.6
Balance plot for the categorical background variables used in the propensity model; Austria, 2004–2011, except for 2009.
Based on 20 replications for each year. The balances for the replications are omitted.

for all the households and for the focal households. The consistent differences in the two sets of results call into question the assumptions of the ordinary regression, but not the assumptions of the matched-pairs analysis.

Figure 9.6 displays the average contrasts of the background variables within the sets of matched pairs for all the years. Twenty replications are used for each year. The long ticks mark the annual balances of the treatment groups in the entire (unmatched) samples and the balances for the annual sets of matched pairs are marked by gray discs. The balance is not as exemplary as in Figure 9.5. Its sufficiency can be established indirectly, by repeating the analysis for 2011 with some of the interactions deleted from the propensity model. As a result, some imbalance appears in the version of Figure 9.5, but the estimate of the average treatment effect is changed only slightly.

For completeness, Table 9.4 lists the diagnostic summaries for the two

TABLE 9.4
Diagnostic summaries for age and the sampling weights as background variables in matched-pairs analysis for Austria in 2004−2011.

| | Age | | | | Weight | | | |
| | Entire sample | | Matched pairs | | Entire sample | | Matched pairs | |
Year	RM	RSD	RM	RSD	RM	RSD	RM	RSD
2004	−0.135	0.891	0.010	1.001	−0.006	1.062	−0.013	1.005
2005	−0.251	0.896	0.003	1.007	0.104	1.088	0.009	0.988
2006	−0.243	0.900	−0.003	0.998	0.008	1.043	−0.006	1.009
2007	−0.250	0.862	−0.003	0.998	0.045	1.300	0.005	1.015
2008	−0.306	0.875	−0.001	0.997	−0.019	1.062	0.006	1.025
2010	−0.299	0.861	−0.002	0.989	−0.094	0.970	−0.004	1.022
2011	−0.267	0.871	−0.010	0.992	−0.105	0.916	−0.010	0.986

Notes: RM: relative mean of the contrasts; RSD: relative standard deviation of the contrasts. Based on 20 replicate sets of matched pairs.

continuous background variables, age and the sampling weight. The balance is exemplary. One might argue that the balance should be checked also for the interactions. However, in this way we may enter a never-ending spiral of diagnostic checking in which the criteria get more and more stringent as we extend the propensity model.

9.5.1   Results for Other Countries

For Czech Republic, the estimates of the treatment effect based on the same set of covariates in the propensity model (see Table 9.1) are 0.326, 0.320, 0.285, 0.268, 0.318 and 0.320 for the respective years 2005−2008, 2010 and 2011. Educational level was not recorded in 2009. The standard errors due to the replication of matching are in the range 0.009−0.014, so the sampling variance dominates the overall variation. The estimates of the subpopulation differences of the two groups are in the range 0.40−0.43, that is, 49%−53%. The estimates of the treatment effect for Czech Republic are substantially higher than for Austria. The balance of the matched groups is somewhat better than for Austria.

The estimates of the average treatment effect for a selection of countries are listed in Table 9.5. A dash in the table indicates that the estimate for the country in the given year is not available. The most frequent reason for this is that the treatment variable is not recorded. In some instances one or several background variables are not recorded.

The standard errors of the related estimators have two components, one related to the replications of matching, which is in all instances smaller than the component due to sampling variation. Its estimate is only slightly smaller than

TABLE 9.5

Estimates of the average treatment effects for a selection of EU countries and years 2004–2011.

	2004	2005	2006	2007	2008	2009	2010	2011
AT	0.247	0.214	0.232	0.237	0.226	—	0.226	0.242
CZ	—	0.326	0.320	0.285	0.268	—	0.318	0.320
DE	—	0.219	—	—	0.234	—	0.274	0.254
EE	—	0.227	0.170	0.197	0.222	—	0.265	0.243
ES	—	—	—	0.418	0.357	0.409	0.358	0.374
FR	0.358	0.340	0.329	0.318	—	—	0.356	0.373
HU	—	0.455	—	0.392	0.355	—	0.350	0.390
LU	0.326	0.304	0.359	0.454	0.464	—	0.368	0.369
PL	—	—	0.486	0.473	0.446	—	0.415	0.425
UK	—	0.187	0.234	0.259	—	0.215	0.288	0.250

the estimate of the compound standard error (due to sampling and matching), and its values for the countries and years range from 0.02 to 0.03. Austria, Germany, Estonia and the UK have the smallest average treatment effects throughout the studied period and Spain, Hungary and Poland have the highest. The standard errors have magnitudes comparable to the estimated inter-year differences, so a trend is difficult to identify for any country, except for Poland. There the average treatment effect has been declining from very high levels, but even in 2011 its estimate was the highest among the studied countries. For Luxembourg, the estimates for 2007 and 2008 differ from the estimates for the other years by much more than what could be attributed to estimation error. For some of the countries not listed in the table, one or more of the background variables are not recorded.

9.5.2   Regression of Outcome on Treatment and Background

An established alternative to the potential outcomes framework is the analysis by weighted regression. We apply it to Austria in each year, except 2009, using the treatment and background variables and the interactions listed in Table 9.1 as the covariates. We do not have to be concerned with including some possibly redundant variables or interactions in the model, because we have a large number of observations for every year. The principal threat to validity of the model is from a bias, and so erring on the side of a richer model (and greater sampling variance) is inconsequential. In the regression, we use the sampling weights.

The fit for Austria in year 2011 is displayed in Table 9.6, using the same covariates and interactions as in the propensity model, with the treatment indicator added and the weights left out. The estimate of the treatment effect

TABLE 9.6
Linear regression fit to log-eHI for Austria in 2011.

Intercept	8.9190	0.2089	Econ.St.-5	−0.4738	0.2060
T.Educ.	0.2343	0.0242	ChangeJob	0.1994	0.0529
HHtype-6	0.5835	0.0345	Age2	−0.0295	0.0056
HHtype-7	0.5932	0.0389	Sex.GHealth-2	−0.0939	0.0566
HHtype-8	0.9161	0.0422	Sex.GHealth-3	−0.1407	0.0634
HHtype-9	0.3393	0.0661	Sex.EconSt.-3	−0.1012	0.0578
HHtype-10	0.7121	0.0435	Sex.EconSt.-5	0.2631	0.1156
HHtype-11	0.7262	0.0447	EcoSt.GHealth	−0.0053	0.0594
HHtype-12	0.8394	0.0625	Sex.Age	0.0034	0.0018
HHtype-13	0.9711	0.0508	Sex.ChJob	−0.0318	0.0965
Sex	−0.0517	0.1108	GHealth2.Age	−0.0025	0.0019
Age	0.0314	0.0047	GHealth3.Age	0.0000	0.0023
GenHealth-2	0.0643	0.0898	EconSt3.Age	0.0056	0.0025
GenHealth-3	−0.1615	0.1107	EconSt5.Age	−0.0064	0.0045
Econ.St.-3	−0.5718	0.1211	*Res. variance*	0.4011	

is 0.234, with standard error 0.024, not substantially different from the corresponding result for matched pairs. However, this is an estimate of a different target, namely, the difference of the mean log-eHI, assuming it is constant, except for the residual variation.

We can relax this assumption by adding to the covariates interactions of postsecondary education with age, since it is, at face value, an important predictor. Now the estimate associated with tertiary education is 0.119 (standard error 0.081), but it refers to age extrapolated to zero, a meaningless target. For estimating a more reasonable target, we have to take into account the (average) age of the focal group, the squared age and the averages of the ages within the sexes. Further, we have to explore other interactions because our estimation should be based on an acceptable model. We omit the details. To derive the sampling variance of the resulting estimator, we have to operate with a 30 × 30 matrix of cross-products. These points are an effective discouragement from pursuing this approach.

The regression we applied assumes that the values of the covariates have been set, and the outcomes observed thereafter. In the survey, all the variables were recorded simultaneously (during an interview), but the background and treatment variables can reasonably be assumed to be precedents for the outcome. The assumptions of normality and linearity (model specification) are relevant for regression; in the potential outcomes framework we do not have to be concerned with them.

9.5.3 Potential Versions of Variables

We defined background variables as variables that have identical potential versions for the two (or more) treatments. We may argue that some of the variables included as covariates in the propensity model are not background. For example, the household representative's economic status or even general health would be different had the person been in another treatment group. Moreover, another member of the household may have become its representative, and so the values of the variable age (derived from the year of birth, "pb040") might also be different. We can argue similarly about all the variables, including sex and household type. In this respect, our analysis lacks validity. On the other hand, variables that would satisfy the implied higher standard for being background are very difficult to find, because education may even affect household formation.

A variable is called *intermediate* if its potential versions with respect to the treatment variable are not identical, but is itself well-defined as a treatment variable. For example, the level of education has a prima facie effect of attainment of a job after completing one's education. Thus, job attainment is a (binary) outcome variable. However, job attainment is a treatment variable for income. In this setting, income has (at least) four potential values, one for each combination of educational level and job attainment. A change in the treatment (education) may result in a different value of the intermediate variable (job attainment), but if the value of this variable were altered, so would be the outcome.

We can define the effect of the treatment reinforced by the resulting value of the intermediate variable, the effect of the treatment assuming a particular value of the intermediate variable, and the effect of the intermediate variable for a given treatment. An intermediate variable is usually assumed to have the property that the outcome and treatment are independent given the value of the intermediate variable.

An important assumption associated with the potential outcomes framework is that the treatment applied to one unit exerts no influence on the outcome of another unit. This assumption is known by its acronym SUTVA—stable unit treatment value assignment. Without this assumption, the outcome would have unmanageably many potential versions, one linked to each set of treatment assignments. In our setting the assumption is problematic, because members of the labour force compete for vacant jobs and, more generally, adjust their behaviour in the labour market according to the information they have about the competition they face. Also, they are informed by the experiences of others, family members and friends in particular. A wholesale change in the educational background of the labour force would result in a structural change of the labour market and the differences between the two treatment groups are likely to change with it. Therefore, our conclusions and comparisons have very strong caveats that cannot be extrapolated to substantial (longer-term) changes in the labour market and the economy in general, and can best be related to an advice or proposal adressed to one or a few individuals, but not to a large fraction of the population.

Suggested Reading

Rubin (2006) is a collection of studies that apply the potential outcomes framework. A method of matching that differs substantially from propensity scoring and related ideas is developed by Iacus, King and Porro (2011). Analysis with intermediate variables is described by Frangakis and Rubin (2002). Rosenbaum (2002) is an established text on the causal analysis of observational studies. Rosenbaum (2010) deals with the design of observational studies. A lot of the advice in Cochran (1965) and Kish (1987) is highly relevant to modern household surveys. The essential role of sensitivity analysis in causal inference is highlighted by Rubin (1991). Other important references on the subject of causal inference are Angrist, Imbens and Rubin (1996) and Pearl (2009).

9.A   Appendix. Programming Notes

The matched-pairs analysis is implemented in the user-defined function `POTF`. For its steps it uses separate functions which we introduce here and give details in the next section.

The arguments of `POTF` are the country acronym, the year, the number of propensity groups (bins) to be formed, the number of replicate sets of matched pairs to be generated and the maximum number of values of a categorical variable. The latter argument is a device to distinguish between categorical and continuous variables.

```
POTF <- function(co, ye, nbn=50, mre=50, mxc=12)
{
### Propensity score analysis -- potential outcomes framework

## co   The country
## ye   The year (digit)
## nbn  The number of bins for matching
## mre  The number of replicate matches
## mxc  The maximum number of categories
```

Data is input from a household- and an individual-level dataset:

```
## Data input
datH <- KHread(co, ye, colsH, "H", "hb030")
datP <- KHread(co, ye, colsP, "P", "pb030")
```

Functions for data input were introduced in previous chapters. Here we use the function `KHread` for inputting data at either individual or household level or, more generally, for inputting several types of datasets. The database in EU-SILC comprises datasets named H, P, R and S for each combination of country and year; H and R are defined for households and P and S for individuals. For example, `AT11P.dat` is one such dataset, for Austria in 2011.

The files are created by Linux shell commands using the **grep** utility, which extracts from a (large) file all records that contain a specific character string. For example, the commands

```
grep PL DATA/cross_sec_H_04-11.csv > DATA/PLallH.dat
grep PL DATA/cross_sec_P_04-11.csv > DATA/PLallP.dat
grep PL DATA/H_merged_04-11.csv > DATA/PLallS.dat
```

extract the records that contain the string 'PL' from the three *.csv files and deposits them to files `DATA/PLall?.dat`. We do not type these commands on the command line; they are written in a file called `Slam1PL` and executed by the shell command

```
sh Slam1PL
```

The three files are processed further to extract records that contain the character strings 2005, 2006, ..., 2011, to 3 × 6 files named `PLyrZ.dat`, where **yr** is the two-digit version of the year (e.g., 05 or 10) and Z is the file type (H, P or S). The files thus created contain some inappropriate records which have the year string in some other fields (variables). They are screened out by the R function `KHread`.

We use only datasets H and P. The records of individuals in these datasets are connected to the records of households through their identification numbers (Id.s). The Id. of an individual's household is obtained by clipping (discarding) the last two digits of the individual's Id.

The arguments of the function `KHread` are the country's acronym, the year, the vector of the names of the variables, and the type and name of the variable that contains the Id. The function returns a dataset in which the Id.s form the rownames and all the variables that have constant values, or all their values are missing, are discarded. Details are given below.

Before using the data matrices `datH` and `datP`, the function `VarIn`, with arguments `vars` and `cols`, is applied to check that the variables in `vars` have been retained in the dataset; the argument `cols` is set to the names of the retained variables. Only the first variable in the vector `vars` that is found not to be among `cols` is indicated and the execution is halted.

```
VarIn(vars=c("pb140", "pl030"), cols=colnames(datP))
VarIn(vars="hx060", cols=colnames(datH))
VarIn(c("pb150", "pb140", "pb040", "ph010"), colnames(datP))
VarIn(c("pl030", "pl160", "pe040"), colnames(datP))
```

The function could be applied to one variable at a time. Also, an argument could be added in `VarIn` to indicate the dataset (`datH` or `datP`).

The households' representatives are selected by the function `KHsel1`. Its arguments are the households' Id.s and a score generated from the employment status and age. The score is a total of the following subscores:

- 1000 points for full-time employment, zero otherwise;

- 200 points for being in working age (less than 65 years);

- one point for every year of age.

The function returns the person-level Id.s of the representatives of the households. The corresponding households are obtained by clipping the last two digits of these Id.s (using function substring). The dataset datP is then reduced to the records of the household representatives.

```
##  The calendar year
ye2 <- 2000+as.numeric(ye)

##  Selection of the households' representatives
sel1r <- KHsel1(rownames(datH),
    sco=ye2-datP[, "pb140"] + 200*((ye2-datP[, "pb140"]) < 65) +
      1000*(datP[, "pl030"]==1))

##  The corresponding households
hhsel <- substring(sel1r, 1, nchar(sel1r)-2)

##  Sort the records in datP
datP <- datP[sel1r, ]
```

Next, the matrix of background variables is compiled in two stages. Household type, sex, age and the sampling weight are joined by the recoded versions of the general health, economic status and the indicator of a change of job in the last twelve months. The treatment variable is defined as a separate vector KHv. Listwise deletion is applied, but a warning is issued if the number of deleted records exceeds five. Some of the background variables are recoded.

```
##  The background variables
datX <- cbind(HHtype=datH[hhsel, "hx060"], Sex=datP[, "pb150"],
        Age=ye2-datP[, "pb140"], Weight=datP[, "pb040"]/1000)

##  General Health  --  recoding
KHx <-  datP[, "ph010"]
KHx[KHx > 3] <- 3
KHx[is.na(KHx)] <- 2

##  Attach the recoded variable to the dataset
datX <- cbind(datX, GenHealth=KHx)

##  Repetitive code deleted
...


##  The treatment variable  (level of education)
KHv <- datP[, "pe040"]
KHv <- (KHv > 3)
```

```
warning("Background variables - ready")

##  Listwise deletion
whi <- apply(is.na(cbind(datX, KHv)), 1, sum) == 0
datX <- datX[whi, ]
KHv <- KHv[whi]

if (sum(!whi) > 5)
warning(paste(sum(!whi), "observations deleted."))
```

The logistic regression is now fitted to the treatment variable (KHv) in terms of the background variables and their interactions. Note that the interactions are 'hard-coded' in the function. The arguments of the function LogitW are the matrix of covariates, vector of the outcomes, vector of the weights (with an option for equal weights), and arguments controlling the convergence. Each covariate has to have a unique name; these names are used in the output.

```
##  Logistic regression
Fpro <- LogitW(cbind(datX, Age2=datX[, 3]^2/100,
###    Optional extension of the model
###        Weight2=datX[, 4]^2,
###        Age.Weight =datX[, 3]*datX[, 4],
    Sex.GHealth=1+(datX[, 2]-1)*(datX[, 5]-1),
    Sex.EconSt.=1+(datX[, 2]-1)*(datX[, 6]-1),
    EcoSt.GHealth=(datX[, 5] > 2) * (datX[, 6] > 1),
    Sex.Age=(datX[, 2]-1)*datX[, 3],
    Sex.ChJob=(datX[, 2]-1)*(datX[, 7]-1),
    GHealth2.Age=(datX[, 5]==2)*datX[, 3],
    GHealth3.Age=(datX[, 5]==3)*datX[, 3],
    EconSt3.Age=(datX[, 6]==3)*datX[, 3],
    EconSt5.Age=(datX[, 6]==5)*datX[, 3]), KHv, 1, maxit=10, tol=8)
```

Weights can be specified, and set to the sampling weights in particular. We prefer to use the weights as a covariate, so that its transformations and interactions could also be used if desired. The function returns a list (Fpro) of several elements. The first, Fpro[[1]], is the table (matrix) of estimates and estimated standard errors. The second (Fpro[[2]]) is a matrix, the columns of which are the values of the treatment variable (coerced from logical FALSE/TRUE to numeric 0/1) and the fitted propensity. The latter is coarsened into the specified number of propensity groups by the function KHbin, and the result is attached to Fpro[[2]] as the third column.

```
##  Add the bin No. to the propensities
Fpro[[2]] <- cbind(Fpro[[2]], Bin=KHbin(Fpro[[2]][, 2], nbn))
```

We refer to Fpro[[2]] as the propensity matrix.
    The values of the outcome variable (log-eHI) are evaluated next.

```
##  The outcome variable
inc <- datH[hhsel, "hy020"][whi]
inc[inc < 0] <- 0
```

```
##  Log-transformation
inc <- log(1+inc)
```

The vector whi, defined earlier for listwise deletion, indicates complete (selected) records.

In the final step, replicate sets of matched pairs are formed, using the function KHmtc and the averages of the outcome variable are evaluated within the pairs of matched groups to form a matrix with two rows. This matrix is the first element of the list that is returned by the function POTF. The second element is the output of function KHbal which evaluates the balance for each covariate; see details below. Further elements of the list contain the results of the logistic regression, the table that describes the representation of the treatment groups within the bins, and the results of trivial analyses (the weighted means and medians within the treatment groups).

```
##  Replications of matching and matched-pairs analysis
list(Estimates = replicate(mre,
{
##  One set of matched pairs
mtcR <- KHmtc(Fpro[[2]])
mtcR <- substring(mtcR, 1, nchar(mtcR)-2)

##  Analysis of the matched pairs
apply(matrix(inc[mtcR], ncol=2), 2, mean)}),

##  Diagnostics  --  check on the balance
Balance=KHbal(datX, Fpro, mxc, mre),

##  The propensity model fit and the fitted probabilities/bins
Lgt.Fit=Fpro[[1]], Propensities=Fpro[[2]],

##  Tabulation of the bins
EducLev45=table(KHv, Fpro[[2]][, 3]),

##  The trivial analysis
AveeHI=tapply(inc*datX[, "Weight"], KHv, sum) /
       tapply(datX[, "Weight"], KHv, sum),

##  The within-treatment group medians
MedeHI=tapply(inc, KHv, median))
}  ##  End of function  POTF
```

The following is an application, with measuring the elapsed time added.

```
Time <- proc.time()

##  Analysis for Austria, 2004, with 50 bins and 20 replicates
AT04pf <- POTF("AT", 4, 50, 20, 12)

Time <- proc.time() - Time
##  Takes  28 sec.
```

9.A.1   Second-Level Functions

This section documents the functions called by the function POTF.

A dataset is read by the function KHread. The dataset to be read is identified by the country's acronym, year and type. Further, the names of the variables in the dataset and of the variable that contains the unique record identifier have to be declared. The location of the dataset (directory) and the extension of the file have default values.

```
KHread <- function(co, ye, cols, tp="H", id="hb030", ext="dat",
                   dire="~/Splus/JHrad0813/DATA")
{
###   Data input for matched-pairs analysis (POTF)

##  co    The country (acronym)
##  ye    The year
##  cols  The names of the variables (columns)
##  tp    The data-type  (H, P or R)
##  id    The variable id
##  ext   The file extension
##  dire  The directory
```

The appropriate value of the argument cols is colsH for type H and colsP for type P. Their values are

```
colsH <- c("hb010", "hb020", "hb030", "hh010", "hh020",
           "hh030", "hh050", "hh060", "hh070", "hy010",
           "hy020", "hy022", "hy023", "hx040", "hx050",
           "hx060", "hx070", "hx090")


colsP <- c("pb010", "pb020", "pb030", "pb040", "pb140",
           "pb150", "pb190", "pe010", "pe040", "ph010",
           "ph020", "ph030", "ph040", "ph060", "pl015",
           "pl020", "pl025", "pl030", "pl040", "pl060",
           "pl140", "pl150", "pl160", "pl200")
```

They indicate the variables that were extracted from the original database. In KHread, the year of the survey is established and then the data is input.

```
##  The various formats for the year
if (nchar(ye) == 1)  ye <- paste(0, ye, sep="")
if (nchar(ye) == 4)  ye <- substring(ye, 3,4)

##  The number of columns
ncl <- length(cols)

##   Data input
dat <- matrix(scan(paste(dire, "/", co,ye,tp,".",ext, sep=""),
       what=character(), sep=";"), ncol=ncl, byrow=T,
       dimnames=list(NULL, cols))
```

```
##  Reduce to the given year
dat <- dat[dat[, 1] == 2000 + as.numeric(ye), ]
```

The dataset is reduced by deleting variables that have constant values or indicate the country and its regions. Variables with all their values missing are also discarded because a vector comprising NA has one unique value. For example, LeUni(rep(NA, 100)) is equal to unity.

```
##  Where characters are used or all the values are the same
whi <- apply(substring(dat, 1, 2) == co, 2, sum, na.rm=T) > 0

whi <- whi | apply(dat, 2, LeUni) == 1
names(whi) <- cols
whi[id] <- T

Dtid <- dat[, id]

##  Delete those columns
dat <- apply(dat[, !whi], 2, as.numeric)

rownames(dat) <- Dtid
dat
}  ##  End of function  KHread
```

The following statement is for forming the individual-level (P) dataset for Czech Republic in 2011.

```
CZ11P <- KHread("CZ", 11, colsP, "P", "pb030")
```

The function **VarIn** checks whether a set of variables is included in a dataset. Recall that a variable may be excluded from a dataset read by **KHread** if its values are all missing or are constant.

```
VarIn <- function(vars, cols)
{
###   Function to check on the presence of variables in the data

##  vars   The names of the variables
##  cols   The variables in the dataset (colnames)

for (vr in vars)
{
if (is.na(match(vr, cols)))
    stop(paste("Variable", vr, "not in the dataset."))
}}  ##  End of function  VarIn
```

The function **KHsel1** selects the representatives of the households. Its arguments are the Id.s of the households and the scores defined for individuals. The scores have to be a vector with the individual-level Id.s as its names. Within the function, the auxiliary function **MaxS** is defined; it returns the name of the element of a vector with the highest value. This function is applied, using **tapply**, within each household.

```
KHsel1  <- function(ind, sco)
{
## ind   The indices  (related to rownames)
## sco   The score

## Function to identify the element with the lowest score
MaxS <- function(scoI)
names(scoI)[sort.list(scoI)][1]

## The person id
rwn <- names(sco)

## The lowest scores of  -sco  within  households
tapply(-sco, substring(rwn, 1, nchar(rwn)-2), MaxS)
}  ## End of function  KHsel1
```

The function LogitW implements the generalised least squares for the special case of logistic regression.

```
LogitW <- function(X, y, w, maxit=100, tol=8, dgt=3, mcx=12, nbin=0,
                ctp=c(), plt=T)
{
###   Logistic regression with sampling weights

## X       The regression matrix (vector if only one variable)
## y       The outcome variable (vector of zeros and ones)
## w       The sampling weights
## maxit   The maximum number of iterations
## tol     The convergence criterion
## dgt     The number of digits in rounding
## mcx     The maximum number of values of a covariate
## nbin    The number of bins for diagnostics
## ctp     The cutpoints for the bins
## plt     Whether to plot  (logical)
```

The initial checks are followed by defining the indicator variables for the categorical variables in X. A categorical variable is recognised as having at most mcx distinct values. The regression matrix is formed from the continuous variables and the indicators of the categories of the categorical variables, with the first category designated as the reference, and left out.

```
## The number of subjects
ns <- length(y)

## Truncation of the outcomes
y[y>1] <- 1
y[y<0] <- 0

## Warning about decimal numbers
if (sum(y==1 | y==0) < ns)
    warning("Some fractional outcomes.")
```

```
##   Deal with categorical variables
Lni <- apply(X, 2, LeUni)

##   The regression design matrix and colnames
XX <- 1
XXn <- 1

##   Loop over the variables
for (cl in colnames(X))
{
if (Lni[cl] > mcx)
{
##   Continuous variable; a single column of XX
XX <- cbind(XX, X[, cl])

##   The names of the variables collected in a vector
XXn <- c(XXn, cl)
}
else
{
##   Categorical variable
Xc <- X[, cl]

##   Missing values are a separate category
Xc[is.na(Xc)] <- max(Xc, na.rm=T) + 1

##   The first category (after sorting) is the reference
srt <- sort(unique(Xc))[-1]

##   The indicator variable
for (xd in srt)
XX <- cbind(XX, Xc==xd)

##   The names of the indicator variables
XXn <- c(XXn, paste(cl, srt, sep="-"))
} ##   End of the condition over conti/categ variables
} ##   End of the loop over the variables  (cl)
```

The generalised least squares is an iterative procedure. It requires an initial solution (estimates) and some other objects related to monitoring the convergence.

```
##   The weights and their total
if (length(w)==1) w <- rep(1, ns)
SW <- sum(w)

##   The probability estimate
ybar <- sum(w*y)/SW

##   The number of columns in  XX
```

```
np <- ncol(XX)

##  Prepare for iterations  --  the initial solution
betN <- matrix(c(Lgt(ybar), rep(0, np-1)))

##  The initial log-likelihood
devN <- ns*(ybar*betN[1] + log(1-ybar))

##  The constant part of the score function:  X'y
Xy <- t(XX) %*% matrix(w*y)

##  Initialisation for convergence monitoring and the iterations
Conver <- c(betN, devN, NA)
iter <- 0
dist <- 0
```

The iterations are organised in a conditional loop. They are stopped when the maximum number of iterations (the argument `maxit`) is reached, or a convergence criterion (controlled by the argument `tol`, see below) is satisfied. In the loop, the iteration number `iter` is updated and the current deviance is stored, so that it could later be compared with its updated value.

```
##  The iterations
while (iter < maxit & dist < tol)
{
##  The next iteration
iter <- iter + 1

##  Store the current deviance
devP <- devN

##  The linear predictors
lp <- XX %*% betN

##  The fitted probabilities
prb <- expl(lp)

##  Warning about extreme probabilities
if (min(prb*(1-prb)) < 10^(-tol))
    warning(paste("An extreme probability in iteration", iter))

##  The score function and the Hessian  (the core of GLS)
ss <- Xy - t(XX) %*% matrix(w*prb)
H <- t(XX) %*% (XX * matrix(w*prb*(1-prb), ns, np))

##  The range of the eigenvalues of  H
reig <- range(eigen(H)[[1]])

##  A zero eigenvalue
if (reig[1] < 10^(-tol))
```

```
{
warning(paste("Singularity reached.
   Stopping in iteration ", iter, ".", sep=""))
dist <- tol+1
}
else
{
## The updating
dlt <- solve(H, ss)

## The new solution
betN <- betN + dlt

## The new deviance
devN <- (sum(w*y*lp) + sum(w*log(1-prb))) / SW * ns

## The convergence criterion
dist <- -log((devN-devP)^2 + mean(dlt^2), base=10)/2
} ## End of the condition re. singularity

## Convergence monitoring -- bookkeeping
Conver <- rbind(Conver, c(betN, devN, dist))
} ## End of the iteration loop
```

The suffix P refers to the previous solution and N to the current (new) value of the deviance, dev). For details of the method, the expressions for lp, prb, ss, H, betN and devN, we refer to standard textbooks for generalised linear models, of which McCullagh and Nelder (1989) probably commands the highest authority. Hosmer and Lemeshow (1989) and Dobson (2001) have a more practical orientation. Aitkin *et al.* (2009) has many in-depth analyses by generalised linear models, with details of the applications in R.

The scalar dist evaluates the distance between two consecutive solutions. It can be interpreted as the number of decimal places of precision. When it exceeds the value of the argument tol, the iterations are stopped; see the expression which sets off the iterations (while ...). See Section 7.B for similar use of dist and Conver.

At the conclusion of the iterations, the standard errors are estimated and suitable objects are formed for an effective list of results. The function includes (optional) diagnostic plots. Details are omitted to conserve space. In any case, the conventional diagnostics are not relevant for propensity modelling because their principal goal is balance of the matched pairs, which is checked directly.

```
## The estimated standard errors
ste <- sqrt(SW * diag(solve(H))) / ns )
names(ste) <- XXn

## Attach the standard errors
Conver <- rbind(Conver, c(ste, NA, NA))
```

```
dimnames(Conver) <- list(c(seq(0,iter), "St.error"),
   c(XXn, "Log-likelihood", "Conv.crit"))

rownames(betN) <- colnames(XX)

if (iter == maxit)
warning(paste("The maximum number of iterations", iter, "reached."))

## Diagnostics  (omitted)
...

## lapply(
list(Estimates=rbind(beta=as.vector(betN), St.err=ste),
     Propensity=cbind(Y=y, P=prb), Conv.table=Conver,
     Convergence=c(Iters=iter, Converge=dist), Diagn=Diagn)
## , round, dgt)
} ## End of function LogitW
```

The propensity scores are coarsened (discretised) by the function KHbin. The cutpoints are defined as quantiles of the propensities. The group to which a propensity is assigned is defined as the number of cutpoints which the propensity exceeds, with one added.

```
KHbin <- function(vec, nbin)
{
### Coarsenend propensity scores

## vec   The vector of propensities
## nbin  The number of bins

## The cutpoints
ctp <- quantile(vec, seq(nbin-1)/nbin)

## Initialisation
Bin <- 1

## Allocation to the bins
for (ct in ctp)
Bin <- Bin + (vec > ct)
} ## End of function  KHbin
```

Matched pairs are formed by the function KHmtc. Its sole argument is the propensity matrix (treatment group, fitted propensity and propensity group for each unit in the sample). The function returns a two-column matrix in which the rows are the Id.s of the matched pairs, the first element (column) for the reference and the second for the focal group. In the function, the treatment group that is in the majority is established for each propensity group.

```
KHmtc <- function(pro)
{
```

```
###    Forming a set of matched pairs
##   pro    The matrix of propensities and binning

##   The treatment-by-bin table
TB <- table(pro[, 1], pro[, 3])

##   Which group is in the majority
Ind <- TB[1, ] <= TB[2, ]
```

Then, in a loop over the propensity groups, a sample is drawn from the treatment subgroup of the propensity group that has a majority. The sample size is equal to the size of the minority subgroup. The sample contains the donors for the units in the minority subgroup. The Id.s of the recipients and donors are then vertically stacked in the result-matrix PRS.

```
PRS <- c()

##   Forming matched pairs for each bin
for (bn in colnames(TB))
{
##   Sample from the majority subgroup
SS <- sample.int(TB[1+Ind[bn], bn], TB[2-Ind[bn], bn])

whi <- list()

##   The Id.s for the two subgroups
for (i in c(0,1))
whi[[i+1]] <- rownames(pro)[pro[, 1]==i & pro[, 3]==bn]

##   Reduction to the matched pairs
whi[[1+Ind[bn]]] <- whi[[1+Ind[bn]]][SS]

##   Collecting the matched pairs
PRS <- rbind(PRS, cbind(whi[[1]], whi[[2]]))
} ##   End of the loop over the bins (bn)

}   ##   End of function   KHmtc
```

The function KHbal compiles the statistics on which the diagnostic plots for checking the balance of the within-treatment means are based. Suppose there are r replicate sets of matched pairs. Then for each categorical variable with K categories an $(r + 1) \times K$ matrix is formed. Its first row contains the contrasts (differences) of the percentages of each category between the two treatment groups. The other rows contain these contrasts for the sets of matched pairs. For a binary variable, the first column is dropped; it is equal to the negative of the retained column. For a continuous variable, the relative contrasts and the ratios of the within-treatment standard deviations are evaluated and collected into an $(r + 1) \times 2$ matrix. The arguments of the function KHbal include the propensity matrix. If it is given as a list that contains the matrix as an element, then this (second) element is extracted.

```
KHbal <- function(Xmat, Pro, MxCat=12, nre)
{
###    Diagnostics for the balance of matched pairs

## Xmat     The matrix of the covariates
## Pro      The propensity matrix   (e.g., KHpro[[2]])
## MxCat    The  maximum number of categories
## nre      The number of replicate matched pairs

## Simplify  Pro  to a matrix
if (is.list(Pro)) Pro <- Pro[[2]]

## Exclude categories that appear in only one treatment group
tb0 <- apply(table(Pro[, 1], Pro[, 3]), 2, min)

tb0 <- names(tb0)[tb0==0]

if (length(tb0)>0)
{
Pro <- Pro[-match(tb0, Pro[, 3]), ]
Xmat <- Xmat[-match(tb0, Pro[, 3]), ]
}
```

The contrasts are first evaluated for the entire sample. A variable is regarded as continuous if it has more than MxCat distinct values.

```
## The numbers of categories
Ncat <- apply(Xmat, 2, LeUni)

## Initialisation for tables (contrasts)
TBL <- list()

for (xm in colnames(Xmat))
{
## Categorical variable  --  contrasts of percentages
if (Ncat[xm] <= MxCat)
    TBL[[xm]] <- apply(apply(table(Pro[, 1], Xmat[, xm]), 1, Perc),
        1, diff)
else
## Continuous variable  --  relative contrast and sd ratio
    TBL[[xm]] <- c(diff(tapply(Xmat[, xm], Pro[, 1], mean)) /
        sd(Xmat[, xm]),
    sd(Xmat[Pro[, 1]==1, xm]) / sd(Xmat[Pro[, 1]==0, xm]))
}
```

Next, similar evaluations are conducted for replicate sets of matched pairs, which are generated in a loop.

```
for (rp in seq(nre))
{
## Generate a set of matched pairs
```

```
Xmtc <- KHmtc(Pro)
Xmtc <- matrix(substring(Xmtc, 1, nchar(Xmtc)-2), ncol=2)

## The number of pairs
npr <- nrow(Xmtc)

## Reduce the dataset to the matched pairs
Pdat <- list()

## The reduced datasets
for (i in seq(2))
Pdat[[i]] <- Xmat[Xmtc[, i], ]

for (xm in colnames(Xmat))
{
## If it is a categorical variable
if (Ncat[xm] <= MxCat)
## Contrast of the percentage
tab <- 100*(table(Pdat[[2]][, xm]) - table(Pdat[[1]][, xm]))/npr

else
tab <- c(DIF=(mean(Pdat[[2]][, xm]) - mean(Pdat[[1]][, xm])) /
             sd(as.vector(sapply(Pdat, ExtrC, xm))),
             STD=sd(Pdat[[2]][, xm]) / sd(Pdat[[1]][, xm]))

## Collect the contrasts
TBL[[xm]] <- rbind(TBL[[xm]], tab)
} ## End of the loop over the variables  (xm)
} ## End of the replication loop  (rp)

## Get rid of one category for binary variables
for (xm in names(TBL)[Ncat==2])
TBL[[xm]] <- matrix(TBL[[xm]][, -1])

TBL
} ## End of function  KHbal
```

Note that different sets of matched pairs are formed in KHbal than in estimation of the average treatment effect.

9.A.2 Graphics for the Balance Diagnostics

The diagram of the sample and matched-pairs contrasts for the categorical background variables (Figure 9.5) is drawn by the function KHblg. Its sole argument is the output list of function KHbal. The horizontal limits and the vertical spacing of the diagram are sorted out first and the empty shell of the plot is drawn. The list argument has an element for each background variable, and each element is a matrix. For a binary variable, the matrix has one column;

for a categorical variable with $K > 2$ categories, it has K columns, and for a continuous variable, it has two columns.

```
KHblg <- function(res)
{
### The balance plot for categorical variables
## res    The list-result of KHbal

## The number of categories
Ncat <- sapply(res, ncol)

## The number of large (inter-variable) and small (inter-categ.) gaps
gaps <- c(sum(Ncat != 2)-1, sum(Ncat[Ncat != 2]))

## The horizontal axis limits
Mxh <- 1.05*max(abs(unlist(res[Ncat != 2])))

## The vertical axis limits
ylm <- -sum(gaps*c(2, 1))

## Empty shell of the plot
plot(Mxh*c(-1,1),  c(ylm, -1), type="n", yaxt="n",
     xlab="Percent", ylab="", main="Categorical variables",
     cex.lab=0.9, cex.axis=0.9, cex.main=0.9)
```

The vertical coordinates of the horizontal segments that correspond to the categories are negative, so that variables and their categories are presented from top to bottom. The vertical axis labels are hidden by the setting `yaxt="n"`. In the code below, the scalar `sta` keeps a tab on the height in the loop over the variables. The inner loop (`nc`) is over the categories of a variable. Continuous variables (two-column elements of `res`) are dropped from plotting, but summaries of the balance are evaluated and returned by the function. Binary variables have to be treated separately because they are represented by a single column which reverts by subsetting to a vector that is no longer a matrix.

```
## The vertical zero  (perfect balance)
abline(v=0, lty=2, lwd=0.2)

## The vertical location initialised
sta <- 0

## Loop over the categorical variables
for (xv in names(res)[Ncat != 2])
{
## The number of columns
Nct <- Ncat[xv]

## Horizontal segments
segments(res[[xv]][1, ], sta - seq(Nct),
```

```
          -res[[xv]][1, ], sta - seq(Nct), lwd=0.5)

## The long vertical ticks
for (sgn in c(-1,1))
segments(sgn*res[[xv]][1, ], sta - (seq(Nct) - 0.4),
         sgn*res[[xv]][1, ], sta - (seq(Nct) + 0.4), lwd=0.8+0.4*sgn)

## The ticks for the replicate contrasts
for (nc in seq(Nct))
segments(res[[xv]][-1, nc], sta - nc+0.28, res[[xv]][-1, nc],
    sta - nc-0.8, lwd=0.125)

## Different treatment for the binary variables
if (Nct > 1)
    mns <- apply(res[[xv]][-1, ], 2, mean)
else
    mns <- mean(res[[xv]][-1])

## The gray discs for the means
points(mns, sta - seq(Nct), pch=16, cex=1.1, col="gray75")

## Print the name of the variable
text(-Mxh*1.025, sta-Nct/2-0.5, xv, cex=0.75, adj=0)

## The categories
nms <- colnames(res[[xv]])

## The fix-up for binary variables
if (length(nms)==0) nms <- 2

## Print the categories of the variable (right-hand margin)
text(Mxh*1.025, sta-seq(Nct), nms, cex=0.75)

## The height for the next variable
sta <- sta - Nct - 2
} ## End of the loop over the variables (xv)

## Summary for the continuous variables
SMR <- list()

## Loop over the continuous variables
for (xv in names(res)[Ncat==2])
SMR[[xv]] <- rbind(Mean=c(All=res[[xv]][1,1],
      St.dev.=c(res[[xv]][1, 2], exp(MeVa(log(res[[xv]][-1, 2]))))) )
SMR
} ## End of function  KHblg

## Example of generating a postscript document
KHblgR <- Figure("bal9", KHblg(KHbalR), 4.75, mgm=0.85)
```

The function `KHblf` draws the balance plot for a set of balance lists (Figure 9.6). It is adapted from `KHblg`. Supplementary materials (R code) contain other functions that are similar to those listed here. They are used for secondary purposes, or were used in the construction of the code but became defunct in the process.

Most of the results quoted or displayed in this chapter are obtained from the results of the expressions

```
##  Results for Austria 2006
ATO6pf <- POTF("AT", 6, 20, 20)
ATO6pg <- KHblg(ATO6pf[[2]])
ATO6pf2 <- REGR("AT", 6, 12)
```

and their versions for other countries and years.

Epilogue

This epilogue collects a few ideas for projects that would expand on the material presented in this volume. Most of them are essentially computational, but they also contain an element of 'detective' and other analytical work, exploration of secondary data sources, such as administrative registers, and some theoretical development.

1. Throughout, the standard for poverty, such as 60% of the median equivalised household income (eHI), is estimated. It would be desirable to account for the uncertainty in its estimation by methods simpler and computationally less demanding than bootstrap.

2. The analysis of a dataset, such as for a country's cross-sectional data for a year, ignores some relevant information available in other datasets. For example, the panel data could contribute to the analysis related to a cross-sectional survey.

3. Related to the previous point is an exploration of discrepancies between the cross-sectional and panel datasets for a country. For example, are the rates of extremely low and high income similar in the two databases (for each country)? More generally, are there any signs that the distributions of some key variables in the two datasets differ? Can they be related to different methods of data collection or (possibly) to nonresponse and its management?

4. The general ideas of trimming and shrinking the weights could be followed up by establishing some rules of thumb for the combinations of types of inferences and variation of the weights. How should the weights be shrunk for the established estimators to have their mean squared errors reduced to minimum?

5. For countries with administrative registers, some analyses could be conducted on the registers instead of the data from the European Union Statistics on Income and Living Conditions (EU-SILC). The feasibility of this could be explored, or how some aggregate-level information from the register, not necessarily in the public domain, could contribute to the analysis of EU-SILC data. Is there any scope for checking whether the estimators used for EU-SILC data are unbiased?

6. Describe the (possible) future of a survey like EU-SILC, in which the sampling frame is constructed from an administrative register

and a sample is drawn from it. Households and individuals are contacted merely to supplement the data extracted from the register by an automated process. Discuss the ethical issues involved and how access to the data should be controlled, and at which stage.

7. Find the details of how the original sampling weights are adjusted and reproduce the adjustment on an EU-SILC survey. Relate the details of the auxiliary information used for the adjustment to the variation of the weights and the need for trimming or shrinking the weights. Apply the adjustment with less auxiliary information. Is more auxiliary information rewarded by more efficient inferences? A simulation study could be conducted in which the processes of sampling, with or without some nonresponse and imperfect sampling frame, and estimation are replicated and the mean squared errors of several common estimators compared. Is the variance of the adusted sampling weights, or their inflation over the original weights a suitable summary for how the weights promote efficiency?

8. The distribution of income and of its various equivalised forms does not change dramatically from one year to the next. The principal causes of the changes are (wage) inflation and changes in the exchange rates (where applicable), and they affect most households uniformly. Exceptions are payments received from outside the country. Given this temporal similarity, how could inferences about a single year be strengthened by using information from a few previous years?

9. Which elements of small-area estimation can be applied to estimating the indices included among the Laeken indicators? How can the advantage of small countries in satisfying certain criteria related to small regional differences (region-level variation) be objectively assessed and adjusted for, so that comparisons with larger and more populous countries would be meaningful?

10. Although the national poverty rates are important headline figures, their comparison is problematic because the median value of eHI in one country is regarded on par with the median in another. How could a scheme for more appropriate comparisons be developed? For example, the percentiles of eHI could be compared across the countries, after an appropriate inflation for the perceived or plausible differences of the two countries. Suppose the median in country A is equivalent to between 110% and 125% of the median in country B. Then we compare the poverty rates (or another index) for the two countries by replacing the median in country B with its 1.1- or 1.25-multiple. Could a consensus about these inflation factors be reached? Would the plausible ranges of these factors be narrow enough to obtain comparisons with not too many instances of impasse?

11. Compare the consistency of the results of the univariate and multi-variate mixture analysis across cohorts and panels. Do some changes in the details of how data (about income) is collected leave a trace in the results? Can improved data quality be detected in the results? Is it plausible that the distribution of income has perceptibly changed in (some) countries over a period of a few years? Study the changes that can be attributed to the changes instituted recently in the social transfer systems is a few countries.

12. Assess the realistic nature of the Europe 2020 goals. That is, compare the changes that have to occur in the next few years with the changes that have occurred in the recent past. Could it happen that the goals are reached by manipulating the relevant indices as surrogates, without altering substantially the underlying latent summaries?

13. How can the methods and software for poverty rate and other indices be adapted for indices based on variables other than shortfall? Measures of social exclusion are an important example.

14. Many European households do not have an obvious (long-term) country of residence because they have homes in more than one country, reside temporarily in one country but consider staying (becoming residents) there. In brief, there is an element of convenience for many European citizens (or residents) as to which country they declare as their own. Find out which variables in EU-SILC hold some clues about these issues, and how EU-SILC could be expanded by (further) questions related to it. How could a survey be conducted about cross-border life, work, consumption, education and retirement?

15. Explore the feasibility of asking the members of the EU-SILC panel in the first year of their participation (year 1) not only about their income and related matters in the previous year 0, but also in year -1, so that for some purposes we would have five-variate longitudinal data.

16. Is the current economic crisis reflected is EU-SILC data? That is, can its onset be detected and the latest trends reported by central banks, governments and the media confirmed by data analysis? Are the cross-sectional or the panel data better suited for this, and for which aspects?

17. Devise an index for the extent to which social transfers (or another source of income) alter the distribution of eHI in a country. This could be done by summarising the distribution of a contrast of the scaled eHI (eHI%) in the A- and R-worlds, but other options could also be explored. Describing the subpopulations that benefit from the social transfers most and least would be of interest.

18. Perform a critical assessment of the pros and cons of the potential outcomes framework (propensity score matching) and regression for comparing the outcomes in two categories of a variable. Is the nature of the analysis (causal or not) a factor in this assessment?

Bibliography

[1] Aitkin, M., Francis, B., Hinde, J., and Darnell, R.: *Statistical Modelling with* R. Oxford University Press, Oxford, UK (2009).

[2] Akaike, H.: A new look at statistical model identification. *IEEE Transactions on Automatic Control* AU–19, 716–722 (1974).

[3] Alfons, A., Kraft, S., Templ, M., and Filzmoser, P.: Simulation of close-to-reality population data for household surveys with application to EU-SILC. *Statistical Methods and Applications* 20, 383–407 (2011).

[4] Alfons, A., and Templ, M.: Estimation of social exclusion indicators from complex surveys. The R package `laeken`. *Journal of Statistical Software* 54, 1–25 (2013).

[5] Angrist, J., Imbens, G., and Rubin, D. B.: Identification of causal effects using instrumental variables. *Journal of the American Statistical Association* 91, 444–455 (1996).

[6] Atkinson, A. B.: On the measurement of economic inequality. *Journal of Economic Theory* 2, 244–263 (1970).

[7] Atkinson, A. B.: On the measurement of poverty. *Econometrica* 44, 749–764 (1987).

[8] Atkinson, A. B.: *Poverty in Europe.* Blackwell Publishers, Oxford, UK (1998).

[9] Atkinson, A. B., Cantillon, B., Marlier, E., and Nolan, B.: *Social indicators. The EU and Social Inclusion.* Oxford University Press, Oxford, UK (2002).

[10] Atkinson, A. B., and Marlier, E. (Eds.): *Income and Living Conditions in Europe.* Publications Office of the European Union, Luxembourg (2010).

[11] Atkinson, A. B., Morrisson, C., and Bourguignon, F.: *Empirical Studies of Earnings Mobility (Fundamentals of Pure and Applied Economics).* Routledge, London, UK (1992).

[12] Banfield, J. D., and Raftery, A. E.: Model-based Gaussian and non-Gaussian clustering. *Biometrics* 49, 716–723 (1993).

[13] Basford, K. E., and McLachlan, G. J.: Estimation of allocation rates in a cluster context. *Journal of the American Statistical Association* **80**, 286–293 (1985).

[14] Bibi, S., and Duclos, J. Y.: A comparison of the poverty impact of transfers, taxes and market income across five OECD countries. *Bulletin of Economic Research* **62**, 387–406 (2010).

[15] Caminada, K., and Goudswaard, K.: Social income transfers and poverty: A cross-country analysis for OECD countries. *International Journal of Social Welfare* **21**, 115–126 (2012).

[16] Carpenter, J. R., and Kenward, M. G.: *Multiple Imputation and Its Application.* Wiley, Chichester, UK (2013).

[17] Claeskens, G., and Hjort, N. L.: *Model Selection and Model Averaging.* Cambridge University Press, Cambridge, UK (2008).

[18] Cleveland, W. S., and Devlin, S. J.: Locally weighted regression: An approach to regression analysis by local weighting. *Journal of the American Statistical Association* **83**, 596–610 (1988).

[19] Cochran, W. G.: The planning of observational studies of human populations. *Journal of the Royal Statistical Society* Series A **128**, 134–155 (1965).

[20] Cochran, W. G.: *Sampling Techniques*, 3rd ed. Wiley, New York (1977).

[21] Crow, E. L., and Shimizu, K. (Eds.): *Lognormal Distributions. Theory and Applications.* M. Dekker, New York (1998).

[22] Dagum, C.: A new approach to the decomposition of the Gini income inequality ratio. *Empirical Economics* **22**, 515–531 (1997).

[23] Dalgaard, P. *Introductory Statistics with* R. Springer-Verlag, New York (2008).

[24] Davison, A. C., and Hinkley, D. V.: *Bootstrap Methods and Their Application.* Cambridge University Press. Cambridge, UK (1997).

[25] Delhausse, B., Lüttgens, A., and Perelman, S.: Comparing measures of poverty and relative deprivation. An example for Belgium. *Journal of Population Economics* **6**, 83–102 (1993).

[26] Dempster, A. P., Laird, N. M., and Rubin, D. B.: Maximum likelihood for incomplete data via the EM algorithm. *Journal of the Royal Statistical Society* Series B **39**, 1–38 (1977).

[27] Deville, J.-C.: Variance estimation for complex statistics and estimators: Linearization and residual techniques. *Survey Methodology* **25**, 193–203.

[28] Deville, J.-C., Särndal, C.-E., and Sautory, O.: Generalized raking procedures in survey sampling. *Journal of the American Statistical Association* **88**, 1013–1020 (1993).

[29] Diggle, P. J., Heagerty, P., Liang, K.-Y., and Zeger, S. L.: *Analysis of Longitudinal Data*, 2nd ed. Oxford University Press, Oxford (2002).

[30] Dobson, A.: *An Introduction to Generalized Linear Models*, 2nd ed. Chapman and Hall/CRC, London, UK (2001).

[31] Duncan, G.J., Gustafsson, B., Hauser, R., Schmauss, G., Messinger, H., Muffels, R., Nolan, B., and Ray, J.-C.: Poverty dynamics in eight countries. *Journal of Population Economics* **6**, 215–234 (1993).

[32] Efron, B., and Morris, C.: Limiting the risk of Bayes and empirical Bayes estimators—part II: the empirical Bayes case. *Journal of the American Statistical Association* **67**, 130–139 (1972).

[33] Efron, B., and Morris, C.: Stein's estimation rule and its competitors—an empirical Bayes approach. *Journal of the American Statistical Association* **68**, 117–130 (1973).

[34] Efron, B., and Morris, C.: Data analysis using Stein's estimator and its generalizations. *Journal of the American Statistical Association* **70**, 311–319 (1975).

[35] Efron, B., and Tibshirani, R.: *An Introduction to the Bootstrap*. Chapman and Hall, London, UK (1993).

[36] Elliot, M. R.: Model averaging methods for weight trimming. *Journal of Official Statistics* **24**, 517–540 (2008).

[37] Eurostat: Continuity of indicators between end-ECHP and start-SILC. Algorithms to compute cross-sectional indicators of poverty and social inclusion adopted under the open method of cooperation. Publications Office of the European Union, Luxembourg (2005).

[38] Eurostat: Combatting poverty and social exclusion. A statistical portrait of the European Union 2010. Publications Office of the European Union, Luxembourg (2010).

[39] Everitt, B. S., and Hand, D. J.: *Finite Mixture Models*. Chapman and Hall, London, UK (1981).

[40] Fay, R. E., and Herriott, R. A.: Estimates of income for small places: An application of James-Stein procedures to census data. *Journal of the American Statistical Association* **74**, 269–277 (1979).

[41] Förster, M. F.: The European social space revisited: Comparing poverty in the enlarged European Union. *Journal of Comparative Policy Analysis: Research and Practice* **7**, 29–48 (2005).

[42] Foster, J., Greer, J., and Thorbecke, E.: A class of decomposable poverty measures. *Econometrica* **81**, 761–766 (1984).

[43] Foster, J., Greer, J., and Thorbecke, E.: The Foster-Greer-Thorbecke (FGT) poverty measures: Twenty-five years later. Working Paper IIEP-WP-2010-14. George Washington University, Washington, DC (2010).

[44] Frangakis, C., and Rubin, D. B.: Principal stratification in causal inference. *Biometrics* **58**, 21–29 (2002).

[45] Gastwirth, J. L.: The estimation of the Lorenz curve and Gini index. *The Review of Economics and Statistics* **54**, 306–316 (1972).

[46] Gelman, A., Carlin, J. B., Stern, H., and Rubin, D. B.: *Bayesian Data Analysis*, 2nd ed. Chapman and Hall/CRC, New York (2003).

[47] Ghosh, M., and Rao, J. N. K.: Small area estimation: An appraisal. *Statistical Science* **9**, 55–93 (1994).

[48] Gill, P. E., Murray, W., and Wright, M. H.: *Practical Optimization.* Academic Press, New York (1981).

[49] Gini, C.: Variabilità e mutabilità: contributo allo studio delle distribuzioni e delle relazioni statistiche. *Studi Economico-Giuridici della R. Università di Cagliari* **3**, 3–159 (1912).

[50] Golub, G. H., and van Loan, C. F.: *Matrix Computations*, 3rd ed. Johns Hopkins University Press, Baltimore, MD (1996).

[51] Gonzalez, M. E., and Hoza, C.: Small-area estimation with application to unemployment and housing estimates. *Journal of the American Statistical Association* **73**, 7–15 (1978).

[52] Good, P. I.: *Permutation, Parametric and Bootstrap Tests of Hypotheses*, 3rd ed. Springer-Verlag, New York (2005).

[53] Hoerl, A. E., and Kennard, R. W.: Ridge regression. *Technometrics* **12**, 55–82, (1970).

[54] Hoeting, J. A., Madigan, D., Raftery, A. E., and Volinsky, C. T.: Bayesian model averaging: A tutorial. *Statistical Science* **14**, 382–417 (1999).

[55] Holland, P. W.: Statistics and causal inference. *Journal of the American Statistical Association* **81**, 945–970 (1986).

[56] Holt, D., and Smith, T. F. M.: Post stratification. *Journal of the Royal Statistical Society* Series A **142**, 33–46 (1979).

[57] Horn, R. A., and Johnson, C. R. *Matrix Analysis.* Cambridge University Press, Cambridge, UK (1985).

[58] Horvitz, D. G., and Thompson, D. J.: A generalization of sampling without replacement from a finite universe. *Journal of the American Statistical Association* **47**, 663–685 (1952).

[59] Hosmer, D. W., and Lemeshow, S.: *Applied Logistic Regression*. Wiley, New York (1989).

[60] Iacus, S. M., King, G., and Porro, G.: Multivariate matching methods that are monotonic imbalance bounding. *Journal of the American Statistical Association* **106**, 345–361 (2011).

[61] James, W., and Stein, C.: Estimation with quadratic loss. *Proceedings of the 4th Berkeley Symposium on Mathematical Statistics and Probability*, Vol. I, pp. 361–379 (1961).

[62] Jenkins, S. P.: Modelling household income dynamics. *Journal of Population Economics* **13**, 529–567 (2000).

[63] Kass, R. E., and Raftery, A. E.: Bayes factors. *Journal of the American Statistical Association* **90**, 773–795 (1995).

[64] Kish, L.: *Survey Sampling*. Wiley, New York (1965).

[65] Kish, L.: *Statistical Design for Research*. Wiley, New York (1987).

[66] Klir, G. J., and Yuan, B. (Eds.): *Fuzzy Sets, Fuzzy Logic and Fuzzy Systems: Selected Papers by Lotfi A. Zadeh (Advances in Fuzzy Systems: Application and Theory)* World Scientific, Singapore (1996).

[67] Kovačević, M. S., and Binder, D. A.: Variance estimation for measures of income inequality and polarization—the estimating equations approach. *Journal of Official Statistics* **13**, 51–58 (1997).

[68] Laird, N. M., and Ware, J. H.: Random-effects models for longitudinal data. *Biometrics* **38**, 963–974 (1982).

[69] Lambert, D.: Zero-inflated Poisson regression, with an application to defects in manufacturing. *Technometrics* **34**, 1–14 (1992).

[70] Lange, K.: *Numerical Analysis for Statisticians*. Springer-Verlag, New York (1999).

[71] Langel, M., and Tillé, Y.: Statistical inference for the quintile share ratio. *Journal of Statistical Planning and Inference* **141**, 2976–2985 (2011).

[72] Lemmi A., and Betti, G.: *Fuzzy Set Approach to Multidimensional Poverty Measurement*. Springer-Verlag, New York (2006).

[73] Little, R. J. A., and Rubin, D. B.: *Statistical Analysis with Missing Data*, 2nd ed. Wiley, New York (2002).

[74] Lohmann, H.: Comparability of EU-SILC survey and register data: The relationship among employment, earnings and poverty. *Journal of European Social Policy* **21**, 37–54 (2011).

[75] Longford, N. T.: Multivariate shrinkage estimation of small area means and proportions. *Journal of the Royal Statistical Society* Series A **162**, 227–245 (1999).

[76] Longford, N. T.: Missing data and small area estimation in the UK Labour Force Survey. *Journal of the Royal Statistical Society* Series A **167**, 341–373 (2004).

[77] Longford, N. T.: *Missing Data and Small-Area Estimation. Modern Analytical Equipment for the Survey Statistician.* Springer-Verlag, New York (2005).

[78] Longford, N. T.: *Studying Human Populations. An Advanced Course in Statistics.* Springer-Verlag, New York (2008).

[79] Longford, N. T.: Inference with the lognormal distribution. *Journal of Statistical Planning and Inference* **139**, 2329–2340 (2009).

[80] Longford, N.T.: Small-sample estimators of the quantiles of the normal, log-normal and Pareto distributions. *Journal of Statistical Computation and Simulation* **82**, 1383–1395 (2012a).

[81] Longford, N. T.: 'Which model?' is the wrong question. *Statistica Neerlandica* **66**, 237–252 (2012b).

[82] Longford, N. T.: Assessment of precision with aversity to overstatment. *South African Journal of Statistics* **47**, 49–59 (2013a).

[83] Longford, N. T.: *Statistical Decision Theory.* Springer-Verlag, Heidelberg, Germany (2013b).

[84] Longford, N. T., and Bartošová, J.: A confusion index for measuring separation and clustering. *Statistical Modelling* **14** (2014); to appear.

[85] Longford, N. T., and D'Urso, P.: Mixture models with an improper component. *Journal of Applied Statistics* **38**, 2511–2521 (2011).

[86] Longford, N. T., and Nicodemo, C.: A sensitivity analysis of poverty definitions. IRISS Working Paper Series 2009–15, CEPS/INSTEAD, Differdange, Luxembourg (2009).

[87] Longford, N. T., and Pittau, M. G.: Stability of household income in European countries in the 1990's. *Computational Statistics and Data Analysis* **51**, 1364–1383 (2006).

[88] Lorenz, M.: Methods of measuring the concentration of wealth. *Publications of the American Statistical Association* **9**, 209–219 (1905).

[89] McCullagh, P., and Nelder, J. A.: *Generalized Linear Models*, 2nd ed. Chapman and Hall, London, UK (1989).

[90] McLachlan, G., and Peel, D.: *Finite Mixture Models*. Wiley, New York (2000).

[91] Mahalanobis, P. C.: On the generalised distance in statistics. *Proceedings of the National Institute of Statistics* **2**, 49–55 (1936).

[92] Maître, B., Nolan, B., and Whelan, C. T.: Reassessing the EU 2020 poverty target. An analysis of EU-SILC 2009. Discussion Paper WP 2012/13, Geary Institute, Dublin, Ireland (2012).

[93] Mardia, K. V., Kent, J. T., and Bibby, J.: *Multivariate Analysis*. Academic Press, London, UK (1979).

[94] Marlier, E., Atkinson, A. B., Cantillon, B., and Nolan, B.: *The EU and Social Inclusion: Facing the Challenges*. Policy Press, Bristol, UK (2007).

[95] Marron, J. S., and Wand, M. P.: Exact mean integrated squared error. *The Annals of Statistics* **20**, 712–736 (1992).

[96] Meng, X.-L., and van Dyk, D.: The EM algorithm—an old folk-song sung to a fast new tune. *Journal of the Royal Statistical Society* Series B **59**, 511–567 (1997).

[97] Molenberghs, G., and Verbeke, G.: *Models for Discrete Longitudinal Data*. Springer-Verlag, New York (2006).

[98] Moskowitz, C. S., Venkatraman, E. S., Riedel, E., and Begg, C. B.: Estimating empirical Lorenz curve and Gini coefficient in the presence of error. Working Paper 12, Memorial Sloan-Kettering Cancer Center, Epidemiology and Biostatistics, New York (2007).

[99] Needham, T.: A visual explanation of Jensen's inequality. *American Mathematical Monthly* **100**, 768–771 (1993).

[100] Osier, G.: Variance estimation for complex indicators of poverty and inequality using linearization techniques. *Survey Research Methods* **3**, 167–195 (2009).

[101] Pearl, J.: Causal inference in statistics: An overview. *Statistical Surveys* **3**, 96–146 (2009).

[102] Potthoff, R. F., Woodbury, M. A., and Manton, K. G.: 'Equivalent sample size' and 'equivalent degrees of freedom' refinements for inference using survey weights under superpopulation models. *Journal of the American Statistical Association* **87**, 383–396 (1992).

[103] R Development Core Team: R: *A Language and Environment for Statistical Computing.* R Foundation for Statistical Computing, Vienna, Austria (2011).

[104] Rao, J. N. K.: *Small Area Estimation.* Wiley, New York (2003).

[105] Robbins, H.: An empirical Bayes approach to statistics. *Proceedings of the 3rd Berkeley Symposium on Mathematical Statistics and Probability*, Vol. I, pp. 157–164 (1955).

[106] Rosenbaum, P. R.: *Observational Studies*, 2nd ed. Springer-Verlag, New York (2002).

[107] Rosenbaum, P. R.: *Design of Observational Studies*, 2nd ed. Springer-Verlag, New York (2010).

[108] Rosenbaum, P. R., and Rubin, D. B.: The central role of the propensity score in observational studies for causal effects. *Biometrika* **70**, 41–55 (1983).

[109] Rubin, D. B.: Estimating causal effects of treatments in randomized and nonrandomized studies. *Journal of Educational Psychology* **66**, 688–701 (1974).

[110] Rubin, D. B.: Practical implications of modes of statistical inference for causal effects and the critical role of the assignment mechanism. *Biometrics* **46**, 1213–1234 (1991).

[111] Rubin, D. B.: Multiple imputation after 18+ years. *Journal of the American Statistical Association* **91**, 473–489 (1996).

[112] Rubin, D. B.: *Multiple Imputation for Nonresponse in Surveys*, 2nd ed. Wiley, New York (2002).

[113] Rubin, D. B.: Causal inference using potential outcomes: Design, modelling, decisions. 2004 Fisher Lecture. *Journal of the American Statistical Association* **100**, 322–331 (2005).

[114] Rubin, D. B.: *Matched Sampling for Causal Effects.* Wiley, New York (2006).

[115] Rubin, D. B.: For objective causal inference, design trumps analysis. *The Annals of Applied Statistics* **2**, 808–840 (2008).

[116] Rubin, D. B., and Thomas, N.: Matching using estimated propensity scores: Relating theory to practice. *Biometrics* **52**, 249–264 (1996).

[117] Saltelli, A., Tarantola, S., Campolongo, F., and Ratto, M.: *Sensitivity Analysis in Practice. A Guide to Assessing Scientific Models.* Wiley, New York (2004).

[118] Salverda, W., Nolan, B., and Smeeding, T. (Eds.): *The Oxford Handbook of Economic Inequality*. Oxford University Press, Oxford, UK (2009).

[119] Särndal, C.-E., Swensson, B., and Wretman, J.: *Model Assisted Survey Sampling*. Springer-Verlag, New York (1992).

[120] Schafer, J. L.: *Analysis of Incomplete Multivariate Data*. Chapman and Hall, London (1996).

[121] Schafer, J. L.: Multiple imputation: A primer. *Statistical Methods in Medical Research* **8**, 3–15 (1999).

[122] Schwartz, G.: Estimating the dimension of a model. *Annals of Statistics* **6**, 461–464 (1978).

[123] Sen, A.: Poverty: An ordinal approach to measurement. *Econometrica* **44**, 219–231 (1976).

[124] Sen, A.: Issues in the measurement of poverty. *Scandinavian Journal of Economics* **81**, 285–307 (1979).

[125] Simonoff, J. S.: *Smoothing Methods in Statistics*. Springer-Verlag, New York (1996).

[126] Stewart, M. B., and Swaffield, J. K.: Low pay dynamics and transition probabilities. *Economica* **66**, 23–42 (1999).

[127] Titterington, D. M., Smith, A. F. M., and Makov, U. E.: *Statistical Analysis of Finite Mixture Distributions*. Wiley, New York (1985).

[128] United Nations Economic Commission for Europe (UNECE): *Canberra Group Handbook on Household Income,* 2nd ed. United Nations, New York (2011).

[129] Valliant, R., Dever, J. A., and Kreuter, F.: *Practical Tools for Designing and Weighting Survey Samples*. Springer-Verlag, New York (2013).

[130] van Buuren, S.: *Flexible Imputation of Missing Data*. Chapman & Hall/CRC Press, Boca Raton, FL (2012).

[131] Venables, W. N., and Ripley, B. D.: *Modern Applied Statistics with S-plus*, 4th ed. Springer-Verlag, New York (2010).

[132] Verbeke, G., and Molenberghs, G.: *Linear Mixed Models for Longitudinal Data*. Springer-Verlag, New York (2000).

[133] Wand, M. P., and Jones, M. C.: *Kernel Smoothing*. Chapman and Hall, London, UK (1994).

[134] Zadeh, L. A.: Fuzzy Sets. *Information and Control* **8**, 338–353 (1965).

[135] Zeger, S. L., Liang, K.-Y., and Albert, P.: Models for longitudinal data: A generalized estimating equation approach. *Biometrics* **44**, 1049–1060 (1988).

[136] Zuur, A., Ieno E. N., and Meesters, E.: *A Beginner's Guide to* R. Springer-Verlag, New York (2009).

Subject Index

p value, 75
Excel, 20
LaTex, 34, 126
MATLAB, xiii
R, xii, 20
Stata, xiii, 20
grep (in Linux), 309
laeken (R package), 77

A-eHI, 266
A-eHI%, 264
A-world, 251, 264, 276, 329
absolute rate, 182, 189
additivity, 54, 62, 233
Akaike information criterion, 226
algebra, linear, xiii, 99
algorithm, 100
allocation rule, 108
analysis of variance, 77
antimode, 113
antisymmetry, 186
approximation, 56, 88, 97, 116,
        194, 217, 226, 264, 293
    linear, 168
    normal, 159
argument (in R)
    logical, 85
    mandatory, 21
    optional, 21
arithmetic mean, 63, 118
aspect (factor), 12
assignment, 98, 117, 221, 227, 299
    at random, 285, 300
    of treatment, 289, 307
    probability, 105, 132, 238
    to bins, 129

Atkinson index, 76
attrition (in a panel), 203
autonomous, 287
auxiliary
    data, 152
    estimate, 154
    information, 146, 158, 173,
        328
    statistics, 146
    variable, 147
average effect, 287
averaging, 144, 154, 290

background, 287
    variable, 307, 310
balance, 264, 293, 300, 303, 312,
        318
    plot, 325
basis, 8, 61, 67, 99, 143, 151, 161,
        215, 257
    estimator, 143
Bayes
    factor, 295
    information criterion, 226
    theorem, 98, 217
bias, 42, 142, 150, 158, 199, 280,
        305
    plausible, 42
    selection, 295
binary variable, 38, 53, 320
bootstrap, 40, 49, 70, 138, 157, 166,
        273, 327
    estimate, 42
    sample, 41
boundary of parameter space, 236

calculus, xiii

calibration (of weights), 149
candidate, 243, 294
capacity of social transfers, 251, 261
causal analysis, 308
cause, 285
census, 147
Cholesky decomposition, 216, 250
closure, 215
clustering, 5, 43, 264
clusters, 97, 116, 235
   jointly, 117
   pairwise, 117
coarseness, 130, 191, 197, 280
coefficient, 2, 39, 45, 143, 153, 157, 169, 173
   optimal, 145
   shrinkage, 19, 157
   vector, 147
coercion (in R), 23, 311
cohort, 176, 182, 197, 203
complete
   -data algorithm, 100
   -data analysis, 291
   dataset, 100, 198, 290
   record, 312
completed
   -data analysis, 198
   dataset, 100, 198, 290
completion, 199, 290
compliance
   full, 179
   rate, 177
component, 97, 234, 236
   distribution, 217
   dominant, 103
   exclusive, 222
   improper, 113, 131, 227, 245
   inclusive, 222
   majority, 120, 220
   minority, 112, 120, 220
   of mixture, 216
   proper, 113, 236
components
   confused, 117

   mutually confused, 117
   number of, 97, 122, 236, 237
   separated, 117
composite
   estimate, 157
      multivariate, 152, 157
      univariate, 157
   estimation, 150, 168
   estimator, 143
composition, 97, 99, 143, 151, 161, 169, 200, 210, 217, 272
   bivariate, 152, 154, 155, 158
   dimension of, 173
   estimated, 108
   multivariate, 147, 158, 170
   of household, 2
   univariate, 147, 155, 173
concave curve or function, 263
condition number, 148, 170
conditioning, 216
confidence interval, 6, 42, 197
confusion, 234
   index, 117, 132, 217, 234, 249
   matrix, 120, 132, 217, 234
   rate, 117
   threshold, 235
constituency, 175
contagion, 286
contour plot, 192
contrast, 8, 287, 320
   relative, 320
convergence, 56, 81, 115, 123, 236, 311, 316
   criterion, 236
   monitor, 131, 238
convex
   function, 56, 76
   kernel, 56
convolution, 97
correlation, 39, 138, 200
   matrix, 218, 241
   negative, 221
   structure, 217
covariance, 148, 160, 168
   matrix, 220

region-level, 160
    structure, 215, 234
CPU time, 51
criterion, 38, 47, 57, 226
cross-sectional study, xiii
currency, 54
curvature, 57

data, xi
    format, 20
    frame (in R), 23
data(set)
    complete, 100, 290
    completed, 100, 290
    incomplete, 100, 290
    inspection, 197, 294
    realised, 100
database, xii
decile, 3, 43
decimal comma, 23
degree of freedom, 102
density, 4, 52, 97, 215, 265
    bimodal, 98
    conditional, 289
    improper, 113, 131, 227, 245
        constant, 113, 227
        scaled, 131
    joint, 215, 266
    marginal, 127
    normal, 71, 113, 238
        multivariate, 240
    of mixture, 98
    scaled, 98, 113, 124, 131, 217,
        238, 250
design, 295
    experimental, 285
deviance, 123, 236, 317
diagnostic plot, 320
diagnostics, 318, 322
direct
    estimate, 150, 157, 162, 170
    estimation, 155, 161, 173
    estimator, 142, 152
dispersion, 16
    measure, 64

distribution, 3, 64, 97
    asymmetric, 7
    binary, 288
    conditional, 216, 288
    continuous, 4, 97
    degenerate, 98, 119
    discrete, 97
    function, 4, 98
        scaled, 98
    hybrid, 98
    joint, 39, 215
    lognormal, 7, 48, 52, 71, 100,
        125, 218
    marginal, 108, 215
    mixture, xiii
    multinomial, 97, 108, 292
    multivariate, 234
    nonnormal, 294
    normal, 4, 49, 71, 97, 121, 265
    null, 75
    of income, 7, 328
    of social transfers, 253
        conditional, 264
    of weights, 14
    plausible, 292
    Poisson, 98
    sampling, 41
    semicontinuous, 98
    skewed, 7
    standard normal, 120
    symmetric, 7
    uniform, 99, 253
        continuous, 41
donor, 291
    group, 293
drop-out, 179

E step (of EM algorithm), 100, 217
ECHP, 233
effect, 285
    average, 287, 322
    constant, 302
    unit-level, 287
    unit-specific, 286
effective

sample size, 266
effectiveness of social transfers,
    261, 276
efficiency, 47, 145, 293, 328
    loss of, 145
eHI, 39, 53, 83, 102, 164, 175, 189,
    208, 215, 251, 264, 299
    log-, 103
    missing, 181
    negative, 54, 77, 112, 233
    profile, 217
    relative, 188
    scaled, 258
eHI%, 189, 198, 208, 214, 232, 247,
    258, 264, 277
eHS, 2, 182
eigenvalue, 147, 170, 173, 239
    negative, 152, 153
element of a list (in R), 23
EM algorithm, 100, 113, 121, 131,
    198, 215, 217, 227, 236
entry to poverty, 175, 211
    rate of, 194
equilibrium threshold, 261
equivalence class, 54, 63
equivalisation, 252, 274
equivalised
    household income (eHI), 3
    household size (eHS), 2, 177
estimate, 6, 38, 135, 273
    auxiliary, 154
    bootstrap, 167
    composite, 157
        multivariate, 152, 157, 173
        univariate, 150, 157
    direct, 150, 157, 162, 170
    initial, 102
    pseudo-, 94
    realised, 95
    replicate, 41
estimation, 1, 38, 216, 273, 327
    composite, 150, 168
    direct, 155, 161, 173
    efficient, 152, 198
    error, 305

maximum likelihood, 100
    small-area, xiii, 328
estimator, 5, 38, 53, 138, 151, 159,
    328
    basis, 143
    biased, 289
    composite, 143
    direct, 142, 152
    efficient, 47, 154
    Horvitz-Thompson, 45
    MI, 202
    minimum-variance unbiased,
        48
    moment-matching, 160
    naive, 48, 160
    ratio, 47
    selected-model based, 294
    shrinkage, 146
    single-model based, 294
    unbiased, 45, 160, 289
    weighted, 47
EU-SILC, xii, xiv, 20, 54, 66, 77,
    102, 135, 178, 197, 215,
    230, 252, 262, 267, 290,
    309, 327
    countries in, 35
    longitudinal part, 177
    panel, 200
European Union (EU), xi
Eurostat, 2, 136
exclusive component, 222
exit from poverty, 175, 192, 211
    rate of, 194
exogeneous, 264
expectation, 4, 52, 53, 71, 98, 146,
    160, 215, 290
    conditional, 101, 264
    partial, 52
    region-level, 147
expenditure, 2
experiment, 300
experimental design, 285
exploratory analysis, 197, 204
exponent, 56, 62, 76, 82
    negative, 114

expression (in R), 10, 50, 79, 164, 203
extrapolation, 5, 306
extreme (of a function), 77

failure, 175, 257
    rate, 257
        absolute, 257
        relative, 257
fixed, 6, 37
focal group, 287, 296, 319
focal unit, 287
focus, 19
fragility, 47
frequency, 4
function
    convex, 76
    indicator, 54
    linear, 57
    linear subscore, 192
    link, 294
    log-concave, 254
    logit, 186
    monotone, 109
    nonlinear, 202
    objective, 81
    power, 76
    quadratic, 66, 90
    smooth, 77, 265
    symmetric, 119
function (in R), xiv, 20, 161, 197
    auxiliary, 27, 210, 241
    system-defined, 22, 50
    user-defined, 27, 164

generalised least squares, 294, 315
generalised linear model, 318
geometric mean, 19, 63, 300
Gini coefficient, xiii, 63, 87, 138, 253
    decomposition, 77
graphics, xiv, 173
    device (in R), 34
    window (in R), 129
gray area, 195

grid (of values), 8, 127, 248, 272, 278

harmonic mean, 118
harmonisation, xii
histogram, 7, 42, 99, 103, 127, 188, 221, 248
homoscedasticity, 264
Horvitz-Thompson estimator, 45
hot deck, 291, 297
household, 1
    income, xiii
    intact, 178, 182, 205
    size, 142
        average, 135, 148, 157
hypothesis, 75, 197
    test, 6, 197, 294

identity matrix, 152, 216
impact of social transfers, 254
impasse, 58, 197
improper
    component, 227, 245
    density, 245
        multivariate, 227
imputation, 25, 197
    multiple, 198, 290
    single, 291
imputations
    number of, 199
imputed
    pattern, 197
    value, 100
inclusive component, 222
income, xiii, 1, 112
    disposable, 252, 274
    distribution, 7
    inequality, 53
    negative, 237
    source of, 3, 251
    stability, 232
    total (HI), 3
incomplete dataset, 100, 290
independence, 37, 102, 161, 197, 216, 264

conditional, 288, 290
   mutual, 143, 151, 160, 168,
      198
index, xiii, 2, 53, 75, 82, 135, 146,
      175, 233, 251, 272
   Atkinson, 76
   confusion, 234
   region-level, 143
indicator, 39, 53, 101, 209, 266, 310
   function, 54, 265
   of inclusion, 45
   of presence, 178
   variable, 315
   vector, 147
individual, 1
inequality, xiii, 13, 53, 63, 73, 140
inertia, 150
inference, xii, 140, 197, 273, 327
inferential
   statement, 6, 38, 197, 294
   target, 6, 38, 225, 286
infiltration, 67, 112
inflation, 60, 145, 170, 200, 220,
      328
influence, 76
influential value, 295
information criterion, 103, 294
initial solution, 122, 227, 237, 316
inspection of data, 197
intact household, 207, 219, 232,
      246
integral, 52, 71, 88, 113
   infinite, 227
integrand, 52, 71, 118
integration (numerical), 66, 72
interaction, 294, 303, 311
intermediate variable, 307
interpolation, 24
   linear, 45, 192
interquartile range, 16, 64
invariance, 118
inverse, 4, 76, 152, 186
iteration, 56, 81, 100, 115, 122, 217,
      236, 316

Jensen's inequality, 76

kernel, 53, 75, 164
   convex, 56
   identity, 54
   linear, 54, 139
   multivariate, 265
   normal, 265, 277
   power, 63
   quadratic, 139
   smoothing, 249, 264, 272
   symmetric, 265
   univariate, 265

labelling, 103, 112, 229
Laeken indicators, 2, 328
late entry, 179
legend (in R), 128
library (in R), 22
likelihood, 236
   ratio, 102
linear
   algebra, xiii, 99
   combination, 58, 77, 215
   function, 57
   model, xiii
   poverty gap, 84, 156
   scale, 60, 73, 114, 185
   space, 99
   transformation, 4, 48
linearity, 264, 306
link function, 294
Linux, 309
listwise deletion, 310
location, 7
log
   -concave function, 254
   -determinant, 240
   -eHI, 7, 103, 215, 294, 311
      distribution of, 220
   -income, 48
   -likelihood, 100, 123, 236
      complete-data, 101
      incomplete-data, 101
   -mean, 17

-odds, 185
-poverty gap, 60, 83, 139, 162
-shortfall, 63
-standard deviation, 17
-weight, 16
    scale, 76, 119, 220
    transformation, 218
logarithm, 7
    decimal, 131
logical (in R)
    argument, 85
    variable, 163, 311
    vector, 21
logistic regression, 293, 311
logit, 186
    scale, 186
lognormal distribution, 7, 48, 52,
        71, 100, 125, 218
longitudinal, 177
    data, xiv, 202, 215, 329
    study, 179
loop (in R), 83, 124, 171, 209, 246,
        281
    conditional, 81, 317
    double, 132
    iteration, 238
    variable, 173
Lorenz curve, xiii, 63, 87, 139, 253
    replicate, 70
    transformed, 69

M step (of EM algorithm), 101,
        124, 218, 238
Mahalanobis distance, 265
majority component, 220
marginal probability, 101, 218, 237
match, 293, 330
matched group, 312
matched pair, 293, 308, 319
maximum likelihood, 100, 218
    estimation, 100
mean
    arithmetic, 118
    harmonic, 118
    population, 7

sample, 7
squared error (MSE), 19, 47,
        143, 327
    inflation of, 146
    minimum, 154
median, 3, 64, 83, 287
    eHI, 50, 135, 188, 214, 258, 327
    income, 2
    population, 39
    weighted, 39, 164
member of a population, 4
method, complete-data, 198
minority component, 220, 234
missing
    data, 290
    value, 6, 80, 100, 181, 197, 288,
        309
mixture, 7, 97, 121, 198, 287, 294,
        329
    component, 216
    distribution, xiii
        multivariate, xiii
    finite, 98, 216
    multivariate, 215, 237
    normal, 237
    of lognormal distributions, 102
    of normal distributions, 98,
        216
    of uniform distributions, 99
mobility, 175
    level, 195
mode (of a distribution), 98, 103,
        270
model, 99
    averaging, 295
    fit, 126
    linear, xiii
    parameter, 100
    selection, 116, 294
moment, 17, 52, 160
    matching, 147, 160
moving, 182
multinomial distribution, 97, 108
multiple imputation (MI), 175,
        198, 212, 290

multiplicative scale, 54, 73, 114,
        149, 184, 262
multivariate analysis, xiii

Newton method, 56, 80
nonlinear transformation, 48, 100
nonparametric regression, 264
nonresponse, 6, 16, 177, 327
normal
    density, 71, 113
    distribution, 4, 71, 97, 121,
        265
        multivariate, 215
        standard, 4, 120, 216
        univariate, 215
    kernel, 265, 277
    mixture, 237
normality, 7, 42, 306
numeric (in R), 311

objects (in R), 21
observational study, 285, 308
odds, 8, 185, 256
    log-, 185
    ratio, 8, 185, 256
offset, 86, 281
operating system, 21, 78
ordinal scale, 287
outcome
    potential, 285
    realised, 286
    variable, 3, 287
    would-be, 286
outlier, 108, 116, 155, 227, 232
    status, 227
overestimation, 146, 154
overstatement, 146

panel, xiv, 203, 215
    data, 175, 214, 234, 246, 327
    EU-SILC, 219
    rotating, 176
parent, 235
partial ordering, 66
partial score, 191, 201
partitioning, 216

pattern
    imputed, 197
    incomplete, 179
    of poverty states, 194
    of presence, 178, 205
    poverty, 195
    stable, 195
percentile, 3, 13, 24, 42, 140
    scaled, 73, 93, 137
permutation, 100
    test, 75, 93
pivot, 144, 235
plausible
    bias, 42
    dataset, 198
    distribution, 292
    estimate, 291
    feature, 19
    model, 225
    poverty pattern, 198
    probability, 292
    range, 121, 230
    value, 2, 57, 98, 199, 227, 291
pooling, 151
population, 1, 54, 97, 175, 266
    artificial, 19, 77
    finite, 4
    infinite, 4
    quantity, xiii, 38
    size, 4, 47, 137, 148, 156, 170
        estimated, 167
positive definite, 152, 215, 241
positive part, 53
postscript file, 33, 82
poststratification, 149
potential
    of social transfers, 261, 276
    outcome, 285, 306
    outcomes framework, xiii, 285,
        330
    version, 287, 307
poverty, xi, 1, 53, 233, 251, 261
    dynamics, 175
    financial, 1

gap, 82, 139, 162, 173, 251, 261, 273
  absolute, 82
  curve, 54, 85
  linear, 54, 84, 156, 261
  log-, 60, 83, 139, 162
  national, 157
  regional, 157
  relative, 60, 82, 257
  squared, 139, 162
index, 53, 82, 161, 175, 272
pattern, 182, 195, 209
  plausible, 198, 212
  rate of, 196
persistent, 190, 201
rate, 1, 40, 46, 50, 54, 137, 150, 184, 254
  curve, 8
  function, 8
  national, 150, 328
  regional, 150
  scaled, 137
  standard, 188, 251, 261, 266, 327
  status, 2, 175, 188, 215, 251, 273
    plausible, 201
  threshold, 2, 135
  transient, 191
  unequivocal, 257
power kernel, 63
precision, 42, 99, 125, 143, 238, 318
  averaged, 144
predictor, 306
pro-rating, 184
probability, 4
  assignment, 113
  fitted, 103
  marginal, 132
  of inclusion, 45
process, 37
profile, 222, 242, 299
  lower, 244
  mean, 223, 244
  upper, 244

propensity, 293, 319
  analysis, 300
  fitted, 311
  group, 293, 308
  matrix, 311, 319
  model, 293, 318
  score, 293, 319, 330
protocol, 37, 273
proxy, 273
pseudo-
  estimate, 94
  quantile, 95
  sample, 75
  value, 75

quadratic function, 56, 66, 90
  multivariate, 147
quantile, 3, 23, 43, 64, 72, 319
  function, 3
  pseudo-, 95
  scaled, 73, 91, 137
    function, 73
quantile-quantile (q-q) plot, 13, 43
quartile, 3, 64
  scaled, 137
questionnaire, 3
quintile, 64
  share ratio, 71

R-cHI, 266
R-eHI%, 264
R-world, 251, 276, 329
  hypothetical, 261
random, 6, 37
  assignment, 285
  mechanism, 201
  noise (in diagram), 106, 130
  sample, 97, 132
  vector, 39
randomisation, 288, 299
ratio estimator, 47
realised dataset, 100
recipient, 293
record
  complete, 177

missing, 177
recycling, 31
reference group, 287, 319
reference unit, 287
region, xiii, 135, 156, 161
   -variable, 161
register, xi, 147, 177, 327
regression, 264, 294, 306, 330
   logistic, 293, 311
   matrix, 315
   nonparametric, 264, 277
   ordinary, 48, 264, 303
   weighted, 305
relative contrast, 300
relative rate, 183, 190
replicate, 41, 95, 159, 166, 308
   completion, 199
   dataset, 109
replication, 12, 37, 75, 116, 132,
      141, 166, 198, 216, 288
representative, 296, 309
residual, 240
   variance, 264, 306
response, incomplete, 177
reversal, 152, 184, 256
ridge regression, 146
robust, 7, 69, 76, 112
root (of a function), 77
rotational group, 177, 203
Rubin's causal model, 285

sample, 5, 135, 175, 266
   bootstrap, 41, 50
   pseudo-, 75
   quantity, 38
   random, 97, 132
   rate, 38
   realised, 6, 40
   replicate, 41, 291
   selection, 159
   simple random, 40, 159
   size, 26, 38, 103, 135, 156, 167,
      178, 196, 218, 226, 249
     effective, 26, 42, 51, 102,
      148, 167, 199, 266

sampling, xiii, 37, 198
   design, xii, 5, 38, 41, 177, 264
     clustered, 6
   distribution, 41, 292
   frame, 6, 179, 328
   process, 159
   simple random, 5, 16
   variance, 142, 199, 305
     MI, 199
   variation, 12, 40, 69, 138, 188,
     220, 293
   weight, xiii, 5, 14, 53, 82, 102,
     164, 213, 218, 237, 265,
     297, 305, 310, 328
   weights
     distribution of, 80
satellite (component), 118, 235
scale, 13, 62
   linear, 73, 114
   log, 72, 76, 119
   multiplicative, 13, 54, 73, 114,
     262
   ordinal, 287
   relative, 233
scaled
   density, 98, 250
   percentile, 73, 93
   quantile, 91
scaling, 16, 75
score, 53, 75
   transformed, 53
sensitivity analysis, xii, 3, 308
separation, 234
separator, 23
sequence, 270
shell command, 309
shortfall, 53, 164, 175, 260, 273
   log-, 63
   relative, 60
shrinkage, 19, 146, 327
   coefficient, 19
   estimator, 146
significance, 75, 297
similarity, 146, 158
simulation, 19, 118, 132, 158

singularity, 112, 153, 232, 236

skew, 7

small-area estimation, xiii, 142, 164, 328

smoothing, 9, 30, 118, 182, 260, 270, 280
   kernel, 249, 264
   neighbourhood, 272

social exclusion, xii, 2, 329

social transfers, xiii, 251, 274, 329
   capacity of, 251
   distribution of, 253, 264
   effectiveness of, 261
   impact of, 254
   potential of, 261
   relative, 270
   scaled, 266, 277

squared error, 38

stability, 152
   curve, 233
   matrix, 248
   of income, 232, 247
   threshold, 233, 248

standard, xi, 1, 53, 60, 135, 182, 188, 251, 261, 266
   deviation, 42, 265, 298
     region-level, 150, 157
     relative, 300, 320
   error, 42, 70, 135, 154, 166, 196, 213, 298
     MI, 199
   national, 135

standardisation, 4, 118, 124, 265

statement, 197, 294
   inferential, 197

statistic, 160

staying, 182

staying poor, 182

stratification, 5

stratum, 5, 37

subject, 5

submodel, 102

subpopulation, 143

subscore, 192

success, 175, 191, 257

rate, 257
   absolute, 257
   relative, 257

sufficiency, 1

sufficient statistics, linear, 100

summary, 6, 25, 53, 95, 100, 105, 142, 198, 232, 253, 269, 274, 291
   diagnostic, 303
   missing-data, 100
   population, 6
   region-evel, 147
   sample, 160

support (of a distribution), 99, 114

surrogate, 273, 329

survey, xi, 1, 38, 135, 156, 170, 176, 298, 308, 327
   protocol, 177
   questionnaire, 3
   realised, 75

SUTVA, 307

switch, 194

symmetric function, 119

symmetry, 7

target, 6, 19, 38, 147, 160, 225, 273, 286, 302
   function, 270
   quantity, 150
   region, 143

ternary plot, 105, 130, 222, 235

tertile, 64

threshold, 2, 8, 40, 49, 54, 82, 135, 182, 251, 276
   equilibrium, 261
   for confusion, 117, 235
   for stability, 233, 248

transformation, 58, 97, 139, 164, 192, 220, 233, 276, 296, 311
   exponential, 48
   linear, 4, 48, 216
   log, 7, 17, 48, 52, 76, 218
   monotone, 118, 287
   nonlinear, 48, 100, 145, 154

nonsingular, 216
power, 211
transition, xiii, 175, 194, 215
accidental, 188
insubstantial, 188, 197
ordinary, 189
partial, 175, 209
rate of, 212
partially scored, 191
rate, 175, 194, 207
absolute, 182
evolution of, 184
relative, 175, 182
score, 175, 192
substantial, 175, 186, 209
zone, 189, 199
centre of, 189, 201, 210
half-width of, 189, 201, 210
treatment, 287, 310
-assignment mechanism, 288
confounded, 289
unconfounded, 288
assignment, 307
group, 288
trimming, 19, 327
truncation, 54, 84, 148, 164, 218,
233, 247, 258

unbiasedness, 145
uncertainty, xii, 40, 118, 154, 198,
218, 235, 290, 327
denial of, xii, 225
source of, 199
underestimation, 146, 154
understatement, 146
unequivocal poverty, 257
uniform distribution, 41, 253
unit, 53
administrative, 6
elementary, 147
intact, 43
sampling, 6
universality, 54, 62
updating, 218, 238, 317

validity, 12

variable
background, 287, 307, 310
binary, 287, 320
discrete, 4
indicator, 315
intermediate, 307
outcome, 3
treatment, 287
variance, 4, 39, 52, 101, 142, 170,
199
between-imputation, 199, 213
conditional, 266
inflation, 152, 170
log-scale, 72
matrix, 148, 170, 215, 236, 250
region-level, 160
region-level, 144, 160, 169
residual, 48, 264
sampling, 48
within-imputation, 199, 213
variation, 40
binomial, 201

weight, xiii, 25, 39, 53, 69, 82, 102,
122, 149, 173, 180, 209,
218, 237, 246, 265, 277,
297, 305, 310, 327
sampling, 5, 14, 50
scaled, 17, 102
weights
distribution of, 14
equal, 237
window, 264
flat, 264
width of, 265
workspace
in the operating system, 78
in R, 20, 79

Index of User-Defined R Functions

CVcor, 241
Dround, 25, 28, 29, 80, 89
EUBoots, 49, 51
EULrzA, 87
EULrzF, 89
EULrzW, 87
EUMgr, 242
EUMmx1, 237, 239, 245
EUMmx2, 229, 245
EUMmxF, 242
EUMmxG, 244
EUMmxU, 248
EUSQua, 91
EUSReF2, 281
EUSReg, 277
EUSTpf, 276
EUSTra0, 274, 275
EUSTra1G, 276
EUSTra1, 275
EUSTra2G, 276
EUYear, 22, 23
EUbts, 50
EUcnfM, 132
EUconM, 249
EUconf, 132, 249
EUdns, 124
EUeff, 26, 167
EUfnr, 82
EUloc, 31, 276
EUmix2, 121, 125, 127, 237
EUmixB, 131
EUmixF, 126, 131
EUmixH2, 127, 128
EUmixT, 130
EUmix, 131
EUneg, 77, 79, 87
EUnewton, 80, 122

EUperm, 93
EUpla, 32, 249
EUpoefG, 277
EUpoef, 276, 277
EUpov2C, 29
EUpovA, 84, 276
EUpovB, 84
EUpovG, 85, 90
EUpovL, 83
EUpovS, 82, 84
EUquaF, 91
EUreadC, 204
EUreadMx, 246
EUreadQ, 204, 246
EUreadR, 161
EUread, 20–22, 25, 78, 161
EUregA, 164, 168, 170
EUregB, 166, 168
EUregC, 168
EUregD, 170
EUregG, 173
EUshapQ, 206
EUssz, 167
EUsta, 247–249
EUtab, 129
EUtraMI, 212
EUtransA, 207, 209
EUtransD, 209, 210
EUtransF, 210
EUtransK, 210
EUtransM, 213
EUwei, 25
Extr*, 164
ExtrC, 27, 164
ExtrE, 28, 123, 164, 275
ExtrL, 27, 80, 164, 244
ExtrR, 27, 164, 275

Figure, 33, 92, 129, 283
HTable, 206, 276
InitMx, 237, 239
KHbal, 312, 320, 322
KHbin, 311, 319
KHblf, 325
KHblg, 322, 325
KHmtc, 312, 319
KHread, 308, 313
KHsel1, 309, 314
KernSq, 30, 280
Kern, 279
LeUni, 27
LogitW, 311, 315
MaxS, 314
MeVa, 27, 51, 123, 167
NiNh, 78
POTF, 308, 312, 313
Perc, 130
Pos10, 210, 212
Prob, 124, 130
PwTra, 210
QuantC, 24, 29, 91
QuantG, 34
QuantW, 23
SNdns, 238, 240
VarIn, 309, 314